计算机技术开发与应用丛书

# 虚拟化KVM进阶实践

陈 涛 ◎ 编著

清华大学出版社
北京

## 内 容 简 介

虚拟化技术是云计算的底层支撑技术之一。作为已经纳入 Linux 内核的虚拟化解决方案，KVM 虚拟化近年来发展迅猛，是很多云供应商默认的虚拟机管理程序。对于 IT 从业者来讲，掌握一些 KVM 虚拟化知识是很有必要的。

本书是《虚拟化 KVM 极速入门》的进阶篇，共分为 7 章。针对有一定 KVM 虚拟化基础的读者，通过全动手的实验学习虚拟机的迁移、高可用群集、嵌套虚拟化、性能监视与优化、P2V 和 V2V 迁移、备份与恢复、oVirt(RHV)等企业级虚拟化技术。

本书内容源自多个产业项目的实践，也是作者多年讲授 KVM 虚拟化实践课程经验的结晶，可以帮助学习者构建企业级虚拟化平台。



**图书在版编目(CIP)数据**

虚拟化 KVM 进阶实践/陈涛编著. —北京：清华大学出版社，2022.1 (2023.11重印)
(计算机技术开发与应用丛书)
ISBN 978-7-302-58992-1

Ⅰ. ①虚…　Ⅱ. ①陈…　Ⅲ. ①虚拟处理机　Ⅳ. ①TP338

中国版本图书馆 CIP 数据核字(2021)第 174820 号

**责任编辑：** 赵佳霓
**封面设计：** 吴　刚
**责任校对：** 时翠兰
**责任印制：** 沈　露

**出版发行：** 清华大学出版社
**网　　址：** https://www.tup.com.cn，https://www.wqxuetang.com
**地　　址：** 北京清华大学学研大厦 A 座　**邮　　编：** 100084
**社 总 机：** 010-83470000　**邮　　购：** 010-62786544
**投稿与读者服务：** 010-62776969，c-service@tup. tsinghua. edu. cn
**质量反馈：** 010-62772015，zhiliang@tup. tsinghua. edu. cn
**课件下载：** https://www.tup.com.cn ,010-83470236
**印 装 者：** 三河市龙大印装有限公司
**经　　销：** 全国新华书店
**开　　本：** 186mm×240mm　**印　　张：** 23　**字　　数：** 517 千字
**版　　次：** 2022 年 3 月第 1 版　**印　　次：** 2023 年11月第 2 次印刷
**印　　数：** 2001～2500
**定　　价：** 89.00 元

产品编号：093344-01

# 前言

PREFACE

## 本书的由来

与 VMware、Microsoft 虚拟化技术相比，KVM 虚拟化对于初学者并不“友好”。作为虚拟化项目的组成部分，笔者从 2011 年开始为客户讲授 KVM 虚拟化的课程，对此感触特别深。为了“不重复发明轮子”，KVM 虚拟化充分利用了 Linux、QEMU 和 libvirt 等开源技术，是一种组合型的解决方案，对初学者要求较高。

根据长期的 KVM 面授课程的经验，笔者总结出这样一种教学方法：针对每个知识点，先学习适当深度的原理，再动手做实验；先通过图形界面的操作，看到大概的轮廓，再通过大量的命令行、脚本的练习强化学习到的知识；先学基本知识，再掌握最佳实践方案。采用这种教学方法，通过 8 天左右的培训，就可以让初学者成为一个合格的 KVM 虚拟化平台的管理员。

2015 年，笔者将面授课程搬到了线上，制作了“开源虚拟化 KVM 入门”和“KVM 虚拟化进阶与提高”两门视频课程，发布在 51CTO 学院上，目前已有约 23 万人参加学习。

随着 RHEL/CentOS 8 的发布，笔者又将这套课程进行更新迭代，形成了《虚拟化 KVM 极速入门》和《虚拟化 KVM 进阶实践》，仍然沿用“原理＋实验”的风格，希望能够帮助到读者。

## 本书内容

本书共有 7 章。读者需要有一定的 KVM 虚拟化基础，通过学习本书，可以掌握企业级虚拟化技术所需要的知识。

第 1 章介绍虚拟机迁移的基本原理与分类，以及如何实现共享存储的迁移和无共享存储的迁移。

第 2 章介绍如何通过构建高可用群集(Corosync＋Pacemaker)以实现虚拟机的高可用，在 NFS、iSCSI 和 DRBD 3 种常见的存储中做了 4 个典型的群集实验。

第 3 章介绍如何构建基于 KVM 的嵌套虚拟化。

第 4 章介绍性能监视与优化的思路和工具，以及如何使用 Tuned 进行优化配置，针对 CPU、内存、网络和存储的多种优化技术。

第 5 章介绍 libguestfs 工具并详细介绍两个用于转换的工具：virt-v2v、virt-p2v，它们可以实现虚拟机到虚拟机、物理机到虚拟机的转换。

第 6 章介绍 RPO、RTO 等备份恢复的基本概念，以及内存快照、磁盘内部快照和磁盘外部快照的特点及应用场景，最后通过一个脚本将这些技术组合起来实现了一个简单的备份功能。

第 7 章介绍 oVirt 管理操作，包括安装、数据中心、存储、虚拟机、高可用、用户与权限及备份与恢复。

## 如何使用本书

本书既是笔者自己学习和使用 KVM 虚拟化的总结，又是讲授 KVM 虚拟化课程的课件。笔者认为学习原理、动手实践、做好记录、细心排错是学习 KVM 虚拟化的关键。

聪明人下笨功夫。在本书的陪伴下，我们一起：

(1) 深入理解原理。

(2) 精读 man 帮助、官方文档等。

(3) 做所有的实验。

(4) 详细记录实验过程。

(5) 使用思维导图等辅助工具。

(6) 享受排错的过程，在寻求帮助之前先尝试自己解决。

## 致谢

开源软件的世界精彩万千，在本书的写作过程中参考了很多开源社区的资料。在此向开源社区所有参与者和无私的代码贡献者致敬。

感谢龙芯中科杨昆、田延辉先生对龙芯 CPU 运行 KVM 虚拟机技术细节的介绍。

感谢陈庭暄先生在 Red Hat Enterprise Linux 8.4 Beta 上对全部实验进行的验证工作。

感谢清华大学出版社的工作人员为本书付出的辛勤劳动。

云计算技术发展很快，加之笔者水平有限，书中难免存在疏漏，敬请读者批评指正。

本书源代码

陈　涛

2022 年 1 月

# 目录

CONTENTS

第1章

# 实现虚拟机迁移

当虚拟机在宿主机上运行时，如果资源分配不均，则宿主机可能会过载或负载不足。此外，还会在宿主机上执行诸如硬件更换、软件升级、网络调整和故障排除之类的操作，因此，重要的是在不中断服务的情况下完成这些操作。虚拟机实时迁移技术在保证业务连续性的前提下，实现了负载均衡或前摄操作，提高了用户体验、工作效率。虚拟机实时迁移时保存了整个虚拟机的运行状态，迁移到新宿主机之后，业务不会中断，对用户没有影响。

**本章要点**

- ❑ 虚拟机迁移的基本原理。
- ❑ 宿主机内部迁移。
- ❑ 连接远程宿主机。
- ❑ 基于共享存储的迁移。
- ❑ 基于非共享存储的迁移。

## 1.1 虚拟机迁移的基本原理

目前流行的虚拟化产品如KVM、VMware、Hyper-V、Xen都提供了各自的迁移工具。

由于KVM虚拟机由XML配置文件和磁盘映像两部分组成，所以迁移就要考虑这两部分。如果虚拟机正在运行，则要考虑其内存中的数据。

### 1.1.1 虚拟机迁移的应用场景

根据目标宿主机的位置来分，虚拟机迁移分为两种：宿主机内部迁移、宿主机之间迁移。

(1) 如果某个存储池的空间或性能不能满足要求，则需要在同一宿主机中的不同存储池之间迁移虚拟机，这是宿主机内部迁移。这种迁移涉及磁盘映像的移动和XML配置文件的修改。

(2) 为了满足性能、管理等需求，将虚拟机从一台宿主机迁移到另外一台宿主机，这是宿主机之间迁移。

根据虚拟机的状态来分，虚拟机迁移分为3种：实时（虚拟机处于运行状态）迁移、非实时（虚拟机处于休眠状态）迁移、离线（虚拟机处于关闭状态）迁移。

（1）在实时迁移中，虚拟机继续在源宿主机上运行，而虚拟机的内存页被传输到目标宿主机上。在迁移期间，libvirt监视源宿主中内存页中的更改（这些页被称脏页），并在所有初始内存页传递完成后开始传输这些更改。这种传输会反复多次。libvirt会估算迁移过程中的传输速度，当需要传输的剩余数据量可以在一个可设置时间内（默认为10ms）一次性传完时，libvirt将挂起源虚拟机，以便传输剩余数据，然后在目标宿主机恢复虚拟机，从而完成迁移操作。

（2）在非实时迁移时，会先挂起源虚拟机，然后将虚拟机的内存复制到目标宿主机。虚拟机在目标宿主机上恢复之后，会释放在源宿主机上使用的虚拟机的内存。完成这种迁移所造成的停机时间要比实时迁移时间长，所需的时间取决于网络带宽和延迟、虚拟机恢复时间等。

（3）在离线迁移时，源虚拟机处于关闭状态。虚拟机的磁盘映像不必在共享存储上。

**提示：**早期版本的qemu-kvm、libvirt对实时迁移有比较多的限制，例如要求虚拟机的磁盘映像必须位于共享存储上，现在已经通过NBD方式的--copy-storage*选项实现了对基于非共享存储的迁移支持。

虚拟机迁移是一种很重要的技术，应用场景十分广泛，它有以下优势：

**1）负载均衡**

如果某台宿主机负载过重而另一台宿主机未充分利用，则可以将虚拟机迁移到使用率较低的宿主机上。

**2）硬件独立性**

当需要升级宿主机物理配置、添加或删除硬件时，可以先安全地将虚拟机迁移到其他宿主机上，然后进行操作，使业务不会因硬件的变更而中断。

**3）绿色节能**

在非业务高峰的时间段，可以将虚拟机重新分配给其他宿主机，然后可以关闭无业务的宿主机的电源，以便节省能源并降低成本。

**4）地理迁移**

可以将虚拟机移至离客户最近（低网络延迟）的宿主机上，或由于其他原因而将其移至适合的地理位置。

### 1.1.2 虚拟机迁移的要求

在迁移之前，需要确保系统满足迁移的要求。尽可能使用共享存储架构，目前支持的共享存储有以下几种：

（1）iSCSI。

（2）NFS。

（3）基于光纤通道的LUN。

(4) GFS2。

(5) SCSI RDMA 协议(SCSI RCP)。

虚拟机的磁盘资源(例如：映像文件、LUN 资源)保存在共享存储中，源宿主机与目标均可以访问，如图 1-1 所示。在迁移时，需要将虚拟机 XML 配置文件从源宿主机迁移到目标宿主机，但是虚拟机的磁盘资源没有发生变化。如果采用实时迁移和非实时迁移的方式，则需要迁移虚拟机内存页数据。

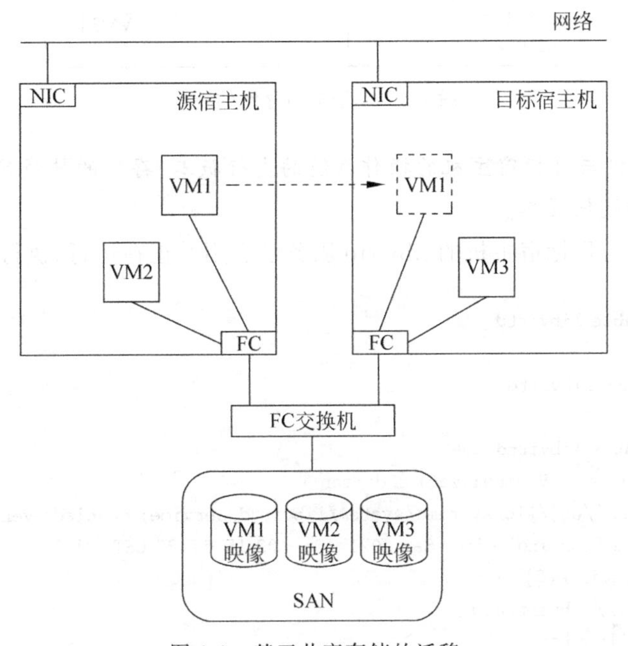


图 1-1　基于共享存储的迁移

实时迁移对宿主机操作系统版本也有一定的要求，RHEL/CentOS 发行版本如表 1-1 所示。

**表 1-1　RHEL/CentOS 发行版本对迁移的支持**

| 迁移方式 | 发行版本号 | 示　例 | 实时迁移支持 |
|---|---|---|---|
| 向前 | 主版本号 | 7.5+→8.x | 完全支持 |
| 向后 | 主版本号 | 8.x→7.y | 不支持 |
| 向前 | 小版本号 | 8.x→8.y(8.0→8.4) | 完全支持 |
| 向后 | 小版本号 | 8.y→8.x(8.4→8.0) | 完全支持 |

如果没有共享存储，则虚拟机的磁盘资源保存在宿主机本地存储中，如图 1-2 所示。早期的 RHEL/CentOS 只支持离线迁移，即虚拟机处于关闭状态。现在的版本已经支持基于非共享存储的虚拟机实时迁移。在迁移时，需要将虚拟机的 XML 配置文件、磁盘映像及内存页从源宿主机迁移到目标宿主机。

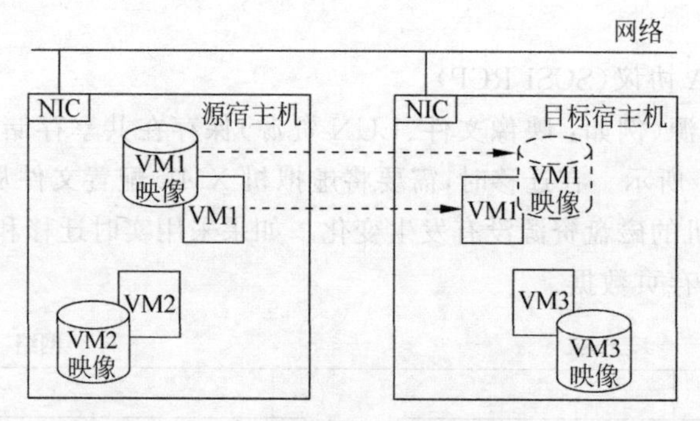

图 1-2 基于本地存储的迁移

**提示**：源宿主机与目标宿主机的操作系统的发行版本、存储的挂载目录应尽可能一致，这是虚拟机迁移的最佳策略。

确保源宿主机与目标宿主机的 libvirtd 服务已启用并正在运行，执行的命令如下：

```
#systemctl enable libvirtd

#systemctl start libvirtd

#systemctl status libvirtd
●libvirtd.service - Virtualization daemon
  Loaded: loaded (/usr/lib/systemd/system/libvirtd.service; enabled; vendor preset: enabled)
  Active: active (running) since Sat 2021-01-02 15:57:33 CST; 1h 38min ago
   Docs: man:libvirtd(8)
         https://libvirt.org
Main PID: 1509 (libvirtd)
   Tasks: 19 (limit: 32768)
  Memory: 89.6M
CGroup: /system.slice/libvirtd.service
 ├─1509 /usr/sbin/libvirtd
 ├─1820 /usr/sbin/dnsmasq --conf-file=/var/lib/libvirt/dnsmasq/default.conf --
leasefile-ro --d>
 └─1821 /usr/sbin/dnsmasq --conf-file=/var/lib/libvirt/dnsmasq/default.conf --
leasefile-ro --d>
...
```

## 1.1.3 虚拟机迁移的限制

具有以下特性的虚拟机不支持迁移：

(1) 拥有直通设备。

(2) 拥有 SR-IOV 设备。

(3) 拥有 vGPU 设备。

(4) 使用非统一内存访问(NUMA)。

RHEL/CentOS 的每个发行版本都有一些新的特性,可能会取消之前的限制。例如 RHEL/Cent OS 8.3 中就引入了两个与虚拟化有关的新特性和增强功能。

(1) JIRA: RHELPLAN-45950 的原文描述如下:

Migrating a virtual machine to a host with incompatible TSC setting now fails faster. Previously, migrating a virtual machine to a host with incompatible Time Stamp Counter (TSC) setting failed late in the process. With this update, attempting such a migration generates an error before the migration process starts.

(2) JIRA: RHELPLAN-45916 的原文描述如下:

Migrating virtual machines with enabled disk cache is now possible.

This update makes the RHEL 8 KVM hypervisor compatible with disk cache live migration. As a result, it is now possible to live-migrate virtual machines with disk cache enabled.

## 1.2 宿主机内部迁移

虚拟机在宿主机内部迁移主要为了解决存储池的空间或性能的问题。目前还没有办法通过 Cockpit、virt-manager 实现,virsh 也没有单个子命令可以完成,所以还需要与 Shell 命令组合实现。

虚拟机在宿主机内部迁移涉及移动磁盘映像和修改 XML 配置文件。下面通过一个示例来演示一下,示例命令如下:

```
1 # virsh pool - list -- details -- all
  Name         State     Autostart  Capacity   Allocation  Available
  ---------------------------------------------------------------------
  default      running   yes        49.98 GiB  47.85 GiB   2.12 GiB
  iso          running   yes        49.98 GiB  47.85 GiB   2.12 GiB
  virtio - win running   yes        49.98 GiB  47.85 GiB   2.12 GiB
  vm           running   yes        99.95 GiB  42.63 GiB   57.31 GiB

2 # virshdomblklist centos6.10
  Target   Source
  ----------------------------------------------------
vda      /var/lib/libvirt/images/centos6.10.qcow2
hda
```

第 1 行命令显示了宿主机所有存储池的详细信息。存储池 default 的可用空间仅有 2.12GiB,假设无法满足业务需要,准备将虚拟机 centos6.10 的磁盘映像移动到存储池 vm 中。

第 2 行命令显示虚拟机 centos6.10 的块设备列表。

接下来执行的示例命令如下：

```
3 #virsh pool-dumpxmlvm
<pool type='dir'>
<name>vm</name>
<uuid>46113e94-2451-4b08-9c9d-7377e26e7f11</uuid>
<capacity unit='Bytes'>107317563392</capacity>
<allocation unit='Bytes'>45777022976</allocation>
<available unit='Bytes'>61540540416</available>
<source>
</source>
<target>
<path>/vm</path>
<permissions>
<mode>0755</mode>
<owner>0</owner>
<group>0</group>
<label>system_u:object_r:unlabeled_t:s0</label>
</permissions>
</target>
</pool>
```

第3行命令显示存储池 vm 的目标路径是/vm。

接下来执行的示例命令如下：

```
4 #mv -b -v /var/lib/libvirt/images/centos6.10.qcow2 /vm

5 #virsh pool-refresh vm

6 #virsh pool-refresh default

7 #virsh pool-list --details --all
  Name         State     Autostart   Capacity    Allocation   Available
  -----------------------------------------------------------------------
  default      running   yes         49.98 GiB   46.53 GiB    3.45 GiB
  iso          running   yes         49.98 GiB   46.53 GiB    3.45 GiB
  virtio-win   running   yes         49.98 GiB   46.53 GiB    3.45 GiB
  vm           running   yes         99.95 GiB   44.54 GiB    55.40 GiB
```

使用第4行命令将映像文件移动到目标存储池中。在生产环境中进行映像文件的操作要谨慎，建议使用-b 选项，它的作用是：当文件存在时，在覆盖前为其创建一个备份。

第5行命令、第6行命令分别刷新存储池 default 和 vm。

第7行命令显示了所有存储池的详细信息。存储池 default 的可用空间增大到了3.45GiB。

下面需要修改虚拟机 centos6.10 的配置文件以反映存储卷的变化，示例命令如下：

```
8 #virsh dumpxml centos6.10 > bak-centos6.10.xml

9 #virsh edit centos6.10
将原有配置：
<disk type='file' device='disk'>
<driver name='qemu' type='qcow2'/>
<source file='/var/lib/libvirt/images/centos6.10.qcow2'/>
<target dev='vda' bus='virtio'/>
<address type='pci' domain='0x0000' bus='0x00' slot='0x07' function='0x0'/>
</disk>

修改为
<disk type='volume' device='disk'>
<driver name='qemu' type='qcow2'/>
<source pool='vm' volume='centos6.10.qcow2'/>
<target dev='vda' bus='virtio'/>
<address type='pci' domain='0x0000' bus='0x00' slot='0x07' function='0x0'/>
</disk>
```

第 8 行命令备份了虚拟机的配置文件以备不时之需。

第 9 行命令修改了配置文件中的 vda 的配置。主要修改了<disk type='volume' device='disk'>和<source pool='vm' volume='centos6.10.qcow2'/>两个属性值。

**提示**：使用存储卷和存储池的标识设置磁盘映像，要比使用映像文件的路径更加灵活。

最后执行的示例命令如下：

```
10 #virsh domblklist centos6.10
    Target    Source
   ------------------------------
    hda       -
    vda       centos6.10.qcow2

11 #virsh start centos6.10
   Domain centos6.10 started
```

第 10 行命令的输出显示了修改后的配置，第 11 行启动了虚拟机以便进行确认。

## 1.3 连接远程宿主机

在宿主机之间迁移虚拟机需要知道如何连接到远程宿主机。不同的管理工具连接到远程宿主机的方法不尽相同，在使用它们之前需要先了解一下 URI。

### 1.3.1 统一资源标识符

统一资源标识符(Uniform Resource Identifier,URI)提供了一种简单且可扩展的方式来标识物理或逻辑资源,例如页面、书籍或文档等。

URI有两种类型,如图1-3所示。

(1) URL:Uniform Resource Locator,指定计算机网络上的位置及其检索技术。例如https://www.Kernel.org。

(2) URN:Uniform Resource Name,指定URN方案的资源名称,而不指定地址。例如书籍的ISBN号。

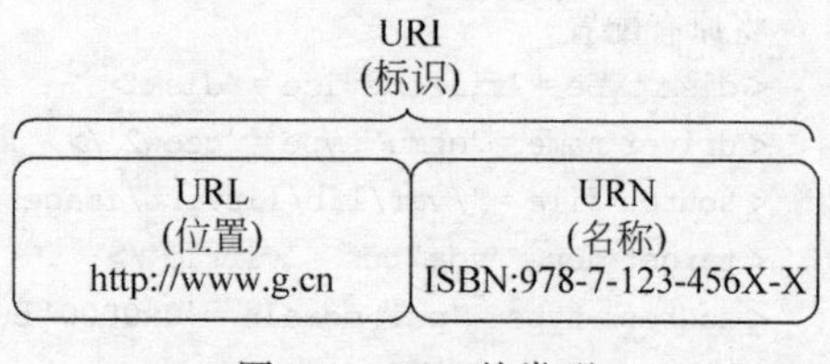


图1-3 URI的类型

人们经常会混淆URI与URL,其关键区别如下:

(1) URL是URI的子集,用于指定资源存在的位置及检索资源的机制,而URI是用于标识资源的URL的超集。

(2) URL的主要目的是获取资源的位置或地址,而URI的主要目的是查找资源。

### 1.3.2 通过virsh连接远程宿主机

默认情况下virsh命令用于连接到本机,可以通过-c选项或connect子命令连接到远程宿主机,这就需要使用URI了。

由于libvirt支持许多类型的虚拟化,所以它的URI采用以下格式:

```
driver[+transport]://[username@][hostname][:port]/path[?extraparameters]
```

其中,KVM虚拟化的驱动程序(driver)就是qemu。libvirt支持的传输模式(transport)有以下几种。

(1) TLS:默认的模式,需要有证书的支持。

(2) SSH:最简单的方式。

(3) UNIX Socket:本地访问。

(4) TCP:非加密的TCP/IP套接字,不推荐使用。

**提示**:libvirt的URI的更详细描述可参见https://libvirt.org/uri.html。

下面,通过在宿主机kvm1(192.168.1.231)上远程访问宿主机kvm2(192.168.1.232)做一个练习,示例命令如下:

```
[root@kvm1 ~]# virsh
Welcome to virsh, the virtualization interactive terminal.

Type:   'help' for help with commands
```

```
        'quit' to quit

virsh # help hostname
  NAME
    hostname - print the hypervisor hostname

  SYNOPSIS
    hostname

virsh # hostname
kvm1

virsh # help uri
  NAME
uri - print the hypervisor canonical URI

  SYNOPSIS
uri

virsh # uri
qemu:///system
```

通过 hostname 子命令获得当前宿主机的主机名。通过 uri 子命令获得所连接 hypervisor 的 URI，qemu:///system 表示 virsh 目录连接到本机。

接下来执行的示例命令如下：

```
virsh # help connect
  NAME
    connect - (re)connect to hypervisor

  SYNOPSIS
    connect [--name <string>] [--readonly]

  DESCRIPTION
    Connect to local hypervisor. This is built-in command after shell start up.

  OPTIONS
    --name <string>   hypervisor connection URI
    --readonly        read-only connection

virsh # connect "qemu+ssh://root@192.168.1.232/system"
The authenticity of host '192.168.1.232 (192.168.1.232)' can't be established.
ECDSA key fingerprint is SHA256:eEf777j0Z+7k2Egd9tvqcsUUW8WhSjWSBb/Ijg4wTPE.
Are you sure you want to continue connecting (yes/no/[fingerprint])? 输入 yes
root@192.168.1.232's password: 输入密码
```

通过 connect 子命令连接到 qemu+ssh://root@192.168.1.232/system 所指定的资源。在这个 URI 中，qemu 表示 KVM 虚拟化，ssh 表示传输模式，root@192.168.1.232 表示远程主机 IP 及用户名，system 表示资源的标识。

在第 1 次通过 SSH 连接远程主机时，需要确认传递过来的 SSH 服务器的公共密钥，然后输入远程宿主机的 root 密码。

最后执行的示例命令如下：

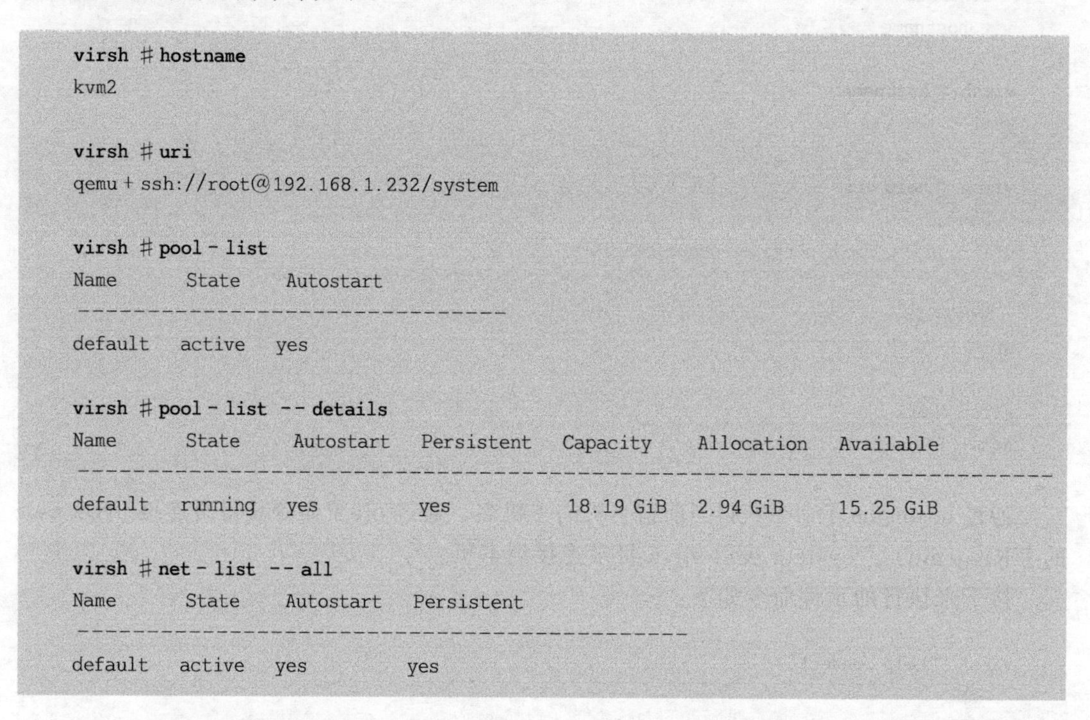

```
virsh # hostname
kvm2

virsh # uri
qemu+ssh://root@192.168.1.232/system

virsh # pool-list
Name       State     Autostart
-----------------------------------
default    active    yes

virsh # pool-list --details
Name       State     Autostart   Persistent   Capacity    Allocation   Available
-------------------------------------------------------------------------------
default    running   yes         yes          18.19 GiB   2.94 GiB     15.25 GiB

virsh # net-list --all
Name       State     Autostart   Persistent
---------------------------------------------
default    active    yes         yes
```

连接成功后，就可以像管理本地宿主机一样对远程宿主机 kvm2 进行管理操作了。

### 1.3.3 通过 virt-manager 连接远程宿主机

默认情况下 virt-manager 仅连接到本机，可以通过添加新连接的方式将远程宿主机添加到管理界面中。

(1) 打开 File 菜单，然后选择 Add Connection 菜单项，此时会出现 Add Connection 窗口，如图 1-4 所示。

(2) 从下拉式 Hypervisor 选择框中选择 QEMU/KVM。本示例使用 SSH 连接到远程宿主机，输入用户名，设置远程宿主机名称或 IP 地址。单击 Connect 按钮进行连接。

(3) 在提示窗口中输入远程宿主机的 root 密码。

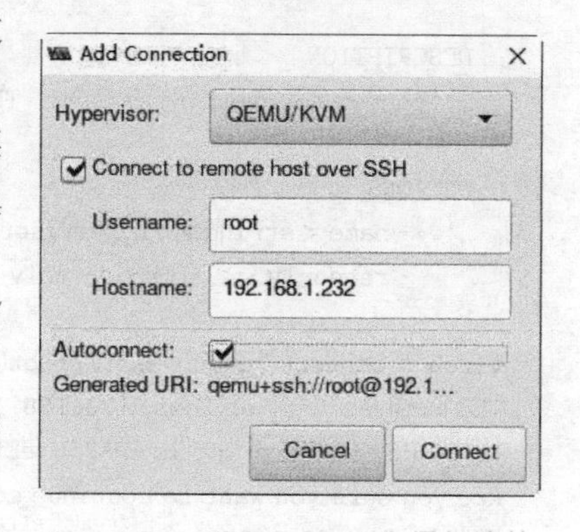


图 1-4 在 virt-manager 中添加连接

(4) 成功连接远程主机后,它会出现在 virt-manager 主窗口中,如图 1-5 所示。

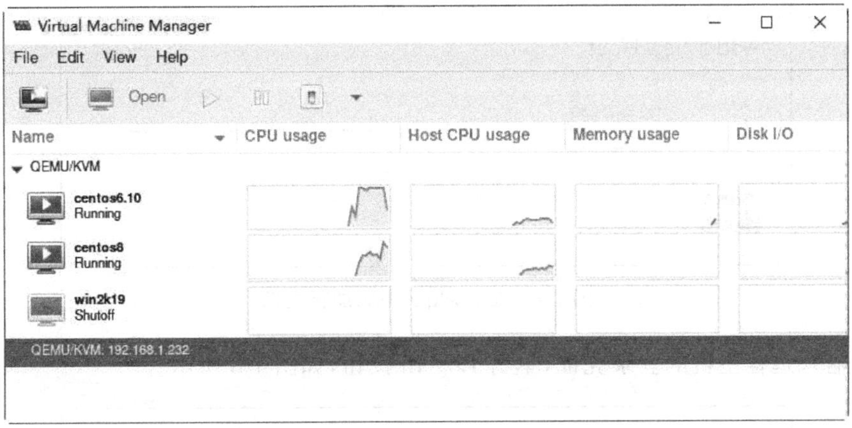

图 1-5　virt-manager 主窗口中的远程主机

**提示:** 在连接远程宿主机时,如果遇到类似"Unable to connect to libvirt…Configure SSH key access for the remote host, or install an SSH askpass package locally."的错误提示,则可以通过 dnf -y install openssh-askpass 来安装 ssh-askpass 软件包解决此问题。

## 1.3.4　通过 Cockpit 连接远程宿主机

虽然当前版本的 Cockpit(cockpit-machines-224.2-1)也可以管理多台宿主机,但是还不能通过它来迁移虚拟机。

(1) 单击左上角的主机名,打开主机列表,单击 Add new host 按钮,如图 1-6 所示。

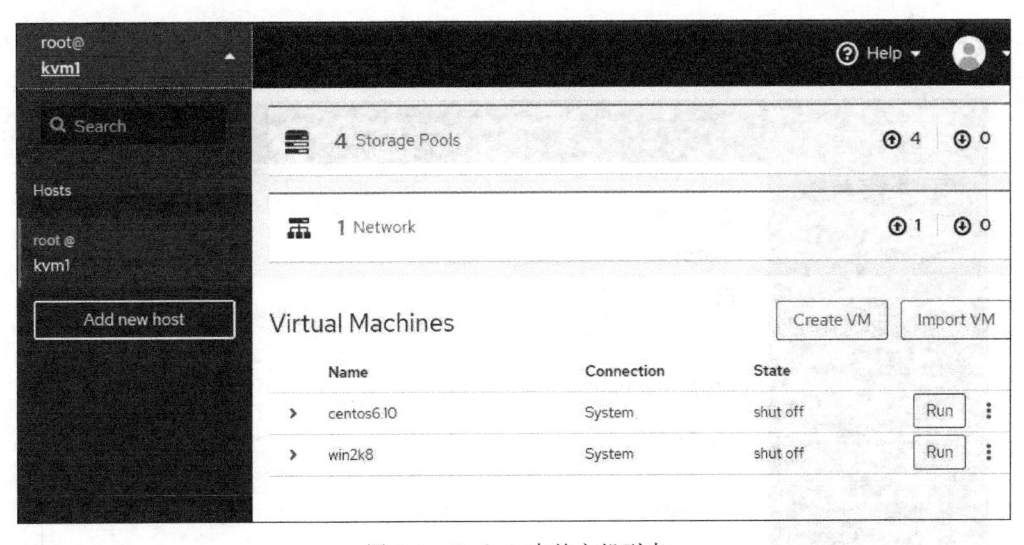


图 1-6　Cockpit 中的主机列表

(2) 输入远程主机的主机名或 IP 地址，单击 Add 按钮，如图 1-7 所示。

Add a New Host ×
Address 192.168.1.232
Color
Add Cancel

图 1-7 在 Cockpit 中添加新主机

(3) 输入远程主机的登录凭证，单击 Log In 按钮，如图 1-8 所示。

Log in to 192.168.1.232 ×
Cockpit was unable to log into **192.168.1.232**. You can change your authentication credentials below. You may prefer to synchronize accounts and passwords.
User name root
Authentication Type a password
Password ······
Log In Cancel

图 1-8 输入登录远程主机的凭证

(4) 登录成功后，远程主机的信息就会出现在主机列表中，如图 1-9 所示。

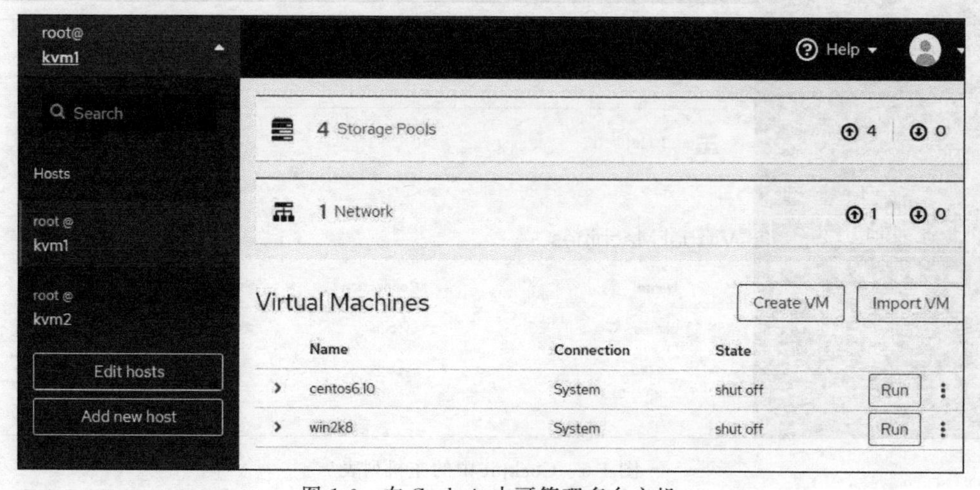


图 1-9 在 Cockpit 中可管理多台主机

# 1.4　基于共享存储的迁移

推荐将虚拟机的磁盘资源保存在共享存储中，如 iSCSI、NFS、GlusterFS 等存储池。在迁移时，磁盘资源不会发生变动，而仅需要处理虚拟机的 XML 文件和内存数据(如果虚拟机处于运行状态)。可以通过 virsh 和 virt-manager 进行基于共享存储的迁移。

下面的实验将使用两台虚拟化宿主机、1 台 NFS 存储主机，详细信息如表 1-2 所示。

表 1-2　实验环境中宿主机的信息

| 主　机　名 | 角　　色 | IP 地址 | 操作系统版本 |
| --- | --- | --- | --- |
| kvm1 | 虚拟机宿主机 | 192.168.1.231 | CentOS 8.3.2011 |
| kvm2 | 虚拟机宿主机 | 192.168.1.232 | CentOS 8.3.2011 |
| stor1 | NFS 存储主机 | 192.168.1.233 | CentOS 8.3.2011 |

## 1.4.1　准备工作

进行虚拟机迁移之前，需要对存储、虚拟化宿主机做一些准备，主要包括：

(1) 存储池准备。

(2) 设置主机名称解析。

(3) 设置 SELinux。

(4) 配置防火墙策略。

### 1. 存储池准备

在 NFS 存储主机中配置共享目录时，需要注意参数的配置。示例命令如下：

```
[root@stor1 ~]# cat /etc/exports
/vmdata *(rw,no_root_squash,sync)
```

在 nfs-utils 1.0.0 之后版本，sync 已经是默认设置。如果采用的是专用 NFS 存储设备，则要设置与"rw，no_root_squash，sync"相同的效果参数。

源宿主机与目标宿主机的存储池名、目标路径要尽可能一致。本实验中存储池名称为 nfs1，目录路径是/vmdata，执行的示例命令如下：

```
[root@kvm1 ~]# virsh pool-dumpxml nfs1
<pool type='netfs'>
<name>nfs1</name>
<uuid>7da63af0-d46b-4802-adfa-7d2bf90638f7</uuid>
<capacity unit='Bytes'>19529728000</capacity>
<allocation unit='Bytes'>3944218624</allocation>
<available unit='Bytes'>15585509376</available>
```

```
<source>
<host name='192.168.1.233'/>
<dir path='/vmdata'/>
<format type='auto'/>
</source>
<target>
<path>/nfs1</path>
<permissions>
<mode>0777</mode>
<owner>0</owner>
<group>0</group>
<label>system_u:object_r:nfs_t:s0</label>
</permissions>
</target>
</pool>
```

**2. 设置主机名称解析**

要保证在所有虚拟化宿主机上都可以将主机名解析成 IP 地址。本实验使用 hosts 文件进行解析，在每台宿主机中的/etc/hosts 文件中添加相应的记录，示例命令如下：

```
[AllNodes~]# cat /etc/hosts
127.0.0.1    localhost localhost.localdomain localhost4
192.168.1.231 kvm1
192.168.1.232 kvm2
```

**注意**：下面凡是有[AllNodes ～]提示符的操作，都需要在所有主机上执行一次。

**3. 设置 SELinux**

在所有虚拟化宿主机上设置 SELinux 的布尔值，允许 libvirt 访问 NFS 资源。示例命令如下：

```
[AllNodes~]# setsebool virt_use_nfs on

[AllNodes~]# getsebool -a | grep virt_use
virt_use_comm --> off
virt_use_execmem --> off
virt_use_fusefs --> off
virt_use_glusterd --> off
virt_use_nfs --> on
virt_use_pcscd --> off
virt_use_rawip --> off
virt_use_samba --> off
virt_use_sanlock --> off
virt_use_usb --> on
virt_use_xserver --> off
```

**4. 配置防火墙策略**

在所有虚拟化宿主机上设置防火墙，允许 libvirt 及 libvirt-tls 协议数据包的入站，允许动态迁移时使用的 TCP 49152～49215 动态端口的入站，示例命令如下：

```
[AllNodes~]#firewall-cmd --add-service={libvirt,libvirt-tls} --permanent

[AllNodes~]#firewall-cmd --add-port=49152-49215/tcp --permanent

[AllNodes~]#firewall-cmd --reload

[AllNodes~]#firewall-cmd --list-all
public (active)
  target: default
icmp-block-inversion: no
  interfaces: ens32
  sources:
  services: cockpit dhcpv6-client libvirt libvirt-tlsrdpsshvnc-server
  ports: 5900-5950/tcp 49152-49215/tcp
  protocols:
  masquerade: no
  forward-ports:
  source-ports:
icmp-blocks:
  rich rules
```

## 1.4.2 使用 virsh 进行实时迁移

可以使用 virsh 的 migrate 子命令进行虚拟机迁移，至少需要提供两个参数：

(1) [--domain]<domain>：虚拟机的名称、ID 或 UUID。

(2) [--desturi]<desturi>：从客户端(常规迁移)或源宿主机(p2p 迁移)看到的目标宿主机的连接 URI。

如果未指定迁移模式，则可使用默认的--live(实时迁移)。如果虚拟机未运行，则需要显式地指定--offline 选项以便实现离线迁移。示例命令如下：

```
1 [root@kvm1 ~]#virsh migrate centos6.10 qemu+ssh://kvm2/system
  root@kvm2's password: *************
  error: Requested operation is not valid: domain is not running

2 [root@kvm1 ~]#virsh start centos6.10
  Domain centos6.10 started

3 [root@kvm1 ~]#ls -lt /etc/libvirt/qemu
  total 24
```

```
  -rw-------. 1 root root 4638 Jan  6 14:38 centos6.10.xml
  -rw-------. 1 root root 4638 Nov 24 17:05 win2k3.xml
  -rw-------. 1 root root 5603 Nov 20 15:55 win2k19.xml
  drwx------. 3 root root   42 Nov  4 09:06 networks

4 [root@kvm1 ~]# virsh list --all
  Id   Name         State
  ------------------------------
  2    centos6.10   running
  -    win2k19      shut off
  -    win2k3       shut off
```

第 1 行命令出错的原因是虚拟机 centos6.10 未运行，而默认的迁移模式是实时迁移。

第 2 行命令启动此虚拟机。这台虚拟机是永久性虚拟机，即在/etc/libvirt/qemu 目录中有相应的 XML 文件。

接下来执行的示例命令如下：

```
5 [root@kvm2 ~]# virsh list --all
  Id   Name   State
  ---------------------
6 [root@kvm2 ~]# ls -lt /etc/libvirt/qemu
  total 0
drwx------. 3 root root 42 Nov  4 09:06 networks
```

在迁移之前先通过第 5 行、第 6 行命令考察目标宿主机 kvm2 的情况：无虚拟机、无 XML 文件。

接下来执行的示例命令如下：

```
7 [root@kvm1 ~]# virsh migrate centos6.10 qemu+ssh://kvm2/system
  root@kvm2's password: *************

8 [root@kvm1 ~]# virsh list --all
  Id    Name         State
  ------------------------------
  -     centos6.10   shut off
  -     win2k19      shut off
  -     win2k3       shut off

9 [root@kvm1 ~]# ls -lt /etc/libvirt/qemu
  total 24
  -rw-------. 1 root root 4638 Jan  6 14:38 centos6.10.xml
  -rw-------. 1 root root 4638 Nov 24 17:05 win2k3.xml
  -rw-------. 1 root root 5603 Nov 20 15:55 win2k19.xml
drwx------. 3 root root   42 Nov  4 09:06 networks
```

第 7 行命令进行实时迁移,输入目标系统的 root 密码。实时迁移所需要的时间取决于网络带宽、系统负载和虚拟机的大小等。迁移成功之后,默认没有任何提示信息,可以通过--verbose 选项来显示迁移的进度。

第 8 行命令的输出显示:在源宿主机上,虚拟机 centos6.10 处于停止状态。

第 9 行命令的输出显示:在源宿主机上,还存在此虚拟机的 XML 文件。

默认情况下,将虚拟机实时迁移到目标宿主机之后,在源宿主机上还保留着虚拟机的配置信息,并处于关闭状态。

接下来执行的示例命令如下:

```
10 [root@kvm2 ~]# virsh list --all
Id   Name        State
------------------------------
1    centos6.10   running

11 [root@kvm2 ~]# ls -lt /etc/libvirt/qemu
  total 0
drwx------. 3 root root 42 Nov  4 09:06 networks
```

在目标宿主机上执行第 10 行命令,会看到虚拟机迁移成功。

默认情况下并不会在目标宿主机上创建 XML 配置文件,所以此虚拟机是临时性的。

**提示**:由于实例迁移会迁移虚拟机内存中的数据,所以虚拟机中的进程不受影响,在目标宿主机上的虚拟机中会继续运行。

最后执行的示例命令如下:

```
12 [root@kvm2 ~]# virsh shutdown centos6.10
  Domain centos6.10 is being shutdown

13 [root@kvm2 ~]# virsh list --all
   Id   Name   State
  ----------------------

14 [root@kvm2 ~]# virsh start centos6.10
  error: failed to get domain 'centos6.10'
```

第 12 行命令用于关闭虚拟机。由于它是无配置文件的临时性虚拟机,所以即使通过--all 选项也无法再看到这台虚拟机了。当然,第 14 行命令尝试启动此虚拟机时会报错。

如果希望迁移到目标宿主机之后的虚拟机是持久性的,就需要使用--persistent 选项,示例命令如下:

```
[root@kvm1 ~]# virsh migrate --persistent centos6.10 qemu+ssh://kvm2/system
```

如果希望迁移成功之后,在源宿主机中取消虚拟机的定义,就需要使用--undefinesource 选项,示例命令如下:

```
[root@kvm1 ~]# virsh migrate --persistent --undefinesource centos6.10 qemu+ssh://kvm2/
system
```

如果希望迁移到目标宿主机后使用新的配置,例如使用不同的存储池,则可以手工创建一个新的配置文件,然后在迁移虚拟机时使用新配置文件中的设置,示例命令如下:

```
[root@kvm1 ~]# virsh dumpxml --migratable centos6.10 > centos6.10v2.xml

[root@kvm1 ~]# vi centos6.10v2.xml
根据需要进行修改
[root@kvm1 ~]# virsh migrate --live --persistent --xml centos6.10v2.xml centos6.10
qemu+ssh://kvm2/system
```

如果在迁移过程中发生错误,则需要清理目标宿主机上的资源,并且继续在源宿主机上运行虚拟机。如果想取消正在进行的迁移操作,则可执行示例命令如下:

```
[root@kvm1 ~]# virsh domjobabort centos6.10
```

如果虚拟机正在运行内存密集型工作负载,则可能需要使用如下选项来确保迁移操作的完成。

(1) --auto-converge:在实时迁移过程中执行强制收敛。

(2) --timeout <number>:如果实时迁移超时(以秒为单位),则执行下面的--timeout-*选项所指定的操作(默认为暂停虚拟机)。

(3) --timeout-suspend:超时后暂停虚拟机。

(4) --timeout-postcopy:超时后执行 post-copy 操作。

可以设置和显示在迁移操作中所使用的网络带宽(单位为 MiB/s),它们的语法格式如下:

```
# virsh migrate-setspeed <domain> <bandwidth>

# virsh migrate-getspeed <domain>
```

### 1.4.3 使用 virsh 进行离线迁移

可以使用--offline 选项来迁移未运行的虚拟机,示例命令如下:

```
[root@kvm1 ~]# virsh list --all
Id      Name            State
--------------------------------
-       centos6.10      shut off
-       centos8         shut off
-       win2k19         shut off
-       win2k3          shut off
-       win2k8          shut off

[root@kvm1 ~]# virsh migrate --offline --persistent --undefinesourcecentos6.10 qemu+
ssh://kvm2/system
```

### 1.4.4　使用 virt-manager 进行实时迁移

在 virt-manager 界面中，右击要迁移的虚拟机，然后单击 Migrate 选项，如图 1-10 所示。

在 New host 字段中，从下拉列表中选择目标宿主机，其他选项采用默认值，然后单击 Migrate 按钮，如图 1-11 所示。

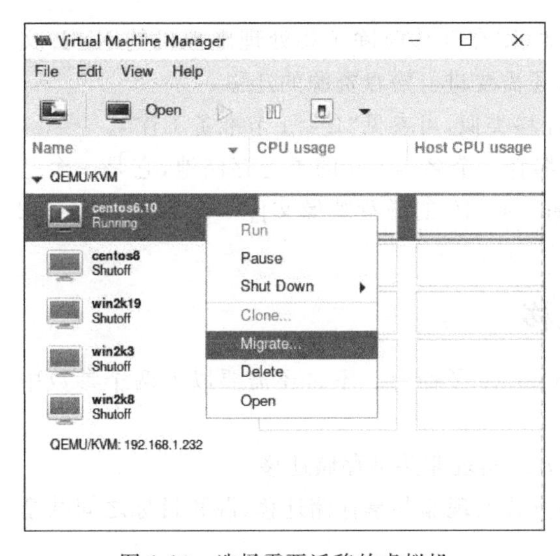


图 1-10　选择需要迁移的虚拟机

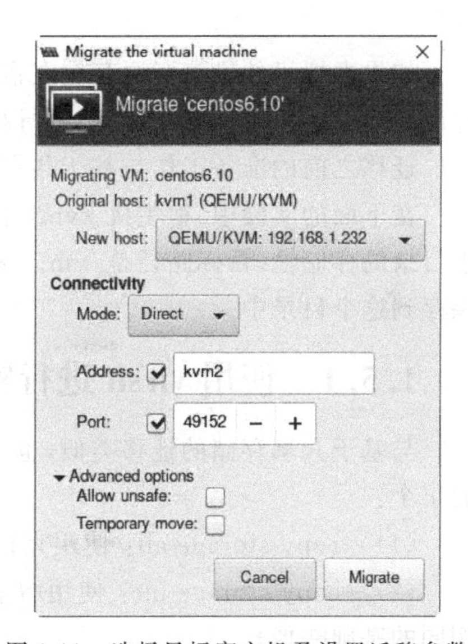


图 1-11　选择目标宿主机及设置迁移参数

这时会出现一个迁移进度窗口。如果迁移顺利完成，则 virt-manager 会在目标宿主机下面显示刚迁移过来的虚拟机，如图 1-12 所示。

**提示**：virt-manager 不支持离线迁移。

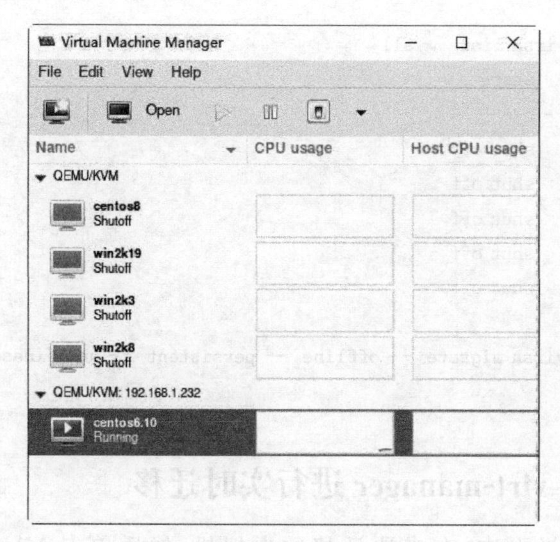


图 1-12 在目标宿主机中运行迁移过来的虚拟机

# 1.5 基于非共享存储的迁移

如果虚拟机的磁盘资源存储在宿主机本地，则在迁移时除了要处理虚拟机的 XML 文件和内存数据（如果虚拟机处于运行状态）外，还需要进行磁盘资源的迁移。

迁移之前的准备工作与基于共享存储的迁移类似，可参见"1.4.1 节准备工作"。

在下面的实验中，宿主机 kvm1 和 kvm2 都有一个名为 vm 的本地存储池，它是一个基于目录的存储池，目标路径是/vm。虚拟机 centos6.10 的磁盘映像文件 centos6.10.qcow2 保存到这个目录中。

## 1.5.1 使用 virsh 进行实时迁移

与基于共享存储的迁移类似，也是执行 migrate 子命令。不过还需要以下两个参数中的 1 个：

(1) --copy-storage-all：使用完整复制的方式实现非共享存储迁移。

(2) --copy-storage-inc：使用增量复制的方式实现非共享存储迁移(源和目标之间共享相同的基础映像)。

示例命令如下：

```
1 [root@kvm1 ~]# virsh list
   Id   Name          State
  --------------------------------
   1    centos6.10    running
```

```
2 [root@kvm1 ~]# virsh migrate --live --persistent --undefinesource --copy-storage-all --verbose centos6.10 qemu+ssh://kvm2/system
  root@kvm2's password: *************
  Migration: [ 12 % ]迁移进度,到 100 %结束

3 [root@kvm1 ~]# virsh list --all
   Id     Name        State
  ----------------------------
   -      centos8     shut off
   -      win2k19     shut off
   -      win2k3      shut off
   -      win2k8      shut off

4 [root@kvm1 ~]# ls /vm/centos6.10.qcow2
  /vm/centos6.10.qcow2
```

第 1 行命令的输出显示了虚拟机 centos6.10 正在运行。

第 2 行命令进行实时迁移，采用--copy-storage-all 选项以实现完整磁盘复制，选项--verbose 会显示迁移的进度。

**提示**：非共享存储的迁移使用的是 NBD 方式。NBD 是 Network Block Device 的缩写，它可以让用户通过网络访问块设备、设备镜像，其工作方式也遵循 C/S 架构。qemu 使用 NBD 进行存储迁移，通过流将磁盘数据发送到本地文件或远程主机。

由于迁移时使用了--undefinesource 选项，所以在第 3 行命令的输出中就不会看到虚拟机 centos6.10 的信息了。

第 4 行命令的输出显示：在源宿主机上，还存在此虚拟机的磁盘映像文件。在确认成功迁移之后，可以手工删除这些磁盘映像文件以释放存储空间。

接下来执行的示例命令如下：

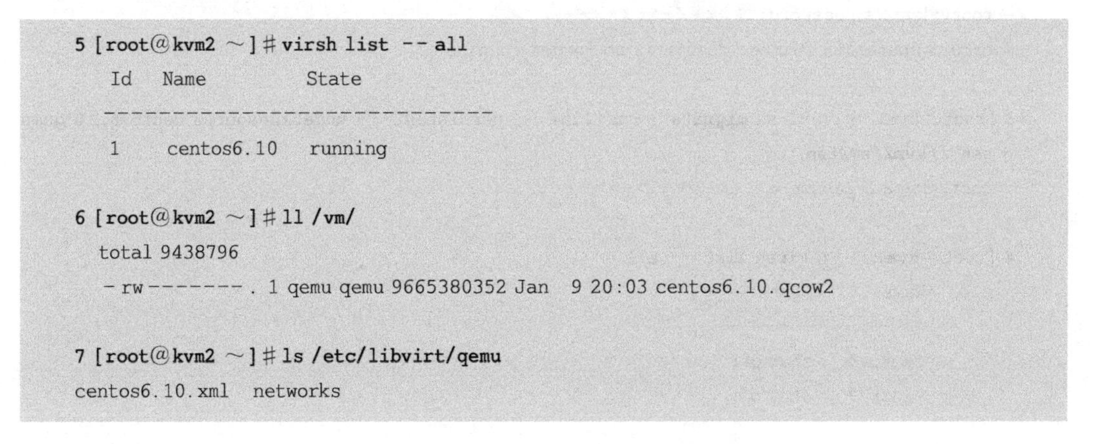

```
5 [root@kvm2 ~]# virsh list --all
   Id   Name           State
  ------------------------------
   1    centos6.10     running

6 [root@kvm2 ~]# ll /vm/
  total 9438796
  -rw-------. 1 qemu qemu 9665380352 Jan  9 20:03 centos6.10.qcow2

7 [root@kvm2 ~]# ls /etc/libvirt/qemu
centos6.10.xml   networks
```

在目标宿主机 kvm2 上执行第 5 行命令，会看到虚拟机迁移成功。

从第 6 行命令的输出中可以看出：在目标宿主机中有复制过来的新磁盘映像文件。

由于迁移时使用了--persistent 选项，所以在第 7 行命令的输出中会看到虚拟机 centos6.10 的 XML 配置文件。

需要说明的是：并非所有的 Linux 发行版本和 qemu-kvm 版本都支持--copy-storage-all 选项。本实验使用的版本信息如下：

```
[root@kvm1 ~]# cat /etc/redhat-release
CentOS Linux release 8.3.2011

[root@kvm1 ~]# /usr/libexec/qemu-kvm -version
QEMU emulator version 4.2.0 (qemu-kvm-4.2.0-34.module_el8.3.0+555+a55c8938)
Copyright (c) 2003-2019 Fabrice Bellard and the QEMU Project developers
```

在 CentOS 7 及更早的版本中，要想使用--copy-storage-all 选项，还需要安装 Red Hat Virtualization 中 qemu-kvm 增强版本，安装方法如下：

```
# yum install CentOS-release-qemu-ev
```

### 1.5.2 使用 virsh 进行离线迁移

对于保存在非共享存储的虚拟机进行离线迁移，仅会传递虚拟机的配置，而不会复制磁盘映像。示例命令如下：

```
1 [root@kvm1 ~]# virsh domstate centos6.10
  shut off

2 [root@kvm1 ~]# virsh migrate --offline --persistent --undefinesource --copy-storage-all --verbose centos6.10 qemu+ssh://kvm2/system
  root@kvm2's password: *************
  error: operation failed: domain is no longer running

3 [root@kvm1 ~]# virsh migrate --offline --persistent --undefinesource centos6.10 qemu+ssh://kvm2/system
  root@kvm2's password: *************

4 [root@kvm1 ~]# virsh list --all
   Id   Name      State
  --------------------------
   -    centos8   shut off
   -    win2k19   shut off
```

由于--copy-storage-all 选项仅支持处于运行状态的虚拟机，所以第 2 行命令会出错。

第 3 行命令不使用--copy-storage-all 选项，会迁移成功而且迁移的速度特别快。

由于迁移时使用了--undefinesource 选项，所以在第 4 命令的输出中已经看不到虚拟机 centos6.10 的信息了。

接下来执行的示例命令如下：

```
5 [root@kvm2 ~]# virsh list --all
  Id   Name          State
  ------------------------------
  -    centos6.10    shut off

6 [root@kvm2 ~]# virsh start centos6.10
  error: Failed to start domain centos6.10
  error: Storage volume not found: no storage vol with matching path '/vm/centos6.10.qcow2'

7 [root@kvm2 ~]# ls /vm/ -lh
  total 0
```

在目标宿主机上执行第 5 行命令，会看到虚拟机的信息，虚拟机处于关闭状态。

第 6 行命令在启动虚拟机时出错，提示没有找到存储卷。第 7 行命令的输出显示：虚拟机的存储卷并没有被复制过来。

最后执行的示例命令如下：

```
8 [root@kvm2 ~]# scp kvm1:/vm/centos6.10.qcow2 /vm

9 [root@kvm2 ~]# virsh start centos6.10
  Domain centos6.10 started
```

可以使用第 8 行命令将磁盘映像复制到目标宿主机上。

第 9 行命令就可以成功启动虚拟机了。

还有一种进行非共享存储的离线迁移的方法：先在源宿主机上执行 virsh dumpxml 生成配置文件，然后将其复制到目标宿主机上，在目标宿主机上执行 virsh define 生成新虚拟机的定义。对比之下，显然使用 virsh migrate 的方法更高效。

### 1.5.3 使用 virt-manager 进行实时迁移

通过 virt-manager 进行非共享存储的虚拟机的实时迁移，如果使用默认配置，则会出现错误提示 Unable to migrate guest：Unsafe migration：Migration without shared storage is unsafe，如图 1-13 所示。

如果想在 virt-manager 中完成迁移，就必须选中 Allow unsafe 选项，如图 1-14 所示。

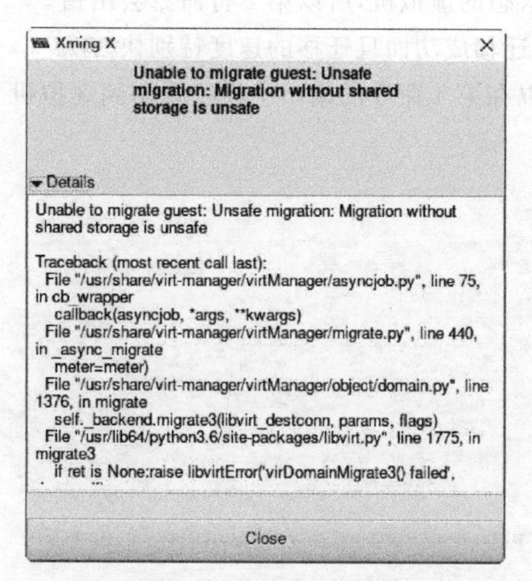


图 1-13 迁移虚拟机时出错

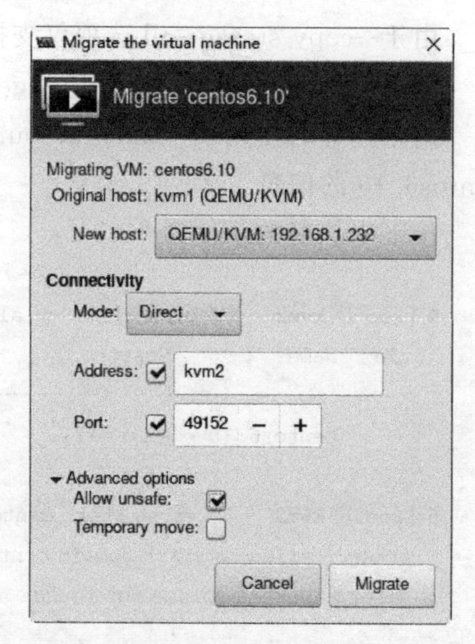


图 1-14 迁移时选中 Allow unsafe 选项

unsafe 是 libvirt 为迁移提供的一个选项。在某些情况下，libvirt 可能会拒绝迁移虚拟机，因为这样做可能会导致潜在的问题，例如数据损坏，因此这种迁移被认为是不安全的。对于 QEMU 虚拟机，如果其使用磁盘时未将缓存模式设置为 none，则可能会发生这种情况。除非将磁盘映像存储在可保证一致性的群集文件系统（例如 GFS2 或 GPFS）中，否则迁移这些虚拟机是不安全的。如果确定这种迁移是安全的或根本不在乎安全性，则可以采用--unsafe 进行强制迁移。

虽然现在 libvirt 已经可以安全地实现非共享存储的实时迁移，但是 virt-manager 软件包的作者并没有进行相应的更新，所以还必须指定 unsafe 模式。这有可能会掩盖迁移时发生的其他错误，所以不推荐使用 virt-manager 进行非共享存储的实时迁移。

## 1.6 本章小结

本章讲解了虚拟机迁移，包括虚拟机迁移的基本原理与分类、应用场景及限制，以及如何实现有共享存储的迁移和无共享存储的迁移。

第 2 章将讲解如何实现虚拟机的高可用。

# 第2章 实现虚拟机高可用

高可用是整个IT基础架构的第一道防线。将多台宿主机构建为高可用群集，当一台宿主机不可用时，虚拟机可自动迁移到其他宿主机上，从而减少业务系统的停机时间。

**本章要点**

- ❑ Linux高可用群集基本原理。
- ❑ Linux高可用群集安装。
- ❑ 基于NFS的KVM群集构建。
- ❑ 基于iSCSI的KVM群集。
- ❑ 基于DRBD的KVM群集构建。

## 2.1 Linux高可用群集基本原理

通过冗余的计算、存储和网络等资源构建高可用群集。高可用性群集软件的主要功能是监视服务运行情况，当检测到服务器中断时，无须手动干预即可进行切换，它是一种成熟的解决方案，提供可靠性和可用性。

### 2.1.1 什么是高可用群集

计算机群集(Computer Cluster)就是一组计算机，它们作为一个整体向用户提供一组计算资源。计算机群集可以充当一台功能更强大的计算机，提供更快的处理速度、更大的存储容量、更好的数据完整性、出众的可靠性和更广泛的资源可用性。有许多类型的计算机群集，包括以下几类。

(1) 高可用性群集：如果正在运行的服务器遇到故障，则可由其他的节点提供服务。

(2) 负载均衡群集：将传入的网络请求分布到各个节点进行处理。

(3) 高性能群集：计算任务分布在多个节点。

高可用群集(High Availability Cluster，HA Cluster)是指以减少服务中断(如因服务器宕机等引起的服务中断)时间为目的服务器群集技术。

高可用群集使群集的整体服务尽可能可用，从而减少由计算机硬件和软件易错性所带来的损失。它通过尽可能地提供不间断的服务，把因软件/硬件/人为造成的故障对业务的影响降到最低程度。

高可用群集中的单个计算机被称为节点(Node)。当某个节点失效时,它的备用节点将在很短的时间内接管它的职责,因此,对于用户而言,群集永远不会停机。高可用群集软件的主要作用就是实现故障检查和业务切换的自动化。

"双机热备"是高可用群集的一种特殊形式,是指由2个节点构建的高可用群集,它们互相备份。当一个节点出现故障时,由另一台节点承担服务任务,不需要人工干预,自动保证系统能持续对外提供服务。

对于关键业务,停机带来的损失是巨大的。计算机系统可用性分类如表2-1所示。

**表2-1 可用性与年停机时间**

| 可 用 性 | 年停机时间 | 可用性分类 |
|---|---|---|
| 99.5 | 3.7d | 常规系统(Conventional) |
| 99.9 | 8.8h | 可用系统(Available) |
| 99.99 | 52.6min | 高可用系统(Highly Available) |
| 99.999 | 5.3min | 故障弹性(Fault Resilient) |
| 99.9999 | 32s | 容错(Fault Tolerant) |

## 2.1.2 开源高可用群集技术选型

开源软件的世界百花齐放,群集是IT系统的底层架构,所以有很多开源的解决方案。据不完全统计,大约有70多种,有高可用的,有负荷平衡的,有通用型的,也有针对特定产品的,如数据库、中间件、Web服务等,如图2-1所示。

HAproxy:Web 负载均衡解决方案
Codis:Redis 集群解决方案
LVS:Linux 虚拟服务器
MySQL Cluster:MySQL 集群
DRBD:文件同步系统
LCMC:Linux 集群管理控制台
Hadoopy:Python 的 Hadoop 扩展
Seesaw:负载均衡系统
ADSG-LBaaS-Driver:OpenStack LBaaS 驱动
Apache Aurora:Mesos 框架
Apache Ranger:Hadoop 集群权限框架
Redis-Migrate-Tool:Redis 集群迁移工具
Nomad:集群管理器和调度器
Pound:反向 HTTP 代理/负载均衡器和 SSL 封装器
Lazy balancer:基于 Nginx 的负载均衡管理系统
Swiftiply:网络集群化代理服务器
Photon Controller:集群管理系统
heartbeat:Linux 集群系统
Keepalived:服务器状态监测
Gearman:任务调度程序
Terracotta:开源群集框架
Galera Load Balancer:负载均衡器
Pen:负载均衡器
OpenAIS:集群框架的应用程序接口规范
Apache Mesos:集群管理器
Pacemaker:集群管理器
Coherence:集群计算部件
synctool:集群配置同步工具
DRBD :管理控制台
Mesosphere Marathon:
mod_backhand:Apache 负载均衡模块
Apache Ambari:Hadoop 管理监控工具
Corosync:Linux 高可用集群
openMosix:Linux 的群集技术
KTCPVS:TCP 负载均衡器
Zen Load Balancer:
JSession:java web 集群软件
Apache Helix:集群管理框架
Perlbal:
norbert:CS 模式的集群的 JAVA API
G5:通讯转发、(负载均衡)通讯分发器
Crossroads:负载均衡器
mod_cluster:
OpenSSI:Linux 集群解决方案
OAQL:负载均衡服务器
SessionContainer:java web 集群软件
TrueCL:集群解决方案
Riverdrums:负载均衡器
ogslb:DNS 轮询系统
SuperMon:高速集群监控系统
GreatTurbo Load Balance Server:
HA-Tools:
Plasma:Map/Redure 框架
Cloudbreak:基于 Hadoop 的 Docker service API
G6:负载均衡器 G5 的第二版
nginx-lua-ds-loadbalancer:
ldirectord:集群服务器管理
izBalancing:负载均衡脚本
pymesos:Mesos 调度器和执行器
GCMT:Gentoo 集群管理工具
Traefik:HTTP 反向代理、负载均衡软件
High Availability Toolkit:高可用技术工具包
JBoss Clustering:
L3DSR:负载均衡服务器
ocelli:交互负载均衡程序
Redis Cluster Unofficial:认证模式下的自动化工具
crmsh:高可用集群管理工具
Minuteman:负载均衡器

图2-1 开源软件中的群集项目列表

高可用群集的核心功能是节点之间的消息传递和资源控制。在众多的开源项目中，Corosync+Pacemaker 的组合是一种流行的解决方案。Corosync 项目实现了节点之间的消息传递，Pacemaker 项目实现对资源的控制。

Corosync 与 Pacemaker 是既独立发展又相互促进的两个项目，它们的发展历程如图 2-2 所示。

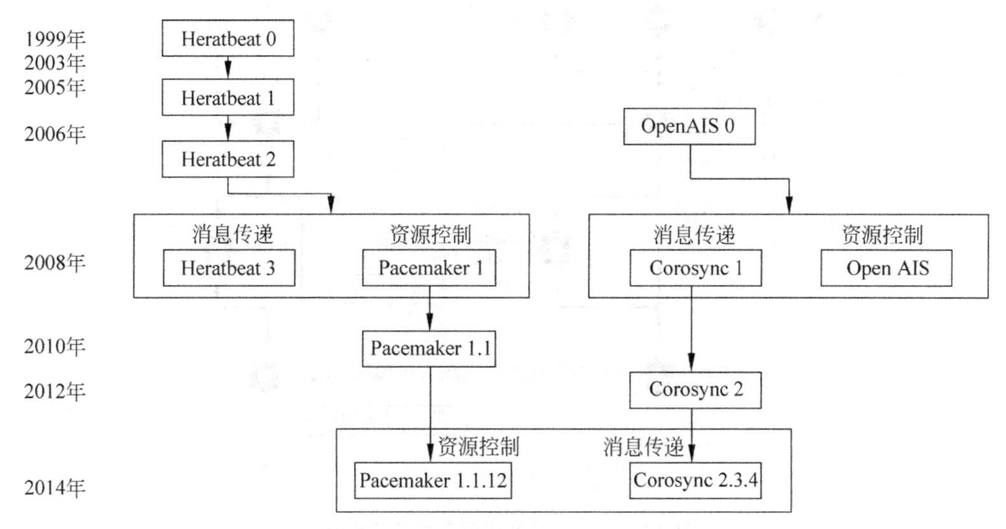


图 2-2 Corosync 与 Pacemaker 的发展历程

**说明**：图 2-2 改编自 http://openstandia.jp/oss_info/corosync/。

Corosync 来源于 OpenAIS(开放式应用接口规范)项目，当前最新版本是 3.10，CentOS 8.3 的软件仓库中的版本是 3.0.3。Pacemaker 来源于 Heartbeat 项目，当前最新版本是 2.0.5，CentOS 8.3 的软件仓库中的版本是 2.0.4。

除了这两个核心组合之外，针对具体的应用还可能用到以下组件。

(1) CLVM：Clustered LVM，群集化的 LVM 是 LVM 的一个群集方面的扩展。允许群集中的计算机通过 LVM 管理共享存储。

(2) DRBD：Distributed Replicated Block Device，是软件实现的、无共享的、服务器之间镜像块设备内容的存储复制解决方案，类似于 RAID1。

(3) GFS2：Global File System 2，群集文件系统使用群集中所有节点并发地通过标准文件系统接口访问存储设备。

(4) OCFS：Oracle Cluster File System，Oracle 公司发起的与 GFS2 类似的群集文件系统。

**提示**：在 RHEL 中 Corosync+Pacemaker 的组合被称为高可用插件。

## 2.1.3 Corosync+Pacemaker 体系结构

可以将通过 Corosync+Pacemaker 组合构建的高可用群集划分为 4 个层次，如图 2-3 所示。

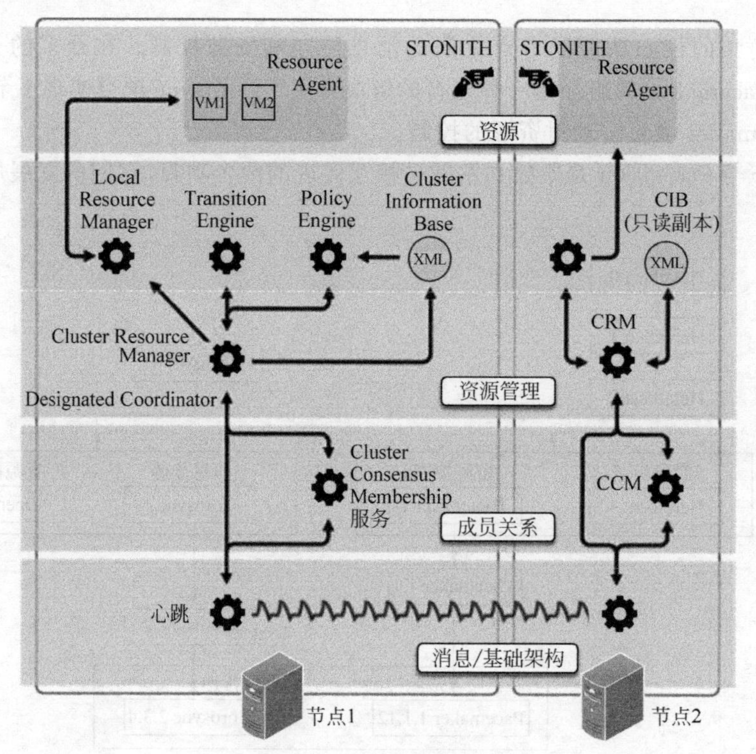


图 2-3 Linux 高可用群集的体系结构

### 1. 消息/基础架构层

这一层主要实现节点之间心跳信息传递。群集中每个节点都不停地将自己的状态信息通告给其他节点,这种信息被称为心跳信息。可以通过串口或网络在节点之间传递心跳信息。如果通过网络传递,则建议使用单独的网络来传递心跳信息,可以采用广播、组播、单播等方式。

### 2. 成员关系层

每个节点运行着一个负责维护群集成员关系的服务,叫作 Cluster Consensus Membership 服务(CCM 服务)。

CCM 服务起着承上启下的作用。它监听底层所收到的心跳信息,当监听不到心跳信息的时候就重新计算整个群集的票数和收敛状态信息,并将结果传递给上层,由上层决定所要采取的措施。CCM 服务还能够生成一个各节点状态的拓扑结构图。

### 3. 资源管理层

这是真正实现群集服务的层,包含以下组件。

(1) CRM(Cluster Resource Manager):群集资源管理器是核心组件,负责资源的分配和管理。所有节点会选举出一个节点的 CRM 成为 DC(Designated Coordinator),也称为主节点,图 2-3 中 DC 是节点 1。DC 的工作是决策和管理群集中的所有资源。如果主节点宕机,则群集中其他的节点会选举出新的 DC。

(2) CIB(Cluster Information Base)：群集信息基库是 XML 格式的配置文件，工作的时候常驻内存。群集的所有信息都会存储在 CIB 中。只有 DC 才能对 CIB 进行修改，其他节点会从 DC 复制一份只读的 CIB。

(3) PE(Policy Engine)：策略引擎负责定义资源转移的方式，它与 DC 是同一个节点。

(4) TE(Transition Engine)：实施引擎根据 PE 做出的策略执行资源迁移，它与 DC 是同一个节点。

(5) LRM(Local Resource Manager)：本地资源管理器接收来自 CRM 的指令，负责本地资源的启动、停止和监控。

**4. 资源层**

在群集中构成一个完整服务的每部分都被称为资源，都需要进行配置和管理。以 KVM 虚拟化为例：虚拟机进程是资源，磁盘映像是资源。以 Web 应用为例：VIP(虚拟 IP 地址)是资源，HTTP 服务器是资源，存储也是资源。不同服务的资源也不尽相同，其中存储资源的选择、配置、管理是高可用群集中的重点和难点。

群集资源代理(Resource Agent)是所管理的资源的启动、停止和获取状态信息的脚本或应用程序。任何资源代理都要提供相同的接口：接收 start、stop、restart、status 等 4 个参数，执行结束后提供符合要求的输出。

有一个特殊的群集资源叫作 STONITH，它是 Shoot the Other Node in the Head(爆另一个节点的头)的缩写。它可以根据要求强行关闭另外一个节点并将其从群集中删除，这样做的目的是为了确保数据完整性。STONITH 是在 CIB 中配置的，可以作为常规群集资源进行监视。STONITH 是群集隔离技术中的一种，可参见“2.1.4 节隔离技术概述”。

群集中不同层次的组件在一起协调工作。群集上的所有信息都会保存到 DC 的 CIB 中，然后同步到其他节点。PE 根据 CIB 获取资源的配置信息，然后做出决策，一旦做出决策就交由 TE 进行资源的管理。PE 借助于本地的 CCM 通知其他节点 CIB 实现对资源信息的传递，例如：通知其他 CRM 要启动某一资源了，CRM 收到信息后并不负责启动，转由 LRM 启动，LRM 又借助 RA 实现资源管理。

在这个架构中，Corosync 完成群集节点之间的心跳通信，并且管理仲裁。其他功能主要由 Pacemaker 实现。

## 2.1.4　隔离技术概述

Pacemaker 使用隔离技术(Fencing)来确保群集中数据的完整性。当群集中某节点失效而无法执行任何操作时，隔离技术可以切断其与共享存储之间的 I/O，从而确保数据的完整性。在失效节点被隔离之前，Pacemaker 不会启动该节点的资源和恢复服务。

如果没有进行隔离，则不能保证共享存储资源上数据的完整性。假设在一个由节点 A、B 和 C 组成的三节点群集中，没有配置任何隔离设备。节点 A 挂载共享存储的一个 ext4 文件系统，并且正在运行数据库服务，该服务器使用该文件系统中的数据文件和日志文件提供服务。如果节点 A 失效或停止响应，则会触发以下一系列事件：

（1）节点B快速执行文件系统的检查后，从共享存储中挂载文件系统。

（2）节点B启动数据库服务。

（3）节点A恢复之后，会继续写入已经被节点B挂载的同一ext4文件系统。

（4）因为存储不支持并发写入，所以有可能造成文件系统的损坏。

如果有隔离技术，则可以防止恢复后节点A访问文件系统，从而避免文件系统的损坏。其基本流程如下：

（1）节点B和节点C从存储中切断节点A。

（2）节点B快速执行文件系统的检查后，从共享存储中挂载文件系统。

（3）节点B启动数据库服务。

（4）节点A恢复之后，还会尝试写入已安装的文件系统。由于节点A无法再访问共享存储资源，因此失败。

（5）当节点A重新启动并重新加入了群集后，根据群集管理的策略来决定可以挂载共享存储上的哪个ext4文件系统。

隔离程序从群集配置文件中确定要使用哪种隔离技术。群集配置文件中的两个关键元素定义了隔离的方法：隔离代理和隔离设备。隔离程序调用群集配置文件中指定的隔离代理。隔离代理又通过隔离设备隔离该节点。隔离完成后，隔离程序将通知群集管理器。

Pacemaker支持多种隔离技术，分为以下几种。

（1）电源隔离：使用智能电源控制器关闭失效节点的电源，如图2-4所示。节点A的电源被关闭。

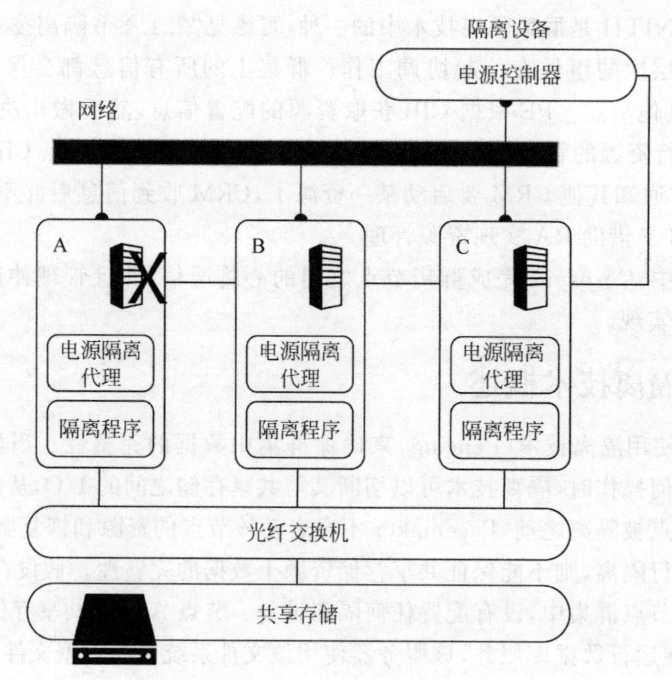


图2-4 电源隔离示例

(2) 存储隔离：禁用失效节点与存储相连的光纤通道端口，如图 2-5 所示。节点 A 的端口被禁用，从而与存储断开连接。

(3) 其他隔离技术：除了传统的带外管理技术如 Dell iDRAC、HP ILO、IPMI 等之外，现在虚拟化产品、云计算服务也提供隔离技术。

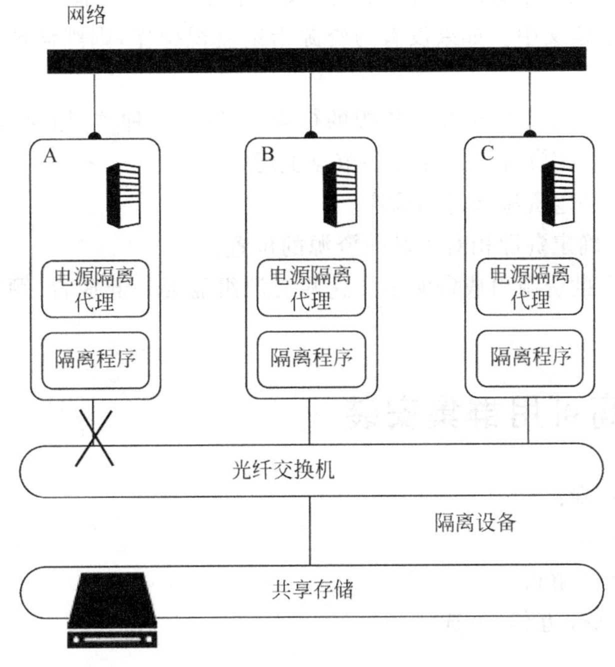


图 2-5　存储隔离示例

可以同时使用多种隔离方法来配置节点。如果某个节点发生故障，则可使用第一种隔离技术对其进行隔离。如果不成功，则可依次使用下一个隔离方法。如果最后仍没有一种隔离方法成功，则将从第 1 个隔离方法重新开始，按顺序循环直到对节点隔离成功为止。

### 2.1.5　法定人数概述

为了维护群集的完整性和可用性，群集系统使用称为法定人数（Quorum）的概念来防止数据损坏和丢失。

当群集节点的一半以上处于联机状态时，即达到法定人数，群集可以正常地工作。例如在 6 节点群集中，当至少 4 个群集节点正常运行时，即达到法定人数。如果大多数节点脱机或不可用，则群集将不再具有法定人数，并且 Pacemaker 将停止群集服务。

法定人数是使用投票系统建立的。当一个群集节点无法正常运行或无法与其余节点通信时，多数正常节点可以投票隔离故障节点。

如果节点之间通信故障，但每个节点还可以独立工作，被称为脑裂（split-brain）现象。

一旦出现了脑裂现象，节点就有可能写入相同的数据，从而导致数据损坏或丢失。Pacemaker中的法定人数功能是为了防止出现这种现象。

### 2.1.6 资源概述

群集资源是由群集服务管理的程序、数据或应用程序。为确保资源保持健康，可以将监视操作添加到资源的定义中。如果没有为资源指定监视操作，则默认情况下将添加一个监视操作。

可以通过配置约束来确定群集中资源的行为，有以下3种类型的约束。

(1) 位置约束：确定资源可以在哪些节点上运行。

(2) 次序约束：确定资源运行的顺序。

(3) 相对约束：确定资源相对于其他资源的位置。

还可以将群集中的资源组成资源组。这些资源组需要一起放置、顺序启动及以相反的顺序停止。

## 2.2 Linux高可用群集安装

安装包括以下步骤：

(1) 群集组件安装。

(2) 配置主机名及解析。

(3) 配置SSH Key互信(可选)。

(4) 配置时钟。

(5) 配置防火墙。

(6) 配置pcs守护程序。

(7) 配置hacluster账号及密码。

实验环境网络及IP地址规划如表2-2所示。

表2-2 IP地址规划

| 主 机 | LAN | Corosync | Storage | 带外管理 |
|---|---|---|---|---|
| kvm1 | 192.168.1.231 | 172.16.1.231 | 10.0.1.231 | 10.0.2.231 |
| kvm2 | 192.168.1.232 | 172.16.1.232 | 10.0.1.232 | 10.0.2.232 |
| stor1 | 192.168.1.235 | | 10.0.1.235 | 10.0.2.235 |

**注意**：下面凡是有[AllNodes ～]提示的操作，都需要在所有节点上执行一次。

### 2.2.1 群集组件安装

将所有的节点上的CentOS 8更新到最新可用版本，示例命令如下：

```
[AllNodes ~]# dnf -y update

[AllNodes ~]# reboot

[AllNodes ~]# cat /etc/redhat-release
CentOS Linux release 8.3.2011

[AllNodes ~]# uname -a
Linux kvm2 4.18.0-240.1.1.el8_3.x86_64 #1 SMP Thu Nov 19 17:20:08 UTC 2020 x86_64 x86_64
x86_64 GNU/Linux
```

在所有的节点上安装 EPEL 仓库，示例命令如下：

```
[AllNodes ~]# dnf -y install epel-release
```

默认情况下，CentOS 8 禁用了 HA 的软件仓库，所以先进行启用操作，示例命令如下：

```
[root@kvm1 ~]# dnf repolist --all | grep ha
ha      CentOS Linux 8 - HighAvailability          disabled

[AllNodes ~]# dnf config-manager --set-enabled ha

[root@kvm1 ~]# dnf repolist
repo id         repo name
appstream       CentOS Linux 8 - AppStream
baseos          CentOS Linux 8 - BaseOS
epel            Extra Packages for Enterprise Linux 8 - x86_64
epel-modular    Extra Packages for Enterprise Linux Modular 8 - x86_64
extras          CentOS Linux 8 - Extras
ha              CentOS Linux 8 - HighAvailability
```

除了使用 dnf 命令来启用之外，还可以修改配置文件/etc/yum.repos.d/CentOS-Linux-HighAvailability.repo 中的 enabled 属性实现。

使用 dnf 命令安装 pacemaker、pcs 和 fence-agents-all 这 3 个 RPM 包，dnf 会根据依赖关系自动安装 corosync 等额外组件，示例命令如下：

```
[AllNodes ~]# dnf -y install pcs pacemaker fence-agents-all
```

### 2.2.2　配置主机名及解析

高可用性群集的所有节点必须能够使用主机名称或 FQDN 名称相互通信，因此可以配置 DNS 服务器或/etc/hosts 进行名称解析。

生产环境中建议使用 DNS，本实验中使用 hosts 文件。

修改所有节点上的/etc/hosts文件，示例命令如下：

```
[AllNodes ~]# vi /etc/hosts
#添加以下内容:
192.168.1.231 kvm1
192.168.1.232 kvm2
172.16.1.231 kvm1-cr
172.16.1.232 kvm2-cr
10.0.1.231 kvm1-stor
10.0.1.232 kvm2-stor
```

建议为不同网络的IP地址设置不同的主机名称，这样清晰明了、不易出错。

配置之后要验证是否可以通过主机名称进行访问，示例命令如下：

```
[AllNodes ~]# ping -c 2 kvm1

[AllNodes ~]# ping -c 2 kvm2

[AllNodes ~]# ping -c 2 kvm1-cr

[AllNodes ~]# ping -c 2 kvm2-cr

[AllNodes ~]# ping -c 2 kvm1-stor

[AllNodes ~]# ping -c 2 kvm2-stor
```

### 2.2.3 配置SSH Key互信

配置SSH Key互信是一个可选的操作，通过配置SSH Key可以免去每次使用SSH访问都需要输入密码的麻烦。示例命令如下：

```
[AllNodes ~]# ssh-keygen -t rsa -P ''
Generating public/private rsa key pair.
Enter file in which to save the key (/root/.ssh/id_rsa):
Your identification has been saved in /root/.ssh/id_rsa.
Your public key has been saved in /root/.ssh/id_rsa.pub.
The key fingerprint is:
SHA256:RoUnU03hIg4XnJW2bUruGKgSe0BfkLDv7qyyf2uACNc allnodes
The key's randomart image is:
+---[RSA 3072]----+
|   .. . ..=++o.  |
|   .o   *o+..    |
|   .. .. +=.o.   |
|. o.E . = .o.o   |
```

```
|o+ ... .So o     |
|o +.. ... o      |
|    =..    +     |
|. oo=      . .   |
|. + oB= .        |
+----[SHA256]-----+

[AllNodes ~]# ls -l -t ~/.ssh/id_rsa*
-rw-r--r--. 1 root root  563 Jan 12 22:20 /root/.ssh/id_rsa.pub
-rw-------. 1 root root 2590 Jan 12 22:20 /root/.ssh/id_rsa
```

在每个节点上通过 ssh-keygen 命令创建一对 RSA 类型的密钥对，默认的保存位置为~/.ssh/，id_rsa.pub 是公共密钥，id_rsa 是私有密钥。

接下来执行的示例命令如下：

```
[root@kvm1 ~]# ssh-copy-id -i ~/.ssh/id_rsa.pub root@kvm2
/usr/bin/ssh-copy-id: INFO: Source of key(s) to be installed: "/root/.ssh/id_rsa.pub"
/usr/bin/ssh-copy-id: INFO: attempting to log in with the new key(s), to filter out any that
are already installed
/usr/bin/ssh-copy-id: INFO: 1 key(s) remain to be installed -- if you are prompted now it is
to install the new keys
root@kvm2's password: *********

Number of key(s) added: 1

Now try logging into the machine, with:   "ssh 'root@kvm2'"
and check to make sure that only the key(s) you wanted were added.
```

在 kvm1 上使用 ssh-copy-id 命令将公共密钥复制到 kvm2 中，这会被添加到文件/root/.ssh/authorized_keys 的尾部。

推荐也将这个公共密钥添加到本机的/root/.ssh/authorized_keys 文件的尾部，示例命令如下：

```
[root@kvm1 ~]# ssh-copy-id -i ~/.ssh/id_rsa.pub root@kvm1
```

类似地，在 kvm2 上也执行类似的操作，示例命令如下：

```
[root@kvm2 ~]# ssh-copy-id -i ~/.ssh/id_rsa.pub root@kvm1

[root@kvm2 ~]# ssh-copy-id -i ~/.ssh/id_rsa.pub root@kvm2
```

查看节点上信任的公共密钥，示例命令如下：

```
[root@kvm1 ~]# cat ~/.ssh/authorized_keys
ssh-rsa
AAAAB3NzaC1yc2EAAAADAQABAAABgQC5m9tKfNdqIQi5iISidr2MGHmPiAEpj2coRE9P0CYVazIpx8z4pE06qJJXhe
5THFddulFCKrkgmSHEd + 21m9BKnOW + zH7az5PR27d9Y6Cdfp2EPqMEASEvWYPDFht/zrjjvgtaTwhTFytRdD
MWSfwZQFE57ni3DwFh/zNw6S7CB2qOkMKAonSSOueYIH0UjGEGbmvbKTdOKckD/W4tdRci/uplWRmPHylvcPyP85TL6
hsCQ8pD5f75lz8EXaOpZEOwkceyxqOH9qJ3rpv70aAu9EriFZJ3U0 + 2mQk5c8rUpAzn7PTiXnFuMWdENd8kNU74U8
rMZ66DRvU8KjlWOyF2gq/0ckJ + LJhDtdMFsPTgA + vqVxQUVtlYV9bMwdfqog51deF2aMSOwb2E8HuL9YEbSu5c
OxwqBEHXT04GDu4aA0StvKGXCzbe3LaIBlJP0ho3NUBi4l4un3nl0gYf7A/WHRtxxy/zblRYb3X7kmVvsaWBP/ZsNo
chbvquX4mVkvs= root@kvm2
ssh-rsa
AAAAB3NzaC1yc2EAAAADAQABAAABgQDQboE8cf + fky3c93J6IYmDgxZiotSITD + foskG6/8ZZkFAvFNwxHYerd0M0bapj/
BY/qDcum2BuaTfyeArihRQXyaLbVhi6OeoEkIsO4s5Rty7clNIkGz2KfQDHkdWiVYVBOgKmNK7SMrOtOk/ICP6Z8bhp
TYNsarH587SdPSmqAsQVPQF582xgPtjSKgpwDXrdm0RbR/1/oNDmv6MaGbkPJGww2gUR4ABYtAmjzFdbPuRh1MdVyr
H4XFcoZykEHHlLi2mHXkoXC46AykMJRJlUzRni0pHSDSPg7kPzB8yyMahyRa4ZtGtxYaRzPidN/01i3FImf6RtIsm4
MKWf2AOL18 + YS877ETF + lhDM43/cyycXZSS9MSOEOxia5efUsH7urzrw1jz2Zaqs/UEP + z + OscQPFK0
ffGExzjx39QCK4qMOqL7pGOxW8eRTd + eoUAxvFj3Ufk36d7i4Bjc1CoOD0LdD2mRS1Zf2119ZdNFOPu3NkOmQ/
Oj6Kge8YlAAec= root@kvm1
```

最终的结果是在每个节点的 authorized_keys 文件中都包含所有节点的 SSH 公共密钥。这样就使拥有对应私有密钥的用户可免密码登录。可以通过 ssh 命令登录远程主机并执行命令进行测试,示例命令如下:

```
[AllNodes ~]# ssh kvm1 "hostname"
kvm1

[AllNodes ~]# ssh kvm2 "hostname"
kvm2
```

### 2.2.4 配置时钟

在高可用性群集中,必须为所有节点设置正确的时钟。根据具体情况,可以通过内网中的时钟服务器来校时,也可以通过互联网中的时钟服务器来校时。在本实验中,将通过互联网中的时钟服务器来校时。

Linux 中常见的时钟服务器守护程序有 2 种类型:chronyd 和 ntpd。应该选择并仅使用一个,以避免同时运行两个 NTP 守护程序导致的不兼容和不稳定。

CentOS 8 的官方软件仓库已经删除了 ntp 和 ntpdate,推荐使用 chrony,它有以下优势:

(1) 高速和准确的时间同步。

(2) 如果无法访问主时钟,则可以保证操作正确(ntpd 需要定期请求)。

(3) 更少的资源消耗。

默认情况下,CentOS 8 会自动安装 chrony。如果没有,则可以通过以下命令来安装:

```
[AllNodes ~]# dnf -y install chrony
```

启动 chronyd 服务并设置为自动启动，示例命令如下：

```
[AllNodes ~]# systemctl start chronyd

[AllNodes ~]# systemctl enable chronyd
```

检查 chronyd 服务的状态是否正常，示例命令如下：

```
[AllNodes ~]# systemctl status chronyd
● chronyd.service - NTP client/server
   Loaded: loaded (/usr/lib/systemd/system/chronyd.service; enabled; vendor preset: enabled)
   Active: active (running) since Wed 2021-01-13 06:53:49 CST; 1h 9min ago
     Docs: man:chronyd(8)
           man:chrony.conf(5)
Main PID: 2330 (chronyd)
    Tasks: 1 (limit: 100937)
   Memory: 2.6M
   CGroup: /system.slice/chronyd.service
   └─2330 /usr/sbin/chronyd

Jan 13 06:53:49 kvm1 systemd[1]: Started NTP client/server.
Jan 13 06:53:57 kvm1 chronyd[2330]: Selected source 202.118.1.130
Jan 13 06:53:57 kvm1 chronyd[2330]: System clock TAI offset set to 37 seconds
Jan 13 06:53:57 kvm1 chronyd[2330]: Source 162.159.200.123 replaced with 78.46.102.180
Jan 13 07:37:30 kvm1 chronyd[2330]: Forward time jump detected!
Jan 13 07:37:30 kvm1 chronyd[2330]: Can't synchronise: no selectable sources
Jan 13 07:40:11 kvm1 chronyd[2330]: Selected source 78.46.102.180
Jan 13 07:40:11 kvm1 chronyd[2330]: System clock wrong by -1.077252 seconds, adjustment started
Jan 13 07:41:51 kvm1 chronyd[2330]: Source 193.182.111.142 replaced with 144.76.76.107
Jan 13 07:42:21 kvm1 chronyd[2330]: Selected source 202.118.1.130
```

通过以下命令检查时间同步是否有效：

```
[AllNodes ~]# timedatectl status
               Local time: Wed 2021-01-13 08:03:26 CST
           Universal time: Wed 2021-01-13 00:03:26 UTC
                 RTC time: Wed 2021-01-13 00:03:26
                Time zone: Asia/Shanghai (CST, +0800)
System clock synchronized: yes
              NTP service: active
          RTC in local TZ: no
```

输出显示 System clock synchronized：yes，NTP service：active，表示工作正常。

chrony 有一个命令行工具 chronyc，可以用于查看时间同步的选项，示例命令如下：

```
[AllNodes ~]# chronyc tracking
Reference ID    : CA760182 (202.118.1.130)
Stratum         : 2
Ref time (UTC)  : Wed Jan 13 00:05:33 2021
System time     : 0.000227154 seconds slow of NTP time
Last offset     : -0.000235339 seconds
RMS offset      : 0.118781812 seconds
Frequency       : 24.653 ppm fast
Residual freq   : -0.021 ppm
Skew            : 2.552 ppm
Root delay      : 0.034480564 seconds
Root dispersion : 0.000546861 seconds
Update interval : 128.5 seconds
Leap status     : Normal
```

输出中的 Reference ID 是当前使用的一台互联网时钟服务器。

查看同步服务器的信息，示例命令如下：

```
[AllNodes ~]# chronyc sources -v
210 Number of sources = 4

  .-- Source mode  '^' = server, '=' = peer, '#' = local clock.
 / .- Source state '*' = current synced, '+' = combined , '-' = not combined,
| /   '?' = unreachable, 'x' = time may be in error, '~' = time too variable.
||                                                 .- xxxx [ yyyy ] +/- zzzz
||      Reachability register (octal) -.           |  xxxx = adjusted offset,
||      Log2(Polling interval) --.      |          |  yyyy = measured offset,
||                                \     |          |  zzzz = estimated error.
||                                 |    |           \
MS Name/IP address         Stratum Poll Reach LastRx Last sample
===============================================================================
^- electrode.felixc.at           3   7   377    32   +2139us[+2139us] +/-  123ms
^- sv1.ggsrv.de                  2   7   377     1   -3334us[-3334us] +/-  114ms
^- de-user.deepinid.deepin.>     3   7   377    68   -5325us[-5035us] +/-  101ms
^* 202.118.1.130                 1   7   177    68   +1928us[+2219us] +/-   19ms
```

如果要使用自定义的时钟器（例如内网的时钟服务器），则可以编辑配置文件/etc/chrony.conf，在其中指定用于同步的 NTP 服务器列表。示例命令如下：

```
[AllNodes ~]# vi /etc/chrony.conf
#在最后添加，例如：
```

```
server 0.CentOS.pool.ntp.org iburst
server 1.CentOS.pool.ntp.org iburst
server 2.CentOS.pool.ntp.org iburst
server 3.CentOS.pool.ntp.org iburst

[AllNodes~]# systemctl restart chronyd
```

**提示**：如果要手动设置日期和时间，则可以继续使用 date 命令，但是使用之前，需要停止或禁用 chronyd 守护程序。

### 2.2.5 配置防火墙

Corosync＋Pacemaker 组合使用的协议比较多，RHEL/CentOS 将其归集为名为 high-availability 的服务。使用以下命令查看这个服务的配置：

```
[root@kvm1 ~]# firewall-cmd --info-service=high-availability
high-availability
  ports: 2224/tcp 3121/tcp 5403/tcp 5404/udp 5405-5412/udp 9929/tcp 9929/udp 21064/tcp
  protocols:
  source-ports:
  modules:
  destination:
  includes:
  helpers:
```

配置防火墙允许使用群集服务，示例命令如下：

```
[AllNodes ~]# firewall-cmd --permanent --add-service=high-availability

[AllNodes ~]# firewall-cmd --reload
```

如果将心跳及存储这两个专用网络认为是可信任的网络，则可以将心跳及存储网络配置为全通过，这样可以降低网络延迟、减少资源消耗，示例命令如下：

```
[AllNodes ~]# firewall-cmd --permanent --zone=trusted \
 --add-source=172.16.1.0/24

[AllNodes ~]# firewall-cmd --permanent --zone=trusted \
 --add-source=10.0.1.0/24

[AllNodes ~]# firewall-cmd --reload
```

允许与虚拟机动态迁移有关的协议，示例命令如下：

```
[AllNodes ~]#firewall-cmd --permanent --add-service={libvirt,libvirt-tls}

[AllNodes ~]#firewall-cmd --permanent --add-port=49152-49215/tcp

[AllNodes ~]#firewall-cmd --reload
```

防火墙的最终设置如下：

```
[AllNodes ~]#firewall-cmd --list-all
public (active)
  target: default
  icmp-block-inversion: no
  interfaces: ens32 ens34 ens35
  sources:
  services: cockpit dhcpv6-client high-availability libvirt libvirt-tls rdp ssh vnc-server
  ports: 5900-5950/tcp 49152-49215/tcp
  protocols:
  masquerade: no
  forward-ports:
  source-ports:
  icmp-blocks:
  rich rules:

[AllNodes ~]#firewall-cmd --list-all --zone=trusted
trusted (active)
  target: ACCEPT
  icmp-block-inversion: no
  interfaces:
  sources: 172.16.1.0/24 10.0.1.0/24
  services:
  ports:
  protocols:
  masquerade: no
  forward-ports:
  source-ports:
  icmp-blocks:
  rich rules:
```

### 2.2.6 配置 pcs 守护程序

pcs 守程序 pcsd 负责同步群集中所有节点的 corosync 配置，所以在配置群集之前，需要启动它并设置为自动启动。示例命令如下：

```
[AllNodes ~]#systemctl start pcsd

[AllNodes ~]#systemctl enable pcsd
```

### 2.2.7 配置 hacluster 账号及密码

安装群集软件包时会创建一个名为 hacluster 的账号，它的密码是禁用的，示例命令如下：

```
[root@kvm1 ~]# grep hacluster /etc/passwd
hacluster:x:189:189:cluster user:/home/hacluster:/sbin/nologin

[root@kvm1 ~]# grep hacluster /etc/shadow
hacluster:!!:18638::::::
```

需要在所有节点为其设置相同的密码，示例命令如下：

```
[AllNodes ~]# echo "P@ssw0rd!" | passwd --stdin hacluster
Changing password for user hacluster.
passwd: all authentication tokens updated successfully.
```

## 2.3 群集配置文件与管理工具

有两个用于 Corosync+Pacemaker 群集部署、监视和管理工具：命令行工具 pcs 和网页界面工具 pcsd。

pcs 的常用子命令如表 2-3 所示。

表 2-3　pcs 的常用子命令

| 子　命　令 | 说　　明 |
|---|---|
| cluster | 配置群集选项和节点 |
| resource | 管理群集资源 |
| stonith | 管理隔离设备 |
| constraint | 管理资源约束 |
| property | 管理 pacemaker 属性 |
| acl | 管理 pacemaker 访问控制列表 |
| qdevice | 管理本机仲裁设备所提供的程序 |
| quorum | 管理群集仲裁设置 |
| booth | 管理群集票务管理器 booth |
| status | 查看群集状态 |
| config | 查看和管理群集配置 |
| pcsd | 管理 pcs 守护程序 |
| host | 管理 pcs/pcsd 的主机 |
| node | 管理群集节点 |
| alert | 管理 pacemaker 警报 |
| client | 管理 pcsd 客户端配置 |
| dr | 管理灾难恢复配置 |
| tag | 管理 pacemaker 标签 |

RHEL/CentOS 8 中还有一个 GUI 的群集管理工具，访问方法为 https://IP 地址：2224，如图 2-6 所示。

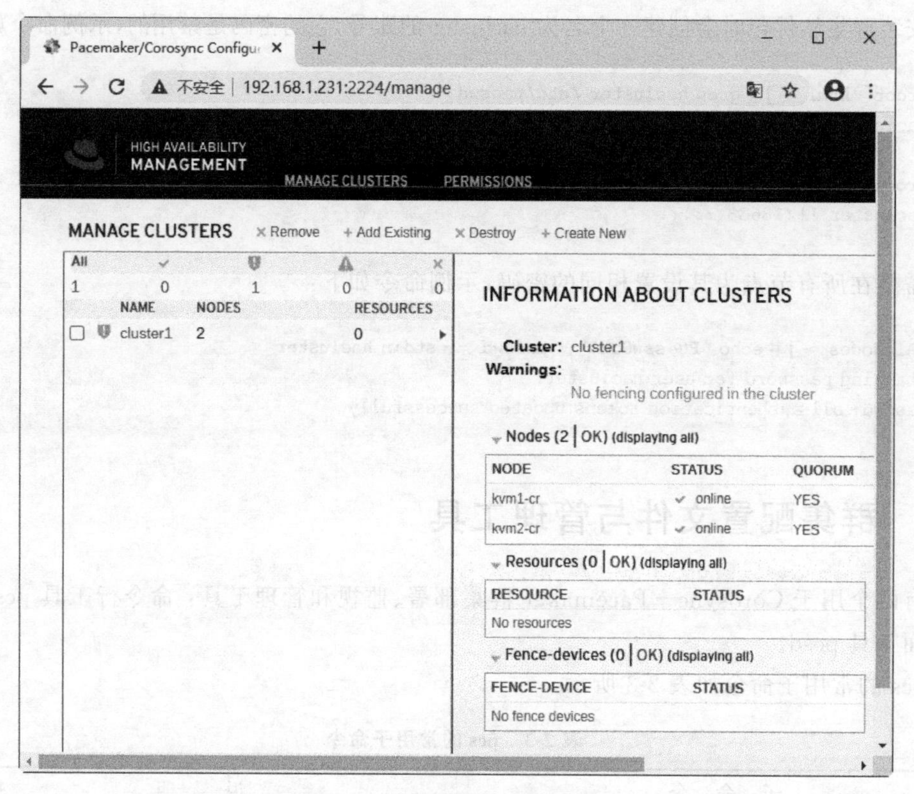


图 2-6 群集管理的 GUI 工具

群集有 2 个重要的配置文件：

(1) corosync.conf：corosync 的选项参数。

(2) cib.xml：Pacemaker 的 Cluster Information Base（CIB），是一个 XML 格式的文件，用于保存群集的配置及群集中所有资源的当前状态。pcsd 守护程序负责在整个群集节点上同步 CIB 的内容。

通常不应直接编辑它们，而应使用 pcs 或 pcsd 进行修改。对于刚安装的新环境，默认为没有这两个配置文件，示例命令如下：

```
[root@kvm1 ~]# tree /etc/corosync/
/etc/corosync/
├── corosync.conf.example
└── uidgid.d

1 directory, 1 file
```

```
[root@kvm1 ~]# tree /var/lib/pacemaker/
/var/lib/pacemaker/
├── blackbox
├── cib
├── cores
└── pengine

4 directories, 0 files
```

# 2.4 创建群集

群集软件安装完毕，就可以按照以下步骤来创建群集了：

(1) 认证组成群集的节点(pcs host auth)。

(2) 配置和同步群集节点(pcs cluster setup)。

(3) 在群集节点中启动群集服务(pcs cluster start)。

(4) 配置隔离设备(pcs stonith create)。

## 2.4.1 认证组成群集的节点

将远程主机上的 pcsd 进程与本地 pcs/ pcsd 进程进行相互认证，此操作仅在一个节点执行即可，示例命令如下：

```
[root@kvm1 ~]# pcs host auth --help
Usage: pcs host auth...
    auth (<host name> [addr=<address>[:<port>]])... [-u <username>]
            [-p <password>]
        Authenticate local pcs/pcsd against pcsd on specified hosts. It is
        possible to specify an address and a port via which pcs/pcsd will
        communicate with each host. If an address is not specified a host name
        will be used. If a port is not specified 2224 will be used.
```

认证完成之后，会在/var/lib/pcsd/目录下创建与身份验证有关的文件。当前目录中没有这些文件，示例命令如下：

```
[root@kvm1 ~]# tree /var/lib/pcsd/
/var/lib/pcsd/
├── pcsd.crt
└── pcsd.key
```

需要通过主机名或 IP 地址指定远程主机，默认使用的是 TCP 2224 端口、hacluster 账号。根据规划，群集节点之间将使用 corosync 专用的链路的主机名，即 kvm1-cr 和 kvm2-cr，示例命令如下：

```
[root@kvm1 ~]# pcs host auth kvm1-cr kvm2-cr
Username: hacluster
Password: ********* 输入 hacluster 的密码
kvm2-cr: Authorized
kvm1-cr: Authorized

[root@kvm1 ~]# tree /var/lib/pcsd/
/var/lib/pcsd/
├── known-hosts
├── pcsd.crt
├── pcsd.key
└── pcs_users.conf

0 directories, 4 files
```

认证完成后会在每个节点生成验证文件 known-hosts 和 pcs_users.conf，其内容如下：

```
[root@kvm1 ~]# cat /var/lib/pcsd/known-hosts
{
  "format_version": 1,
  "data_version": 1,
  "known_hosts": {
    "kvm1-cr": {
      "dest_list": [
        {
          "addr": "kvm1-cr",
          "port": 2224
        }
      ],
      "token": "dd0b0365-4571-4b27-8f3b-84a84ffaf540"
    },
    "kvm2-cr": {
      "dest_list": [
        {
          "addr": "kvm2-cr",
          "port": 2224
        }
      ],
      "token": "66524a78-e9f6-4977-aaaf-509cb2a507dd"
    }
  }
}
```

```
[root@kvm1 ~]# cat /var/lib/pcsd/pcs_users.conf
[
  {
    "username": "hacluster",
    "token": "dd0b0365-4571-4b27-8f3b-84a84ffaf540",
    "creation_date": "2021-01-13 23:42:44 +0800"
  }
]
```

## 2.4.2 配置和同步群集节点

配置群集需要在每个节点上创建配置文件/etc/corosync/corosync.conf，有两种做法。

(1) 手工创建：创建时可参考示例文件/etc/corosync/corosync.conf.example。

(2) 自动生成：通过 pcs setup 命令生成配置文件。

下面我们在一个节点上通过 pcs setup 命令创建群集，并将此节点加入群集，这样就会生成配置文件，示例命令如下：

```
[root@kvm1 ~]# pcs cluster setup cluster1 kvm1-cr kvm2-cr
No addresses specified for host 'kvm1-cr', using 'kvm1-cr'
No addresses specified for host 'kvm2-cr', using 'kvm2-cr'
Destroying cluster on hosts: 'kvm1-cr', 'kvm2-cr'...
kvm2-cr: Successfully destroyed cluster
kvm1-cr: Successfully destroyed cluster
Requesting remove 'pcsd settings' from 'kvm1-cr', 'kvm2-cr'
kvm1-cr: successful removal of the file 'pcsd settings'
kvm2-cr: successful removal of the file 'pcsd settings'
Sending 'corosync authkey', 'pacemaker authkey' to 'kvm1-cr', 'kvm2-cr'
kvm1-cr: successful distribution of the file 'corosync authkey'
kvm1-cr: successful distribution of the file 'pacemaker authkey'
kvm2-cr: successful distribution of the file 'corosync authkey'
kvm2-cr: successful distribution of the file 'pacemaker authkey'
Sending 'corosync.conf' to 'kvm1-cr', 'kvm2-cr'
kvm1-cr: successful distribution of the file 'corosync.conf'
kvm2-cr: successful distribution of the file 'corosync.conf'
Cluster has been successfully set up.
```

创建群集时将群集名字指定为 cluster1，群集中有两个节点：kvm1-cr、kvm2-cr。

查看生成配置文件的配置文件，示例命令如下：

```
[root@kvm1 ~]# ls -lt /etc/corosync/
total 12
-rw-r--r--. 1 root root  482 Jan 14 00:06 corosync.conf
-r--------. 1 root root  256 Jan 14 00:06 authkey
```

```
-rw-r--r--. 1 root root 1917 Jun 16  2020 corosync.conf.example
drwxr-xr-x. 2 root root    6 Jun 16  2020 uidgid.d
```

查看配置文件 corosync.conf 的内容，示例命令如下：

```
[root@kvm1 ~]# cat /etc/corosync/corosync.conf
    totem {                        #图腾，是 totem protocol 项目的代号，开源世界的风格
    version: 2
    cluster_name: cluster1         #设置集群名称
    transport: knet                #传输协议
    crypto_cipher: aes256          #crypto_cipher、crypto_hash 用于相互节点认证
    crypto_hash: sha256
}

nodelist {                         #节点列表
    node {                         #ID 是 1 是的节点
        ring0_addr: kvm1-cr
        name: kvm1-cr
        nodeid: 1
    }

    node {                         #ID 是 2 是的节点
        ring0_addr: kvm2-cr
        name: kvm2-cr
        nodeid: 2
    }
}

quorum {                           #法定人数
    provider: corosync_votequorum
    two_node: 1
}

logging {                          #日志
    to_logfile: yes
    logfile: /var/log/cluster/corosync.log
    to_syslog: yes
    timestamp: on
}
```

### 2.4.3 在群集节点中启动群集服务

启动每个节点上的群集服务，包括 corosync 和 pacemaker 两个服务。

在一个节点使用--all 选项，可以启动所有节点上的群集服务，示例命令如下：

```
[root@kvm1 ~]# pcs cluster start --all
kvm1-cr: Starting Cluster...
kvm2-cr: Starting Cluster...
```

也可以单独在每个节点上独立启动这 2 个服务,示例命令如下:

```
[root@kvm1 ~]# systemctl start corosync.service

[root@kvm1 ~]# systemctl start pacemaker.service

[root@kvm2 ~]# systemctl start corosync.service

[root@kvm2 ~]# systemctl start pacemaker.service
```

查询群集的状态,示例命令如下:

```
[root@kvm1 ~]# pcs status
Cluster name: cluster1

WARNINGS:
No stonith devices and stonith-enabled is not false

Cluster Summary:
  * Stack: corosync
  * Current DC: kvm1-cr (version 2.0.4-6.el8_3.1-2deceaa3ae) - partition with quorum
  * Last updated: Thu Jan 14 00:11:01 2021
  * Last change:  Thu Jan 14 00:10:44 2021 by hacluster via crmd on kvm1-cr
  * 2 nodes configured
  * 0 resource instances configured

Node List:
  * Online: [ kvm1-cr kvm2-cr ]

Full List of Resources:
  * No resources

Daemon Status:
  corosync: active/disabled
  pacemaker: active/disabled
  pcsd: active/enabled
```

如果正处于启动的过程中,则看到信息可能会如下:

```
...
Stack: unknown
```

```
Current DC: NONE
2 nodes and 0 resources configured

Node kvm1 - cr: UNCLEAN (offline)
Node kvm2 - cr: UNCLEAN (offline)
...
```

可以将群集服务设置为自动启动，示例命令如下：

```
[root@kvm1 ~]# pcs cluster enable -- all
kvm1 - cr: Cluster Enabled
kvm2 - cr: Cluster Enabled
```

**提示**：将 corosync、pacemaker 设置为自动启动，这是有争议的。有经验的管理员可能不会将群集服务即 corosync、pacemaker 设置为自动启动。因为如果系统崩溃后又自动重启服务器，这些服务也会自动启动，这样管理员有可能不太容易发现哪里出现了问题。

下面停止节点 2 的群集服务或将其关机，查看群集状态的变化，示例命令如下：

```
[root@kvm1 ~]# pcs cluster stop kvm2 - cr
kvm2 - cr: Stopping Cluster (pacemaker)...
kvm2 - cr: Stopping Cluster (corosync)...
```

或

```
[root@kvm2 ~]# shutdown now - h
```

查看群集的状态，会发现 kvm2-cr 处于 OFFLINE 状态，示例命令如下：

```
[root@kvm1 ~]# pcs status
Cluster name: cluster1

WARNINGS:
No stonith devices and stonith - enabled is not false

Cluster Summary:
  * Stack: corosync
  * Current DC: kvm1 - cr (version 2.0.4 - 6.el8_3.1 - 2deceaa3ae) - partition with quorum
  * Last updated: Thu Jan 14 00:27:08 2021
  * Last change:  Thu Jan 14 00:10:44 2021 by hacluster via crmd on kvm1 - cr
  * 2 nodes configured
  * 0 resource instances configured

Node List:
  * Online: [ kvm1 - cr ]
```

```
  * OFFLINE: [ kvm2-cr ]

Full List of Resources:
  * No resources

Daemon Status:
  corosync: active/enabled
  pacemaker: active/enabled
  pcsd: active/enabled
```

测试完毕之后，将节点 2 的群集服务恢复为正常状态。

### 2.4.4 配置隔离设备

必须为群集中的每个节点配置隔离设备(Fencing Device)，这是解决脑裂现象的必要手段。如果没有配置隔离设备，则会在 pcsstatus 的输出中看到警告信息：No stonith devices and stonith-enabled is not false。

在安装群集组件时会安装了一个名为 fence-agents-all 的软件，它其实是一个“空”包，安装它会自动安装 30 多种隔离设备的代理程序，示例命令如下：

```
[root@kvm1 ~]# rpm -qa | grep fence-agent | sort
fence-agents-all-4.2.1-53.el8_3.1.x86_64
fence-agents-amt-ws-4.2.1-53.el8_3.1.noarch
fence-agents-apc-4.2.1-53.el8_3.1.noarch
...
fence-agents-ipmilan-4.2.1-53.el8_3.1.noarch
...
```

在生产环境中，针对具体的环境可以选择采用不同的隔离设备。下面介绍两种常见的隔离设备的使用。

**提示：** 在纯实验环境，如果条件不允许，则可以不配置隔离设备，然后使用命令 pcs property set stonith-enabled=false 来禁用隔离功能。

1) IPMI

IPMI 设备是最常见的物理隔离设备。IPMI 是智能平台管理界面(Intelligent Platform Management Interface)的缩写，是一种通用的远程管理协议，它有时被称为带外管理(Out-of-Band 或 Lights-Out)。通常由服务器主板上的一个芯片或一组芯片实现。虽然各个硬件厂商使用的名称不同，但其管理方式大同小异。

(1) HP：HPE Integrated Lights Out (iLO)。

(2) Dell：Integrated Dell Remote Access (iDRAC)。

(3) IBM：IBM Integrated Management Module(IMM)。

**提示：** 带外管理通常会提供 KVM 功能，此处的 KVM 是 Keyboard Video Mouse 的缩

写。管理员可以像操作本地计算机一样，通过键盘、鼠标访问和控制远程服务器。

通过以下命令查看 IPMI 代理软件包的信息：

```
[root@kvm1 ~]# rpm -qi fence-agents-ipmilan
Name        : fence-agents-ipmilan
Version     : 4.2.1
Release     : 53.el8_3.1
Architecture: noarch
...
Vendor      : CentOS
URL         : https://github.com/ClusterLabs/fence-agents
Summary     : Fence agents for devices with IPMI interface
Description : Fence agents for devices with IPMI interface.
```

查看 fence-agents-ipmilan 软件包中的文件，可以看出 HP 的 ilo、Dell 的 idrac 和 IBM 的 imm，都是通过 IPMI 实现的，示例命令如下：

```
[root@kvm1 ~]# rpm -ql fence-agents-ipmilan
/usr/sbin/fence_idrac
/usr/sbin/fence_ilo3
/usr/sbin/fence_ilo4
/usr/sbin/fence_ilo5
/usr/sbin/fence_imm
/usr/sbin/fence_ipmilan
/usr/share/man/man8/fence_idrac.8.gz
/usr/share/man/man8/fence_ilo3.8.gz
/usr/share/man/man8/fence_ilo4.8.gz
/usr/share/man/man8/fence_ilo5.8.gz
/usr/share/man/man8/fence_imm.8.gz
/usr/share/man/man8/fence_ipmilan.8.gz
```

IPMI 通常由在 UDP 端口 623 上运行的网络服务实现，并且在服务器上的专用以太网端口上运行（有时标记为“管理”或类似名称），如图 2-7 所示。

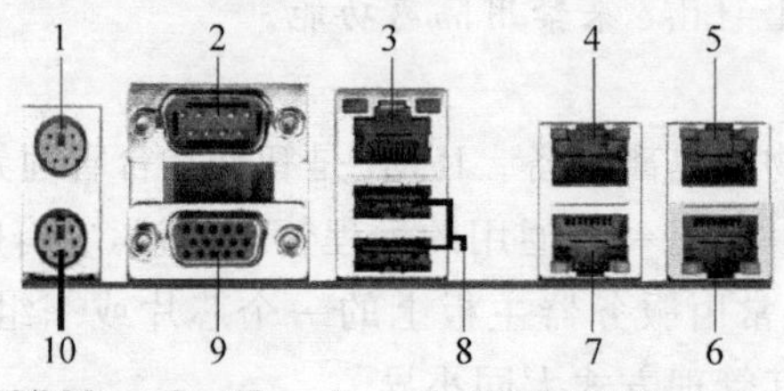


1 PS/2 Mouse Port(Green)
2 Serial Port(COM1)
*3 LAN RJ-45 Port(LAN2)
*4 LAN RJ-45 Port(LAN3)
*5 LAN RJ-45 Port(LAN1)
*6 IPMI LAN Port
*7 LAN RJ-45 Port(LAN4)
8 USB 2.0 Ports(USB01)
9 D-Sub Port(VGA)
10 PS/2 Keyboard Port(Purple)

图 2-7 X86 服务器后面板接口示意图

在所有节点上为 IPMI 配置防火墙，示例命令如下：

```
[AllNodes ~]#firewall-cmd --permanent --add-port=623/udp

[AllNodes ~]#firewall-cmd  --reload
```

配置 IPMI 隔离设备的命令是 fence_ipmilan，可以通过--help 选项查看联机帮助：

```
[root@kvm1 ~]#fence_ipmilan --help
```

测试所有节点的隔离设备是否可以正常工作，示例命令如下：

```
[root@kvm1 ~]#fence_ipmilan -a 10.0.2.231 -P -l admin -p P@ssw0rd -o status -v
2021-01-14 22:10:28,679 INFO: Executing: /usr/bin/ipmitool -I lanplus -H 10.0.2.231 -p
623 -U admin -P [set] -L ADMINISTRATOR chassis power status

0 Chassis Power is on
Status: ON

[root@kvm1 ~]#fence_ipmilan -a 10.0.2.232 -P -l admin -p P@ssw0rd -o status -v
2021-01-14 22:11:01,946 INFO: Executing: /usr/bin/ipmitool -I lanplus -H 10.0.2.232 -p
623 -U admin -P [set] -L ADMINISTRATOR chassis power status

0 Chassis Power is on
Status: ON
```

在节点 1 下达指令以便节点 2 重新启动，示例命令如下：

```
[root@kvm1 ~]#fence_ipmilan -a 10.0.2.232 -P -l admin -p P@ssw0rd -o reboot
Success: Rebooted
```

在节点 2 下达指令以便节点 1 重新启动，示例命令如下：

```
[root@kvm2 ~]#fence_ipmilan -a 10.0.2.231 -P -l admin -p P@ssw0rd -o reboot
Success: Rebooted
```

测试成功后，就可以将隔离设备以资源的方式加入群集了，示例命令如下：

```
[root@kvm1 ~]#pcs stonith create ilo-fence-kvm1-cr fence_ipmilan pcmk_host_list=
"kvm1-cr" ipaddr="10.0.2.231" login="admin" passwd="P@ssw0rd" lanplus=1  power_wait=4
delay=20

[root@kvm1 ~]#pcs stonith create ilo-fence-kvm2-cr fence_ipmilan pcmk_host_list="kvm2
-cr" ipaddr="10.0.2.232"  login="admin" passwd="P@ssw0rd" lanplus=1  power_wait=4
```

**注意**：两节点主机还须在其中一台设置 delay=N 秒，但三节点以上主机不需要设置。

查看群集的 stonith，示例命令如下：

```
[root@kvm1 ~]#pcs stonith show
ilo-fence-kvm1-cr        (stonith:fence_ipmilan):        Started kvm2-cr
ilo-fence-kvm2-cr        (stonith:fence_ipmilan):        Started kvm1-cr
```

查看群集当前的状态，示例命令如下：

```
[root@kvm1 ~]#pcs status
Cluster name: cluster1
Cluster Summary:
  * Stack: corosync
  * Current DC: kvm1-cr (version 2.0.4-6.el8_3.1-2deceaa3ae) - partition with quorum
  * Last updated: Thu Jan 14 22:12:53 2021
  * Last change:  Thu Jan 14 22:12:46 2021 by root via cibadmin on kvm1-cr
  * 2 nodes configured
  * 2 resource instances configured

Node List:
  * Online: [ kvm1-cr kvm2-cr ]

Full List of Resources:
  * ilo-fence-kvm1-cr        (stonith:fence_ipmilan):        Started kvm2-cr
  * ilo-fence-kvm2-cr        (stonith:fence_ipmilan):        Started kvm1-cr

Daemon Status:
  corosync: active/enabled
  pacemaker: active/enabled
  pcsd: active/enabled
```

2）fence-virt

如果群集节点是 KVM 虚拟机，就可以在宿主机上安装 fence-virt 服务器端套件，从而担任隔离设备的角色，如图 2-8 所示。当某个群集节点（虚拟机）出现故障时，为了避免出现脑裂现象，其他的群集节点（也是一台虚拟机）可以向隔离设备（宿主机）发出指令。隔离设备（宿主机）通过验证（通过 key 文件）后，根据指令关闭或重新启动指定的虚拟机。

在下面的实验中，tomkvm1 是 KMV 宿主机，kvm1 和 kvm2 是群集中的两个节点。

（1）在宿主机上安装 Stonith 设备 fence-virt 服务器端套件。

安装软件包的示例命令如下：

```
[root@tomkvm1 ~]#dnf -y install fence-virt fence-virtd fence-virtd-libvirt fence-virtd-multicast fence-virtd-serial
```

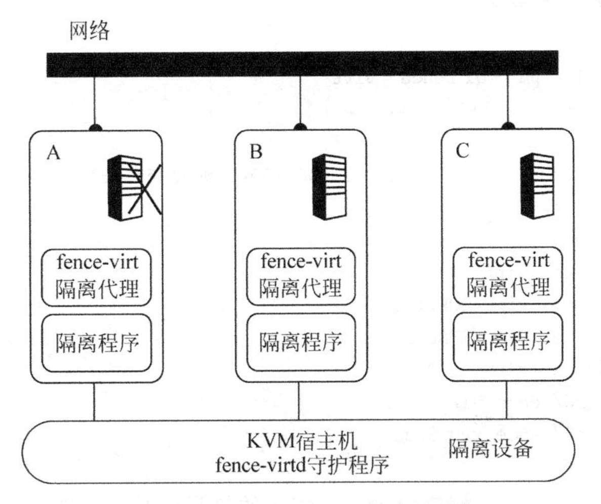


图 2-8　由 KVM 虚拟机构建的群集

其中 fence-virtd 是宿主机上的一个守护程序，软件包的信息如下：

```
[root@tomkvm1 ~]# rpm - qi fence - virtd
Name          : fence - virtd
Version       : 1.0.0
Release       : 1.el8
...
Summary       : Daemon which handles requests from fence - virt
Description :
This package provides the host server framework, fence_virtd,for fence_virt.  The fence_virtd
host daemon is resposible forprocessing fencing requests from virtual machines and routingthe
requests to the appropriate physical machine for action.
```

宿主机上的 fence_virtd 守护程序可用于处理来自虚拟机的隔离请求，并将请求路由到适当的物理机宿主上进行操作。

fence-virt 是隔离设备，需要在宿主机及虚拟机上都进行安装，示例命令如下：

```
[root@tomkvm1 ~]# rpm - qi fence - virt
Name          : fence - virt
Version       : 1.0.0
Release       : 1.el8
...
Summary       : A pluggable fencing framework for virtual machines
Description   :Fencing agent for virtual machines.
```

fence-virt 软件包中的 fence_virt 与将 fence_virtd 守护程序进行交互，可以发出的隔离指令有 null、off、on、reboot、status、list、list-status、monitor、validate-all 和 metadata。

接下来执行以下示例命令：

```
[root@tomkvm1 ~]# rpm -ql fence-virt
/usr/lib/.build-id
/usr/lib/.build-id/3a
/usr/lib/.build-id/3a/40f63989301ab152e029f69b81d67175f668fb
/usr/sbin/fence_virt
/usr/sbin/fence_xvm
/usr/share/doc/fence-virt
/usr/share/doc/fence-virt/COPYING
/usr/share/doc/fence-virt/README
/usr/share/doc/fence-virt/TODO
/usr/share/man/man8/fence_virt.8.gz
/usr/share/man/man8/fence_xvm.8.gz
```

另外一个命令 fence_xvm 其实是 fence_virt 的符号链接，示例命令如下：

```
[root@tomkvm1 ~]# ls -l /usr/sbin/fence_
fence_virt   fence_virtd   fence_xvm
[root@tomkvm1 ~]# ls -l /usr/sbin/fence_*
-rwxr-xr-x. 1 root root 64704 Apr 27  2020 /usr/sbin/fence_virt
-rwxr-xr-x. 1 root root 54664 Apr 27  2020 /usr/sbin/fence_virtd
lrwxrwxrwx. 1 root root    10 Apr 27  2020 /usr/sbin/fence_xvm -> fence_virt
```

(2) 创建 fence-virt 的密钥并复制到所有节点。

在宿主机与虚拟机上 fence-virt 组件需要共享相同的密钥，密钥需要保存在/etc/cluster 目录中，示例命令如下：

```
[root@tomkvm1 ~]# mkdir -p /etc/cluster/

[root@kvm1 ~]# mkdir -p /etc/cluster/

[root@kvm2 ~]# mkdir -p /etc/cluster/
```

接下来，使用 dd 命令创建 fence-virt 密钥，文件名为 fence_xvm.key，示例命令如下：

```
[root@tomkvm1 ~]# dd if=/dev/urandom of=/etc/cluster/fence_xvm.key bs=4k count=1
1+0 records in
1+0 records out
4096 Bytes (4.1 KB, 4.0 KiB) copied, 7.6448e-05 s, 53.6 MB/s

[root@tomkvm1 ~]# file /etc/cluster/fence_xvm.key
/etc/cluster/fence_xvm.key: COM executable for DOS
```

把生成的密钥复制到所有节点的/etc/cluster 目录下，示例命令如下：

```
[root@tomkvm1 ~]# scp /etc/cluster/fence_xvm.key kvm1:/etc/cluster/

[root@tomkvm1 ~]# scp /etc/cluster/fence_xvm.key kvm2:/etc/cluster/
```

(3) 在宿主机上配置 fence-virtd 守护程序。

宿主机上 fence_virtd 守护程序的配置文件是/etc/fence_virt.conf。在安装 fence_virtd 时,会自动创建默认的配置文件。我们既可以直接编辑它,也可以通过交互式对它进行修改,示例命令如下:

```
[root@tomkvm1 ~]# fence_virtd -c
#命令会提交当前的值(或默认值),如果不修改,则可以直接按回车键
Module search path [/usr/lib64/fence-virt/]:        <-保留默认值

Available backends:
    libvirt 0.3
Available listeners:
    vsock 0.1
    multicast 1.2
    serial 0.4

Listener modules are responsible for accepting requests
from fencing clients.

Listener module [multicast]:                        <-监听模式,保留默认值

The multicast listener module is designed for use environments
where the guests and hosts may communicate over a network using
multicast.

The multicast address is the address that a client will use to
send fencing requests to fence_virtd.

Multicast IP Address [225.0.0.12]:                  <-组播 IP 地址,保留默认值

Using ipv4 as family.

Multicast IP Port [1229]:              <-如果更改此端口,则需要记住在防火墙中允许该端口

Setting a preferred interface causes fence_virtd to listen only
on that interface.  Normally, it listens on all interfaces.
In environments where the virtual machines are using the host
machine as a gateway, this *must* be set (typically to virbr0).
```

```
Set to 'none' for no interface.

Interface [virbr0]: virbr1              <-我正在使用 virbr1,可根据集群节点的网卡进行更改

The key file is the shared key information which is used to
authenticate fencing requests.  The contents of this file must
be distributed to each physical host and virtual machine within
a cluster.

Key File [/etc/cluster/fence_xvm.key]:  <-密钥文件名,保留默认值

Backend modules are responsible for routing requests to
the appropriate hypervisor or management layer.

Backend module [libvirt]:                          <-保留默认值

The libvirt backend module is designed for single desktops or
servers.  Do not use in environments where virtual machines
may be migrated between hosts.

Libvirt URI [qemu:///system]:                      <-保留默认值

Configuration complete.

=== Begin Configuration ===                        <-从此处开始是生成配置文件的内容
backends {
    libvirt {
        uri = "qemu:///system";
    }

}

listeners {
    multicast {
        port = "1229";
        family = "ipv4";
        interface = "virbr1";
        address = "225.0.0.12";
        key_file = "/etc/cluster/fence_xvm.key";
    }

}

fence_virtd {
    module_path = "/usr/lib64/fence-virt/";
    backend = "libvirt";
```

```
    listener = "multicast";
  }

  === End Configuration ===
  Replace /etc/fence_virt.conf with the above [y/N]? y              <-确认
```

(4) 启动宿主机上 fence-virtd 守护程序。

示例命令如下：

```
[root@tomkvm1 ~]# systemctl enable fence_virtd --now

[root@tomkvm1 ~]# systemctl status fence_virtd
● fence_virtd.service - Fence-Virt system host daemon
   Loaded: loaded (/usr/lib/systemd/system/fence_virtd.service; enabled; vendor preset:
disabled)
   Active: active (running) since Sun 2021-01-17 08:40:37 CST; 7h ago
     Docs: man:fence_virtd(8)
           man:fence_virt.con(5)
Main PID: 1753 (fence_virtd)
    Tasks: 1 (limit: 203267)
   Memory: 7.0M
   CGroup: /system.slice/fence_virtd.service
 └─1753 /usr/sbin/fence_virtd -w
...

[root@tomkvm1 ~]# netstat -an | grep 1229
tcp        0      0 192.168.1.23:59252       192.168.1.231:1229       TIME_WAIT
tcp        0      0 192.168.1.23:51552       192.168.1.232:1229       TIME_WAIT
udp        0      0 0.0.0.0:1229             0.0.0.0:*
```

启动 fence_virtd 守护程序，并设置为自动启动。通过 netstat 检查监听的端口 1229。

(5) 配置宿主机上的防火墙以允许 fence-virt 的入站数据包。

由于 fence_virtd 使用的端口是默认的 1229，因此必须在防火墙中允许该 TCP、UDP 端口的入站。

首先查看 firewall 的活动区域列表，示例命令如下：

```
[root@tomkvm1 ~]# firewall-cmd --get-active-zones
libvirt
  interfaces: virbr0 virbr3 virbr2
public
  interfaces: virbr1 eno1
```

实验环境中需要开放 public(default)和 libvirt 区域的接口，将规则添加为永久生效，并重新加载防火墙规则以激活更改，示例命令如下：

```
[root@tomkvm1 ~]# firewall-cmd --permanent --add-port=1229/udp

[root@tomkvm1 ~]# firewall-cmd --permanent --add-port=1229/tcp

[root@tomkvm1 ~]# firewall-cmd --permanent --add-port=1229/tcp \
  --zone=libvirt
[root@tomkvm1 ~]# firewall-cmd --permanent --add-port=1229/udp \
  --zone=libvirt

[root@tomkvm1 ~]# firewall-cmd --reload
```

（6）检查 fence-virt 是否工作正常。

在后续的实验中，会用到 fence_xvm 命令的两个选项：

```
-o <operation>       Fencing action (null, off, on, [reboot], status, list,
                     list-status, monitor, validate-all, metadata)

  -H <domain>          Virtual Machine (domain name) to fence
```

在宿主机上使用 fence_xvm 命令来向 fence-virtd 发送 list 请求。如果正常，则会输出宿主机上所有虚拟机的运行状态，示例命令如下：

```
[root@tomkvm1 ~]# fence_xvm -o list
_centos8.3         377937b6-db51-4749-a83c-c3f6bb989e46 off
_centos8.3-KVM     76133e4a-5a46-47df-b1b1-c29535912561 off
centos8.3          9c6a2f70-9738-4591-892d-1275dbc2ad1f off
esxi6.7            768bcc7e-e8a5-465d-84be-896bde1fa145 off
esxi6.7u3          1261bfc6-ce77-4b35-83a7-999589acc23e off
kvm1               25b265b0-aaf5-475a-8080-d0694bbb3288 on
kvm2               71219bf6-062e-4735-b031-021afd7f15c3 on
stor1              1dd9ed69-20a0-476a-b014-3bf730325983 on
win2019            2b8f0015-b42c-4813-a117-2092f424ea50 off
```

在所有的群集节点执行相同的命令，也会获得相同的结果，示例命令如下：

```
[root@kvm1 ~]# fence_xvm -o list
...

[root@kvm2 ~]# fence_xvm -o list
...
```

如果出现以下错误，就需要检查防火墙的状态和 fence_virtd 守护程序的状态，示例命令如下：

```
[root@kvm1 ~]#fence_xvm -o list
#等了一会,才出现
Timed out waiting for response
Operation failed
```

下面再使用 fence_xvm 命令验证群集隔离是否可以正常工作。要实际隔离节点,必须使用 list 命令列出的 UUID 而不是虚拟机名称。可以在 kvm1 上对 kvm2 进行关机、开机操作,示例命令如下:

```
[root@kvm1 ~]#fence_xvm -o list | grep kvm
kvm1    25b265b0-aaf5-475a-8080-d0694bbb3288 on
kvm2    71219bf6-062e-4735-b031-021afd7f15c3 on

[root@kvm1 ~]#fence_xvm -o off -H 71219bf6-062e-4735-b031-021afd7f15c3

[root@kvm1 ~]#fence_xvm -o list | grep kvm
kvm1    25b265b0-aaf5-475a-8080-d0694bbb3288 on
kvm2    71219bf6-062e-4735-b031-021afd7f15c3 off          <-已关闭

[root@kvm1 ~]#fence_xvm -o on -H 71219bf6-062e-4735-b031-021afd7f15c3

[root@kvm1 ~]#fence_xvm -o list | grep kvm
kvm1    25b265b0-aaf5-475a-8080-d0694bbb3288 on
kvm2    71219bf6-062e-4735-b031-021afd7f15c3 on           <-处于运行状态
```

(7) 为群集创建 STONITH 资源。

在确认 fence-virt 可以正常工作之后,就可以为群集创建 STONITH 资源了。可以通过以下命令获得 fence-virt 有关选项参数:

```
[root@kvm1 ~]#pcs stonith describe --help

Usage: pcs stonith describe...
    describe <stonith agent> [--full]
        Show options for specified stonith agent. If --full is specified, all
        options including advanced and deprecated ones are shown.

[root@kvm1~]#pcs stonith describe fence_virt
fence_virt - Fence agent for virtual machines

fence_virt is an I/O Fencing agent which can be used withvirtual machines.

Stonith options:
...
  port: Virtual Machine (domain name) to fence
```

```
...
  pcmk_host_list: A list of machines controlled by this device
...
```

在任意一个节点上创建资源 fence-kvm1 和 fence-kvm2，它们类型都是 fence_xvm，示例命令如下：

```
[root@kvm1 ~]# pcs stonith create fence-kvm1 fence_xvm port=kvm1 \
pcmk_host_list=kvm1-cr

[root@kvm1 ~]# pcs stonith create fence-kvm2 fence_xvm port=kvm2 \
pcmk_host_list=kvm2-cr
```

创建时使用 port 选项来指定虚拟机的名称。pcmk_host_list 是由空格、逗号或分号分隔的节点列表，名称应与 pacemaker 设置的名称完全匹配，可以查看配置文件/etc/corosync/corosync.conf 或 pcs status 输出。

查看 STONITH 资源的配置，示例命令如下：

```
[root@kvm1 ~]# pcs stonith config
 Resource: fence-kvm1 (class=stonith type=fence_xvm)
  Attributes: pcmk_host_list=kvm1-cr port=kvm1
  Operations: monitor interval=60s (fence-kvm1-monitor-interval-60s)
 Resource: fence-kvm2 (class=stonith type=fence_xvm)
  Attributes: pcmk_host_list=kvm2-cr port=kvm2
  Operations: monitor interval=60s (fence-kvm2-monitor-interval-60s)
```

查看群集的状态，示例命令如下：

```
[root@kvm1 ~]# pcs status
Cluster name: cluster1
Cluster Summary:
  * Stack: corosync
  * Current DC: kvm1-cr (version 2.0.4-6.el8_3.1-2deceaa3ae) - partition with quorum
  * Last updated: Sun Jan 17 21:11:38 2021
  * Last change:  Sun Jan 17 21:00:18 2021 by root via cibadmin on kvm1-cr
  * 2 nodes configured
  * 2 resource instances configured

Node List:
  * Online: [ kvm1-cr kvm2-cr ]

Full List of Resources:
  * fence-kvm1    (stonith:fence_xvm):    Started kvm1-cr
```

```
  * fence-kvm2     (stonith:fence_xvm):     Started kvm2-cr

Daemon Status:
  corosync: active/enabled
  pacemaker: active/enabled
  pcsd: active/enabled
```

验证群集是否正确地使用隔离设备，示例命令如下：

```
[root@kvm1 ~]# pcs property --all | grep stonith
 stonith-action: reboot
 stonith-enabled: true
 stonith-max-attempts: 10
 stonith-timeout: 60s
 stonith-watchdog-timeout: (null)
```

输出中的 stonith-action：reboot 表示群集对节点的隔离措施是重新启动。

检查是否可以隔离某个节点，示例命令如下：

```
[root@kvm1 ~]# pcs stonith fence kvm2-cr
Node: kvm2-cr fenced
```

可以在隔离之后，新开一个会话对 kvm2-cr 进行 ping 测试，查看其中断时间，示例命令如下：

```
[root@kvm1 ~]# ping kvm2-cr
PING kvm2-cr (172.16.1.232) 56(84) Bytes of data.
64 Bytes from kvm2-cr (172.16.1.232): icmp_seq=1 ttl=64 time=0.419 ms
...
64 Bytes from kvm2-cr (172.16.1.232): icmp_seq=17 ttl=64 time=0.309 ms
From kvm1-cr (172.16.1.231) icmp_seq=22 Destination Host Unreachable
From kvm1-cr (172.16.1.231) icmp_seq=23 Destination Host Unreachable
...
From kvm1-cr (172.16.1.231) icmp_seq=44 Destination Host Unreachable
From kvm1-cr (172.16.1.231) icmp_seq=45 Destination Host Unreachable
64 Bytes from kvm2-cr (172.16.1.232): icmp_seq=46 ttl=64 time=1200 ms
...
64 Bytes from kvm2-cr (172.16.1.232): icmp_seq=50 ttl=64 time=0.240 ms
^C
--- kvm2-cr ping statistics ---
50 packets transmitted, 22 received, +24 errors, 56% packet loss, time 242ms
rtt min/avg/max/mdev = 0.100/62.921/1200.198/250.859 ms, pipe 4
```

Linux 中的 ping 命令发送数据包的时间间隔为 1s，由于丢了 24 个包，所以本次实验

kvm2 重新启动耗时 24s。

## 2.5 基于 NFS 的 KVM 群集构建

在本实验中，NFS 存储是独立的一个网络，其 IP 地址规划如表 2-4 所示。

表 2-4 基于 NFS 的 KVM 群集实验的 IP 地址规划

| 主 机 | IP 地址 |
|---|---|
| kvm1 | 10.0.1.231 |
| kvm2 | 10.0.1.232 |
| stor1 | 10.0.1.235 |

### 2.5.1 准备 NFS 存储服务器

部署一台 NFS 服务器，示例命令如下：

```
[root@stor1 ~]# cat /etc/redhat-release
CentOS Linux release 8.3.2011

[root@stor1 ~]# dnf -y install nfs-utils

[root@stor1 ~]# systemctl enable nfs-server

[root@stor1 ~]# mkdir /vmdata

[root@stor1 ~]# chmod a+w /vmdata

[root@stor1 ~]# echo "/vmdata *(rw,no_root_squash,sync)" >> /etc/exports

[root@stor1 ~]# systemctl restart nfs-server

[root@stor1 ~]# firewall-cmd --permanent --add-service={nfs,nfs3,rpc-bind,mountd}

[root@stor1 ~]# firewall-cmd --reload
```

在 NFS 服务器上检查配置是否正确，示例命令如下：

```
[root@stor1 ~]# showmount -e localhost
   Export list for localhost:
   /vmdata *
```

在所有节点上测试到 NFS 服务器的连接及文件的读写，示例命令如下：

```
[root@kvm1 ~]# dnf -y install nfs-utils

[root@kvm1 ~]# mkdir /vm

[root@kvm1 ~]# echo "10.0.1.235:/vmdata /vm nfs defaults 0 0" >> /etc/fstab

[root@kvm1 ~]# mount /vm

[root@kvm1 ~]# hostname > /vm/testkvm1.txt

[root@kvm1 ~]# ls -l /vm
total 4
-rw-r--r--. 1 root root 5 Jan 17 22:35 testkvm1.txt
```

在节点 kvm2 执行的示例命令如下：

```
[root@kvm2 ~]# dnf -y install nfs-utils

[root@kvm2 ~]# mkdir /vm

[root@kvm2 ~]# echo "10.0.1.235:/vmdata /vm nfs defaults 0 0" >> /etc/fstab

[root@kvm2 ~]# mount /vm

[root@kvm2 ~]# hostname > /vm/testkvm2.txt

[root@kvm2 ~]# ls -l /vm
total 8
-rw-r--r--. 1 root root 5 Jan 17 22:40 testkvm1.txt
-rw-r--r--. 1 root root 5 Jan 17 22:40 testkvm2.txt
```

使用 setsebool 命令设置所有节点的 SELinux 的布尔值，允许 KVM 使用 nfs 的功能，示例命令如下：

```
[AllNodes ~]# setsebool -P virt_use_nfs 1

[AllNodes ~]# getsebool -a | grep virt_use_nfs
virt_use_nfs --> on
```

### 2.5.2 准备测试用的虚拟机

将之前已经安装好的 centos6.10 的磁盘映像文件复制到/vm 目录下。查看映像文件的信息，示例命令如下：

```
[root@kvm1 ~]# qemu-img info /vm/centos6.10a.qcow2
image: /vm/centos6.10a.qcow2
file format: qcow2
virtual size: 80 GiB (85899345920 Bytes)
disk size: 2.31 GiB
cluster_size: 65536
Format specific information:
    compat: 1.1
    lazy refcounts: false
    refcount bits: 16
    corrupt: false
```

通过 virt-install 命令利用已有的磁盘映像文件创建新的虚拟机,示例命令如下：

```
[root@kvm1 ~]# virt-install --name=centos6.10a  \
   --disk device=disk,bus=virtio,path='/vm/centos6.10a.qcow2'  \
   --vcpus=1 --ram=512  \
   --network network=default,model=virtio  \
   --graphics spice,listen=127.0.0.1 \
   --graphics vnc --video qxl \
   --boot hd \
   --noautoconsole

WARNING  No operating system detected, VM performance may suffer. Specify an OS with --os-
variant for optimal results.

Starting install...
Domain creation completed.

[root@kvm1 ~]# virsh list
Id   Name           State
------------------------------
2    centos6.10a    running
```

**提示**：需要在虚拟机中安装 qemu-ga 驱动程序或服务,保证可以通过 virsh shutdown 正常地关闭虚拟机。否则会导致停止资源超时。

### 2.5.3 测试实时迁移

在创建群集资源之前,要保证虚拟机可以在群集的节点间进行实时迁移。

将虚拟机从 kvm1 迁移到 kvm2,示例命令如下：

```
[root@kvm1 ~]# virsh migrate --persistent --undefinesource centos6.10a \
  qemu+ssh://kvm2/system

[root@kvm1 ~]# virsh list --all
```

```
Id   Name   State
---------------------

[root@kvm1 ~]# virsh --connect=qemu+ssh://kvm2/system list
Id   Name           State
--------------------------------
1    centos6.10a    running
```

将虚拟机从 kvm2 迁回到 kvm1,示例命令如下:

```
[root@kvm1 ~]# virsh --connect=qemu+ssh://kvm2/system migrate --persistent \
    --undefinesource centos6.10a qemu+ssh://kvm1/system

[root@kvm1 ~]# virsh list
Id   Name           State
--------------------------------
3    centos6.10a    running
```

与实时迁移相关的知识,可参见本书的“第1章 实现虚拟机迁移”。

### 2.5.4 创建虚拟机资源

从存储的层面来看,KVM 虚拟机是由配置文件和磁盘映像文件组成的。这两部分都必须放置在共享存储上,而且需保证每个节点都可以正常访问。

首先,将配置文件复制到共享存储上,示例命令如下:

```
[root@kvm1 ~]# mkdir /vm/qemu_config

[root@kvm1 ~]# cp /etc/libvirt/qemu/centos6.10a.xml /vm/qemu_config/
```

取消虚拟机的定义,这只是删除配置文件,磁盘映像文件不受影响。示例命令如下:

```
[root@kvm1 ~]# virsh shutdown centos6.10a

[root@kvm1 ~]# virsh undefine centos6.10a

[root@kvm1 ~]# virsh list --all
Id     Name                                 State
----------------------------------------------------------
```

获得群集的虚拟机资源的参数,示例命令如下:

```
[root@kvm1 ~]# pcs resource list | grep VirtualDomain
ocf:heartbeat:VirtualDomain - Manages virtual domains through the libvirt
```

需要创建 ocf：heartbeat：VirtualDomain 类的资源，下面查看这类资源的使用方法，示例命令如下：

```
[root@kvm1 ~]# pcs resource describe ocf:heartbeat:VirtualDomain
ocf:heartbeat:VirtualDomain - Manages virtual domains through the libvirt virtualization framework

Resource agent for a virtual domain (a.k.a. domU, virtual machine,
virtual environment etc., depending on context) managed by libvirtd.

...
Default operations:
  start: interval = 0s timeout = 90s
  stop: interval = 0s timeout = 90s
  monitor: interval = 10s timeout = 30s
  migrate_from: interval = 0s timeout = 60s
  migrate_to: interval = 0s timeout = 120s
```

了解一个创建资源的命令，示例命令如下：

```
[root@kvm1 ~]# pcs resource create --help

Usage: pcs resource create...
    create <resource id> [<standard>:[<provider>:]]<type> [resource options]
           [op <operation action><operation options> [<operation action>
<operation options>]...] [meta <meta options>...]
           [clone [<clone options>] | promotable [<promotable options>] |
           --group <group id> [--before <resource id> | --after <resource id>] |
           bundle <bundle id>] [--disabled] [--no-default-ops] [--wait[=n]]
Create specified resource. If clone is used a clone resource is
created. If promotable is used a promotable clone resource is created.
If --group is specified the resource is added to the group named. You
...
        Example: Create a new resource called 'VirtualIP' with IP address
           192.168.0.99, netmask of 32, monitored everything 30 seconds,
           on eth2:
           pcs resource create VirtualIP ocf:heartbeat:IPaddr2 \
              ip = 192.168.0.99 cidr_netmask = 32 nic = eth2 \
              op monitor interval = 30s
```

创建虚拟机资源，其中最重要的是为 VirtualDomain 类型的资源指定配置文件/vm/qemu_config/centos6.10a.xml，示例命令如下：

```
[root@kvm1 ~]# pcs resource create centos6.10a_res VirtualDomain  \
  hypervisor = "qemu:///system"  \
```

```
  config = "/vm/qemu_config/centos6.10a.xml"   \
  migration_transport = ssh   \
  op start timeout = "120s"   \
  op stop timeout = "120s"   \
  op monitor timeout = "30"   \
    interval = "10"   meta allow - migrate = "true"  priority = "100"   \
  op migrate_from interval = "0"  timeout = "120s"   \
  op migrate_to interval = "0"  timeout = "120"

Assumed agent name 'ocf:heartbeat:VirtualDomain' (deduced from 'VirtualDomain')
```

VirtualDomain 类型资源的完整标识符是 ocf：heartbeat：VirtualDomain。

查看群集状态，最终会看到 centos6.10a_res 的状态为 Started，示例命令如下：

```
[root@kvm1 ~]# pcs status
Cluster name: cluster1
Cluster Summary:
  * Stack: corosync
  * Current DC: kvm1 - cr (version 2.0.4 - 6.el8_3.1 - 2deceaa3ae)  -  partition with quorum
  * Last updated: Mon Jan 18 12:30:51 2021
  * Last change:   Mon Jan 18 12:30:33 2021 by root via crm_resource on kvm1 - cr
  * 2 nodes configured
  * 3 resource instances configured

Node List:
  * Online: [ kvm1 - cr kvm2 - cr ]

Full List of Resources:
  * centos6.10a_res     (ocf::heartbeat:VirtualDomain):     Started kvm1 - cr
  * fence - kvm1        (stonith:fence_xvm):                Started kvm2 - cr
  * fence - kvm2        (stonith:fence_xvm):                Started kvm1 - cr

Daemon Status:
  corosync: active/enabled
  pacemaker: active/enabled
  pcsd: active/enabled
```

查看当前虚拟机的列表，示例命令如下：

```
[root@kvm1 ~]# virsh list -- all
Id   Name          State
------------------------------
5    centos6.10a   running
```

可以在 libvirt 中看到新的虚拟机。这个虚拟机没有 XML 配置文件，所以是临时型的虚拟机。

### 2.5.5 群集测试

为了检查高可用的效果，我们先做一些准备工作。登录到用于测试的虚拟机 centos6.10a 的控制台，在其中使用 vi 编辑器创建一个新文件，输入一些内容但不保存，如图 2-9 所示。这些数据仅仅在虚拟机的内存中。下面通过 4 个实验来检查是否达到高可用。

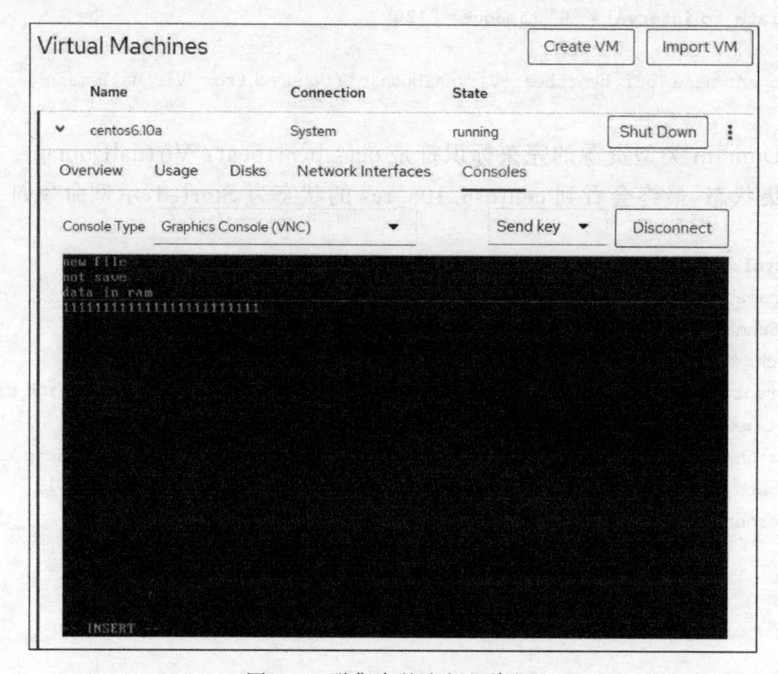


图 2-9 群集中的虚拟机资源

#### 1. pcs resource move

查看群集当前的状态，示例命令如下：

```
[root@kvm1 ~]# pcs status
Cluster name: cluster1
Cluster Summary:
  * Stack: corosync
  * Current DC: kvm2-cr (version 2.0.4-6.el8_3.1-2deceaa3ae) - partition with quorum
  * Last updated: Mon Jan 18 15:20:25 2021
  * Last change:  Mon Jan 18 15:19:33 2021 by root via crm_resource on kvm1-cr
  * 2 nodes configured
  * 3 resource instances configured

Node List:
  * Online: [ kvm1-cr kvm2-cr ]
```

```
Full List of Resources:
  * centos6.10a_res    (ocf::heartbeat:VirtualDomain):    Started kvm1 - cr
  * fence - kvm1       (stonith:fence_xvm):               Started kvm2 - cr
  * fence - kvm2       (stonith:fence_xvm):               Started kvm2 - cr

Daemon Status:
  corosync: active/enabled
  pacemaker: active/enabled
  pcsd: active/enabled
```

群集资源 centos6.10a_res 运行在节点 kvm1-cr 上。

查看当前的约束,示例命令如下:

```
[root@kvm1 ~]#pcs constraint
Location Constraints:
Ordering Constraints:
Colocation Constraints:
Ticket Constraints:
```

当前群集环境中没有资源约束。

从 pcs resource move 的联机帮助可以知道:迁移资源时会自动创建一个-INFINITY 类型的位置约束,这样可以避免资源在节点之间反复迁移。习惯将这种技术称为"冷却时间",示例命令如下:

```
[root@kvm1 ~]#pcs resource move --help

Usage: pcs resource move...
    move <resource id> [destination node] [--master] [lifetime=<lifetime>]
           [--wait[=n]]
Move the resource off the node it is currently running on by creating
a -INFINITY location constraint to ban the node. If destination node isspecified the resource
will be moved to that node by creatingan INFINITY location constraint to prefer the destination
node. If --master is used the scope of the command is limited to the master roleand you must use
the promotable clone id (instead of the resource id).
If lifetime is specified then the constraint will expire after thattime, otherwise it defaults
to infinity and the constraint can becleared manually with 'pcs resource clear' or 'pcs
constraint delete'.
If --wait is specified, pcs will wait up to 'n' seconds for theresource to move and then return
0 on success or 1 on error.   If 'n' is not specified it defaults to 60 minutes.
If you want the resource to preferably avoid running on some nodes but be able to failover to them
use 'pcs constraint location avoids'.
```

如果未指定目标节点,则 pacemaker 会自动选择 1 个节点。本次迁移的目标是 kvm2-cr,示例命令如下:

```
[root@kvm1 ~]# pcs resource move centos6.10a_res
Warning: Creating location constraint 'cli-ban-centos6.10a_res-on-kvm1-cr' with a score of -INFINITY for resource centos6.10a_res on kvm1-cr.
        This will prevent centos6.10a_res from running on kvm1-cr until the constraint is removed
        This will be the case even if kvm1-cr is the last node in the cluster
```

Pacemaker 会发出一个警告提示：为 kvm1-cr 上的资源 centos6.10a_res 创建位置约束 cli-ban-centos6.10a_res-on-kvm1-cr。在删除约束之前，这将阻止 centos6.10a_res 在 kvm1-cr 上运行，即使 kvm1-cr 是集群中的最后一个节点，也是如此，示例命令如下：

```
[root@kvm1 ~]# pcs constraint
Location Constraints:
  Resource: centos6.10a_res
    Disabled on:
      Node: kvm1-cr (score:-INFINITY) (role:Started)
Ordering Constraints:
Colocation Constraints:
Ticket Constraints:
```

查看群集的状态，示例命令如下：

```
[root@kvm1 ~]# pcs status
Cluster name: cluster1
Cluster Summary:
  * Stack: corosync
  * Current DC: kvm2-cr (version 2.0.4-6.el8_3.1-2deceaa3ae) - partition with quorum
  * Last updated: Mon Jan 18 15:27:13 2021
  * Last change:  Mon Jan 18 15:26:38 2021 by root via crm_resource on kvm1-cr
  * 2 nodes configured
  * 3 resource instances configured

Node List:
  * Online: [ kvm1-cr kvm2-cr ]

Full List of Resources:
  * centos6.10a_res     (ocf::heartbeat:VirtualDomain):     Started kvm2-cr
  * fence-kvm1          (stonith:fence_xvm):                Started kvm1-cr
  * fence-kvm2          (stonith:fence_xvm):                Started kvm2-cr

Daemon Status:
  corosync: active/enabled
  pacemaker: active/enabled
  pcsd: active/enabled
```

群集资源 centos6.10a_res 运行在节点 kvm2-cr 上，也就是说虚拟机现在已经运行在 kvm2 上了。

查看宿主机 kvm1 上虚拟机列表，已经没有虚拟机的定义了，示例命令如下：

```
[root@kvm1 ~]# virsh list --all
Id   Name   State
--------------------

[root@kvm1 ~]# virsh -c qemu+ssh://kvm2/system list --all
Id   Name          State
------------------------------
4    centos6.10a   running
```

在宿主机 kvm2 上打开虚拟机的控制台，还会看到 vi 编辑器正处于运行状态，而且正打开那个未保存的新文件。这说明本次迁移是实时迁移，业务系统不会中断。

下面我们再将虚拟机迁移回宿主机 kvm1。在迁移之前，需要先删除约束，可以使用 pcs resource clear 命令清除与某个资源有关的所有约束，示例命令如下：

```
[root@kvm1 ~]# pcs constraint
Location Constraints:
  Resource: centos6.10a_res
    Disabled on:
      Node: kvm1-cr (score:-INFINITY) (role:Started)
Ordering Constraints:
Colocation Constraints:
Ticket Constraints:

[root@kvm1 ~]# pcs resource clear centos6.10a_res
Removing constraint: cli-ban-centos6.10a_res-on-kvm1-cr

[root@kvm1 ~]# pcs constraint
Location Constraints:
Ordering Constraints:
Colocation Constraints:
Ticket Constraints:
```

迁移群集资源 centos6.10a_res，示例命令如下：

```
[root@kvm1 ~]# pcs resource move centos6.10a_res
Warning: Creating location constraint 'cli-ban-centos6.10a_res-on-kvm2-cr' with a score
of -INFINITY for resource centos6.10a_res on kvm2-cr.
        This will prevent centos6.10a_res from running on kvm2-cr until the constraint is removed
        This will be the case even if kvm2-cr is the last node in the cluster

[root@kvm1 ~]# pcs resource
  * centos6.10a_res      (ocf::heartbeat:VirtualDomain):     Started kvm1-cr
```

清除本次迁移生成的约束。这次使用另外一种清除方法，即使用约束的 ID，示例命令如下：

```
[root@kvm1 ~]# pcs constraint delete cli-ban-centos6.10a_res-on-kvm2-cr
```

**提示**：可以通过 pcs constraint --full 查询约束的详细信息从而获得约束的 ID。

2. pcs node standby

如果将指定的群集节点设置为待机(standby)模式，这样它就无法运行所有的资源了，这将触发资源的迁移，示例命令如下：

```
[root@kvm1 ~]# pcs node standby --help

Usage: pcs node <command>
    standby [--all | <node>...] [--wait[=n]]
Put specified node(s) into standby mode (the node specified will nolonger be able to host
resources), if no nodes or options are specified the current node will be put into standby mode,
if --all is specifie dall nodes will be put into standby mode.
If --wait is specified, pcs will wait up to 'n' seconds for the node(s)to be put into standby
mode and then return 0 on success or 1 if the operation not succeeded yet. If 'n' is not specified
it defaults to 60 minutes.
```

查看当前群集的状态，示例命令如下：

```
[root@kvm1 ~]# pcs status
Cluster name: cluster1
Cluster Summary:
  * Stack: corosync
  * Current DC: kvm2-cr (version 2.0.4-6.el8_3.1-2deceaa3ae) - partition with quorum
  * Last updated: Mon Jan 18 16:36:12 2021
  * Last change:  Mon Jan 18 16:17:22 2021 by root via crm_resource on kvm1-cr
  * 2 nodes configured
  * 3 resource instances configured

Node List:
  * Online: [ kvm1-cr kvm2-cr ]

Full List of Resources:
  * centos6.10a_res    (ocf::heartbeat:VirtualDomain):    Started kvm1-cr
  * fence-kvm1         (stonith:fence_xvm):               Started kvm2-cr
  * fence-kvm2         (stonith:fence_xvm):               Started kvm2-cr

Daemon Status:
  corosync: active/enabled
  pacemaker: active/enabled
```

```
  pcsd: active/enabled

[root@kvm1 ~]# pcs constraint
Location Constraints:
Ordering Constraints:
Colocation Constraints:
Ticket Constraints:
```

目前资源 centos6.10a_res 在节点 kvm1-cr 上，而且也没有约束。将 kvm1-cr 设置为待机状态，示例命令如下：

```
[root@kvm1 ~]# pcs node standby kvm1-cr

[root@kvm1 ~]# pcs status
Cluster name: cluster1
Cluster Summary:
  * Stack: corosync
  * Current DC: kvm2-cr (version 2.0.4-6.el8_3.1-2deceaa3ae) - partition with quorum
  * Last updated: Mon Jan 18 16:37:47 2021
  * Last change:  Mon Jan 18 16:37:44 2021 by root via cibadmin on kvm1-cr
  * 2 nodes configured
  * 3 resource instances configured

Node List:
  * Node kvm1-cr: standby
  * Online: [ kvm2-cr ]

Full List of Resources:
  * centos6.10a_res     (ocf::heartbeat:VirtualDomain):     Started kvm2-cr
  * fence-kvm1          (stonith:fence_xvm):                Started kvm2-cr
  * fence-kvm2          (stonith:fence_xvm):                Started kvm2-cr

Daemon Status:
  corosync: active/enabled
  pacemaker: active/enabled
  pcsd: active/enabled
```

当节点 kvm1-cr 处于待机状态时，pcs status 的输出显示为 Node kvm1-cr: standby。通过待机触发的迁移不会生成约束。查看群集的约束，示例命令如下：

```
[root@kvm1 ~]# pcs constraint
Location Constraints:
Ordering Constraints:
Colocation Constraints:
Ticket Constraints:
```

查看所有节点虚拟机的运行情况，示例命令如下：

```
[root@kvm1 ~]# virsh list --all
 Id   Name   State
--------------------

[root@kvm1 ~]# virsh -c qemu+ssh://kvm2/system list --all
 Id   Name          State
------------------------------
 7    centos6.10a   running
```

虚拟机会被迁移至宿主机 kvm2 上，观察虚拟机控制台，会发现原有内存的数据还在，说明业务没有中断。

通过 pcs node unstandby 命令将节点 kvm1-cr 恢复为正常联机状态，示例命令如下：

```
[root@kvm1 ~]# pcs node unstandby kvm1-cr

[root@kvm1 ~]# pcs status
Cluster name: cluster1
Cluster Summary:
  * Stack: corosync
  * Current DC: kvm2-cr (version 2.0.4-6.el8_3.1-2deceaa3ae) - partition with quorum
  * Last updated: Mon Jan 18 16:42:55 2021
  * Last change:  Mon Jan 18 16:42:49 2021 by root via cibadmin on kvm1-cr
  * 2 nodes configured
  * 3 resource instances configured

Node List:
  * Online: [ kvm1-cr kvm2-cr ]

Full List of Resources:
  * centos6.10a_res     (ocf::heartbeat:VirtualDomain):  Started kvm2-cr
  * fence-kvm1          (stonith:fence_xvm):             Started kvm1-cr
  * fence-kvm2          (stonith:fence_xvm):             Started kvm2-cr

Daemon Status:
  corosync: active/enabled
  pacemaker: active/enabled
  pcsd: active/enabled
```

默认情况下，这不会导致虚拟机资源 centos6.10a_res 的回迁，除非通过 pcs constraint location 将 kvm1-cr 指定为首选节点。

**提示：** Pacemaker 早期版本将模式设置为待机模式的命令是 pcs cluster standby。

### 3. pcs cluster stop

pcs cluster stop 命令可以停止某个或全部节点的群集服务：

```
[root@kvm1 ~]# pcs cluster stop --help

Usage: pcs cluster stop...
    stop [--all | <node>... ] [--request-timeout=<seconds>]
Stop a cluster on specified node(s). If no nodes are specified then stop a cluster on the local
node. If --all is specified then stop a cluster on all nodes. If the cluster is running
resources which take long time to stop then the stop request may time out before the cluster
actually stops. In that case you should consider setting --request-timeout to a suitable
value.
```

如果停止某个节点的群集服务，则会先将其上所有的资源迁移到其他的节点上，然后停止服务。

查看当前群集的状态，示例命令如下：

```
[root@kvm1 ~]# pcs status
Cluster name: cluster1
Cluster Summary:
  * Stack: corosync
  * Current DC: kvm2-cr (version 2.0.4-6.el8_3.1-2deceaa3ae) - partition with quorum
  * Last updated: Mon Jan 18 16:56:36 2021
  * Last change:  Mon Jan 18 16:42:49 2021 by root via cibadmin on kvm1-cr
  * 2 nodes configured
  * 3 resource instances configured

Node List:
  * Online: [ kvm1-cr kvm2-cr ]

Full List of Resources:
  * centos6.10a_res     (ocf::heartbeat:VirtualDomain):     Started kvm2-cr
  * fence-kvm1          (stonith:fence_xvm):                Started kvm1-cr
  * fence-kvm2          (stonith:fence_xvm):                Started kvm2-cr

Daemon Status:
  corosync: active/enabled
  pacemaker: active/enabled
  pcsd: active/enabled
```

群集资源 centos6.10a_res 运行在节点 kvm2-cr 上。停止这个节点的群集服务，示例命令如下：

```
[root@kvm1 ~]# pcs cluster stop kvm2-cr
kvm2-cr: Stopping Cluster (pacemaker)...
kvm2-cr: Stopping Cluster (corosync)...
```

再次查看当前群集的状态，示例命令如下：

```
[root@kvm1 ~]# pcs status
Cluster name: cluster1
Cluster Summary:
  * Stack: corosync
  * Current DC: kvm1-cr (version 2.0.4-6.el8_3.1-2deceaa3ae) - partition with quorum
  * Last updated: Mon Jan 18 16:57:14 2021
  * Last change:  Mon Jan 18 16:42:49 2021 by root via cibadmin on kvm1-cr
  * 2 nodes configured
  * 3 resource instances configured

Node List:
  * Online: [ kvm1-cr ]
  * OFFLINE: [ kvm2-cr ]

Full List of Resources:
  * centos6.10a_res    (ocf::heartbeat:VirtualDomain):    Started kvm1-cr
  * fence-kvm1         (stonith:fence_xvm):               Started kvm1-cr
  * fence-kvm2         (stonith:fence_xvm):               Started kvm1-cr

Daemon Status:
  corosync: active/enabled
  pacemaker: active/enabled
  pcsd: active/enabled
```

群集资源 centos6.10a_res 会被迁移到节点 kvm1-cr 上。观察虚拟机控制台，会发现原有内存的数据还在，说明业务没有中断。

恢复实验环境，示例命令如下：

```
[root@kvm1 ~]# pcs cluster start kvm2-cr
kvm2-cr: Starting Cluster...
```

#### 4. 强制关闭一个节点

通过 init 命令模拟节点 kvm1 意外宕机，示例命令如下：

```
[root@kvm1 ~]# init 0
```

查看群集的状态，示例命令如下：

```
[root@kvm2 ~]# pcs status
Cluster name: cluster1
Cluster Summary:
  * Stack: corosync
  * Current DC: kvm2-cr (version 2.0.4-6.el8_3.1-2deceaa3ae) - partition with quorum
  * Last updated: Mon Jan 18 17:06:03 2021
  * Last change:  Mon Jan 18 16:42:49 2021 by root via cibadmin on kvm1-cr
  * 2 nodes configured
  * 3 resource instances configured

Node List:
  * Online: [ kvm2-cr ]
  * OFFLINE: [ kvm1-cr ]

Full List of Resources:
  * centos6.10a_res     (ocf::heartbeat:VirtualDomain):     Started kvm2-cr
  * fence-kvm1          (stonith:fence_xvm):                Started kvm2-cr
  * fence-kvm2          (stonith:fence_xvm):                Started kvm2-cr

Daemon Status:
 corosync: active/enabled
 pacemaker: active/enabled
 pcsd: active/enabled
```

群集资源 centos6.10a_res 会运行在节点 kvm2-cr。

查看虚拟机的状态,示例命令如下:

```
[root@kvm2 ~]# virsh list
Id   Name          State
------------------------------
8    centos6.10a   running
```

虚拟机会自动在宿主机 kvm2 上启动。由于虚拟机是意外关闭的,所以不会触发虚拟机动态迁移,当然内存中的数据会丢失,业务会发生中断,中断的时间取决于虚拟机启动和业务软件启动的时间。

## 2.5.6 删除群集资源

如果不希望由群集管理资源,则可以对其进行禁用或删除操作。被禁用的资源在需要的时候还可以再启用,而删除的资源则无法进行逆操作。

下面我们将删除资源为后续的实验清理环境,示例命令如下:

```
[root@kvm1 ~]# pcs status
Cluster name: cluster1
```

```
Cluster Summary:
  * Stack: corosync
  * Current DC: kvm2-cr (version 2.0.4-6.el8_3.1-2deceaa3ae) - partition with quorum
  * Last updated: Mon Jan 18 17:09:36 2021
  * Last change:  Mon Jan 18 16:42:49 2021 by root via cibadmin on kvm1-cr
  * 2 nodes configured
  * 3 resource instances configured

Node List:
  * Online: [ kvm1-cr kvm2-cr ]

Full List of Resources:
  * centos6.10a_res    (ocf::heartbeat:VirtualDomain):    Started kvm2-cr
  * fence-kvm1         (stonith:fence_xvm):               Started kvm1-cr
  * fence-kvm2         (stonith:fence_xvm):               Started kvm2-cr

Daemon Status:
  corosync: active/enabled
  pacemaker: active/enabled
  pcsd: active/enabled

[root@kvm1 ~]# pcs resource delete centos6.10a_res
Attempting to stop: centos6.10a_res... Stopped
```

pcs resource delete 命令可以删除资源、组、捆绑或克隆。

**提示**：删除虚拟机资源的时间有可能会比较长，可以先使用 virsh destroy 命令终止虚拟机，然后删除群集资源。

## 2.6 基于 iSCSI 的 KVM 群集 1

如果群集的后端存储是SAN，则不管是iSCSI还是FC-SAN，配置Linux群集的方法大同小异。存储设备将LUN资源分配给所有群集节点，节点通常使用LVM进行管理，这比使用传统的分区管理更加灵活。

RHEL/CentOS 8为LVM卷提供了两种不同的集群配置模式。

(1) 主动/被动(Active/Passive)模式：使用故障转移配置中的高可用性LVM卷(HA-LVM)，在该配置中，群集中只有一个节点可以随时访问存储。

(2) 主动/主动(Active/Active)模式：使用lvmlockd守护程序来管理主动/主动配置中的存储设备的LVM卷，在该配置中，群集的多个节点可以同时访问存储。

究竟使用何种模式，取决于应用程序或服务的需求。

如果群集的多个节点需要同时对活动/活动系统中的LVM卷进行读/写访问，则必须使用lvmlockd守护程序并将卷配置为共享卷。lvmlockd守护程序为集群的节点对LVM

卷并发的操作请求进行协调。当节点与卷交互并更改其布局时，lvmlockd守护程序的锁定服务为LVM元数据提供保护。

如果一次只有一个节点需要访问给定的LVM卷，则采用HA-LVM这种主动/被动模式是最佳策略，因为无须使用lvmlockd锁定服务。

大多数应用程序都不是“群集感知”的，并没有针对群集进行专门设计或优化，无法与其他实例同时运行，所以推荐采用主动/被动模式。另外，选择在主动/被动的共享逻辑卷上运行这些应用程序还可能会导致性能下降。这是因为在这些实例中，逻辑卷本身存在群集通信的开销。“群集感知”应用程序是经过专门设计的，可以通过使用支持群集的文件系统和群集逻辑卷来提升性能，可以抵消一部分群集通信的开销。

HA-LVM和使用的共享逻辑卷lvmlockd很相似，可以防止多个节点进行重叠更改，防止LVM元数据及其逻辑卷的损坏。HA-LVM限制了逻辑卷只能以独占方式激活，也就是说，一次仅在一个节点上处于活动状态，避免集群的协调开销，从而提高了性能。

对于KVM虚拟化来讲，采用HA-LVM这种主动/被动模式比较简单，但是存储资源不是共享的，所以虚拟机迁移不是热迁移，迁移时虚拟机中的业务会中断。采用lvmlockd实现的主动/主动模式LVM虽然比较复杂，但是由于提供的是共享存储，可以实现虚拟机的热迁移，虚拟机中的业务不会中断，所以是一个比较好的选择。

在本实验中先采用HA-LVM模式，它的环境与上次实验的环境类似。iSCSI存储是独立的一个网络。IP地址及iSCSI启动器的规划如表2-5所示。

**表2-5 基于iSCSI的KVM群集实验的IP地址规划**

| 主　机 | IP地址 | 启动器名称 |
|---|---|---|
| kvm1 | 10.0.1.231 | iqn.1994-05.com.redhat：kvm1 |
| kvm2 | 10.0.1.232 | iqn.1994-05.com.redhat：kvm2 |
| stor1 | 10.0.1.235 | |

## 2.6.1 准备iSCSI存储服务器

在本实验中，我们在一台运行CentOS 8的主机stor1上安装targetcli软件包，通过LinuxIO实现iSCSI目标。

通过targetcli配置后端、设置ACL之后，最终的配置如下：

```
[root@stor1 ~]#targetcli ls
o- / ......................................................... [...]
  o- backstores .............................................. [...]
  | o- block .................................. [Storage Objects: 0]
  | o- fileio ................................. [Storage Objects: 2]
  | | o- file1  [/iscsifiles/file1.dat (20.0GiB) write-back deactivated]
  | | | o- alua ...................................... [ALUA Groups: 1]
  | | |   o- default_tg_pt_gp .......... [ALUA state: Active/optimized]
```

```
| | o- iscsifiles-file1.dat  [/iscsifiles/file1.dat (20.0GiB) write-thru activated]
| |   o- alua ..................................... [ALUA Groups: 1]
| |     o- default_tg_pt_gp .......... [ALUA state: Active/optimized]
| o- pscsi .................................... [Storage Objects: 0]
| o- ramdisk .................................. [Storage Objects: 0]
o- iscsi ............................................ [Targets: 1]
| o- iqn.2003-01.org.Linux-iscsi.stor1.x8664:sn.a97bf0f9a33d  [TPGs: 1]
|   o- tpg1 ................................ [no-gen-acls, no-auth]
|     o- acls ........................................... [ACLs: 2]
|     | o- iqn.1994-05.com.redhat:kvm1 ............. [Mapped LUNs: 1]
|     | | o- mapped_lun0 ..... [lun0 fileio/iscsifiles-file1.dat (rw)]
|     | o- iqn.1994-05.com.redhat:kvm2 ............. [Mapped LUNs: 1]
|     |   o- mapped_lun0 ..... [lun0 fileio/iscsifiles-file1.dat (rw)]
|     o- luns ........................................... [LUNs: 1]
|     | o- lun0  [fileio/iscsifiles-file1.dat (/iscsifiles/file1.dat) (default_tg_pt_gp)]
|     o- portals ...................................... [Portals: 1]
|       o- 0.0.0.0:3260 ......................................... [OK]
o- loopback ......................................... [Targets: 0]
```

其中有两个核心配置选项。

(1) 后端存储：使用/iscsifiles/file1.dat 文件当作后端存储，大小为 20.0GiB。

(2) ACL：允许群集中的两个节点访问。

为了清晰地标识群集节点，修改每个节点的 iSCSI 启动器的名称，示例命令如下：

```
[root@kvm1 ~]# cat /etc/iscsi/initiatorname.iscsi
InitiatorName=iqn.1994-05.com.redhat:kvm1
```

```
[root@kvm2 ~]# cat /etc/iscsi/initiatorname.iscsi
InitiatorName=iqn.1994-05.com.redhat:kvm2
```

## 2.6.2 为群集准备 LVM 逻辑卷和文件系统

下面在存储上创建 LVM 逻辑卷，这需要在两个节点都进行操作。

### 1. 在节点 kvm1 上执行的操作

通过 iscsiadm 命令进行发现、登录操作，示例命令如下：

```
[root@kvm1 ~]# iscsiadm --mode discovery --type sendtargets \
    --portal 10.0.1.235
10.0.1.235:3260,1 iqn.2003-01.org.Linux-iscsi.stor1.x8664:sn.a97bf0f9a33d

[root@kvm1 ~]# iscsiadm -m node --login
Logging in to [iface: default, target: iqn.2003-01.org.Linux-iscsi.stor1.x8664:sn.
a97bf0f9a33d, portal: 10.0.1.235,3260]
```

```
Login to [iface: default, target: iqn.2003-01.org.Linux-iscsi.storl.x8664:sn.a97bf0f9a33d, portal: 10.0.1.235,3260] successful.

[root@kvm1 ~]# systemctl enable iscsid

[root@kvm1 ~]# systemctl restart iscsid
```

通过 fdisk 命令会查看到新磁盘，示例命令如下：

```
[root@kvm1 ~]# fdisk -l
...
Disk /dev/sda: 20 GiB, 21474836480 Bytes, 41943040 sectors
Units: sectors of 1 * 512 = 512 Bytes
Sector size (logical/physical): 512 Bytes / 512 Bytes
I/O size (minimum/optimal): 512 Bytes / 8388608 Bytes
```

在新磁盘/dev/sda 上创建一个基本分区，使用全部空间，并将分区类型设置为 LVM，示例命令如下：

```
[root@kvm1 ~]# fdisk /dev/sda
...

[root@kvm1 ~]# fdisk -l /dev/sda
Disk /dev/sda: 20 GiB, 21474836480 Bytes, 41943040 sectors
Units: sectors of 1 * 512 = 512 Bytes
Sector size (logical/physical): 512 Bytes / 512 Bytes
I/O size (minimum/optimal): 512 Bytes / 8388608 Bytes
Disklabel type: dos
Disk identifier: 0xa3b9c687

Device     Boot Start      End  Sectors Size Id Type
/dev/sda1       16384 41943039 41926656  20G 8e Linux LVM
```

配置主动/被动的 HA-LVM 最关键的配置就是启用 LVM 的系统 ID 功能。有了系统 ID，就可以将卷组(VG)限制为仅允许一台主机访问。当 VG 资源需要切换到另外一个节点时，Pacemaker 会修改 VG 的系统 ID，从而由另外一个节点接管卷组。

默认情况下未启用 LVM 的系统 ID 功能。下面通过修改文件/etc/lvm/lvm.conf 中的 system_id_source 实现，示例命令如下：

```
[root@kvm1 ~]# lvm systemid
  system ID:

[root@kvm1 ~]# uname -n
kvm1
```

```
[root@kvm1 ~]# vi /etc/lvm/lvm.conf
...
        # Configuration option global/system_id_source.
        # The method LVM uses to set the local system ID.
        # Volume Groups can also be given a system ID (by vgcreate, vgchange,
        # or vgimport.) A VG on shared storage devices is accessible only to
        # the host with a matching system ID. See 'man lvmsystemid' for
        # information on limitations and correct usage.
        #
        # Accepted values:
        # none
        #   The host has no system ID.
        # lvmlocal
        #   Obtain the system ID from the system_id setting in the 'local'
        #   section of an lvm configuration file, e.g. lvmlocal.conf.
        # uname
        #   Set the system ID from the hostname (uname) of the system.
        #   System IDs beginning localhost are not permitted.
        # machineid
        #   Use the contents of the machine-id file to set the system ID.
        #   Some systems create this file at installation time.
        #   See 'man machine-id' and global/etc.
        # file
        #   Use the contents of another file (system_id_file) to set the
        #   system ID.
        #
        # system_id_source = "none"
        system_id_source = "uname"
...
```

再次查看 LVM 的系统 ID，示例命令如下：

```
[root@kvm1 ~]# lvm systemid
  system ID: kvm1
```

现在 LVM 的系统 ID 值为主机名。

修改 LVM 配置之后，还需要重建 initramfs，建议先进行备份再重建。重建之后需要启动主机以便生效，示例命令如下：

```
[root@kvm1 ~]# cp /boot/initramfs-$(uname -r).img /boot/initramfs-$(uname -r).img.$(date +%m-%d-%H%M%S).bak

[root@kvm1 ~]# ls -t /boot/*.bak
/boot/initramfs-4.18.0-240.1.1.el8_3.x86_64.img.01-18-215013.bak
```

```
[root@kvm1 ~]# dracut -f -v
dracut: Executing: /usr/bin/dracut -f -v
...
dracut: *** Creating image file '/boot/initramfs-4.18.0-240.1.1.el8_3.x86_64.img' ***
dracut: *** Creating initramfs image file '/boot/initramfs-4.18.0-240.1.1.el8_3.x86_64.
img' done ***

[root@kvm1 ~]# reboot
```

重新启动完成后，进行 LVM 卷组创建，示例命令如下：

```
[root@kvm1 ~]# pvcreate /dev/sda1
  Physical volume "/dev/sda1" successfully created.

[root@kvm1 ~]# vgcreate vgkvm1 /dev/sda1
  Volume group "vgkvm1" successfully created with system ID kvm1

[root@kvm1 ~]# vgs -o+systemid
  VG     #PV #LV #SN Attr   VSize   VFree  System ID
  cl       1   3   0 wz--n- <79.00g     0
  vgkvm1   1   0   0 wz--n-  19.98g 19.98g kvm1
```

通过使用-o+systemid 选项来执行 vgs 命令，可以显示卷组的系统 ID。

在其上创建 LVM 逻辑卷，使用卷组的全部自由空间，然后在该卷上创建 XFS 文件系统以用于群集。示例命令如下：

```
[root@kvm1 ~]# lvcreate -n lvkvm1 -l 100%FREE vgkvm1
Logical volume "lvkvm1" created.

[root@kvm1 ~]# mkfs.xfs /dev/vgkvm1/lvkvm1
```

### 2. 在节点 kvm2 上执行的操作

类似地，也需要在 kvm2 上启用 LVM 的系统 ID 功能并重建 initramfs，示例命令如下：

```
[root@kvm2 ~]# vi /etc/lvm/lvm.conf
...
        # system_id_source = "none"
        system_id_source = "uname"
...

[root@kvm2 ~]# cp /boot/initramfs-$(uname -r).img /boot/initramfs-$(uname -r).img.
$(date +%m-%d-%H%M%S).bak

[root@kvm2 ~]# dracut -f -v

[root@kvm2 ~]# reboot
```

通过 iscsiadm 命令进行发现、登录操作，示例命令如下：

```
[root@kvm2 ~]# iscsiadm --mode discovery --type sendtargets \
    --portal 10.0.1.235

[root@kvm2 ~]# iscsiadm -m node --login

[root@kvm2 ~]# systemctl enable iscsid

[root@kvm2 ~]# systemctl restart iscsid
```

通过 fdisk 命令会查到新增加的磁盘/dv/sda，其中有一个基本分区，示例命令如下：

```
[root@kvm2 ~]# fdisk -l
...
Disk /dev/sda: 20 GiB, 21474836480 Bytes, 41943040 sectors
Units: sectors of 1 * 512 = 512 Bytes
Sector size (logical/physical): 512 Bytes / 512 Bytes
I/O size (minimum/optimal): 512 Bytes / 8388608 Bytes
Disklabel type: dos
Disk identifier: 0xa3b9c687

Device     Boot Start      End  Sectors Size Id Type
/dev/sda1       16384 41943039 41926656  20G 8e Linux LVM
```

由于启用了 LVM 的系统 ID 功能，所以在非属主(owner)节点上看不到对应物理卷、卷组和逻辑卷的信息，示例命令如下：

```
[root@kvm2 ~]# pvs
  PV         VG Fmt  Attr PSize   PFree
  /dev/vda2  cl lvm2 a--  <79.00g    0

[root@kvm2 ~]## vgs -o+systemid
  VG #PV #LV #SN Attr   VSize   VFree System ID
  cl   1   3   0 wz--n- <79.00g    0

[root@kvm2 ~]# lvs
  LV   VG Attr       LSize   Pool Origin Data%  Meta%  Move Log Cpy%Sync Convert
  home cl -wi-ao----  24.61g
  root cl -wi-ao---- <50.42g
  swap cl -wi-ao----   3.96g
```

### 2.6.3 创建卷组和文件系统资源

在群集中创建虚拟机资源之前，需要先创建所依赖的 LVM 卷组和文件系统资源。

在所有节点上创建逻辑卷的挂载目录，示例命令如下：

```
[AllNodes ~]#mkdir /lvmcluster
```

查看群集当前的状态，示例命令如下：

```
[root@kvm1 ~]#pcs status
Cluster name: cluster1
Cluster Summary:
  * Stack: corosync
  * Current DC: kvm1-cr (version 2.0.4-6.el8_3.1-2deceaa3ae) - partition with quorum
  * Last updated: Mon Jan 18 22:10:13 2021
  * Last change:  Mon Jan 18 17:31:14 2021 by root via cibadmin on kvm1-cr
  * 2 nodes configured
  * 2 resource instances configured

Node List:
  * Online: [ kvm1-cr kvm2-cr ]

Full List of Resources:
  * fence-kvm1    (stonith:fence_xvm):    Started kvm1-cr
  * fence-kvm2    (stonith:fence_xvm):    Started kvm2-cr

Daemon Status:
  corosync: active/enabled
  pacemaker: active/enabled
  pcsd: active/enabled
```

查看群集当前版本所支持的与LVM有关的资源，示例命令如下：

```
[root@kvm1 ~]#pcs resource list | grep -i lvm
ocf:heartbeat:LVM-activate - This agent activates/deactivates logical volumes.
ocf:heartbeat:lvmlockd - This agent manages the lvmlockd daemon
service:lvm2-lvmpolld - systemd unit file for lvm2-lvmpolld
service:lvm2-lvmpolld.socket - systemd unit file for lvm2-lvmpolld.socket
service:lvm2-monitor - systemd unit file for lvm2-monitor
systemd:lvm2-lvmpolld - systemd unit file for lvm2-lvmpolld
systemd:lvm2-lvmpolld.socket - systemd unit file for lvm2-lvmpolld.socket
systemd:lvm2-monitor - systemd unit file for lvm2-monitor
systemd:lvm2-pvscan@ - systemd unit file for lvm2-pvscan@
```

本实验中虚拟机使用的是磁盘映像文件，所以在创建群集资源时要考虑依赖关系：虚拟机依赖文件系统，而文件系统又依赖LVM卷组。

在管理群集资源时，可以将类似这样的依赖性、相关性的资源组成资源组，从而简化管理，例如：可以将资源组中的资源作为整体进行迁移。

先创建 LVM-activate 资源 my-vg，并将其指定为资源组 HA-LVM。由于资源组 HA-LVM 不存在，所以会先创建资源组，示例命令如下：

```
[root@kvm1 ~]# pcs resource create my-vg ocf:heartbeat:LVM-activate \
vgname=vgkvm1 \
activation_mode=exclusive \
vg_access_mode=system_id \
--group HA-LVM
```

创建资源时，资源会自动启动。查看群集状态、观察发生的变化，示例命令如下：

```
[root@kvm1 ~]# pcs status
Cluster name: cluster1
Cluster Summary:
  * Stack: corosync
  * Current DC: kvm1-cr (version 2.0.4-6.el8_3.1-2deceaa3ae) - partition with quorum
  * Last updated: Tue Jan 19 16:05:09 2021
  * Last change:  Tue Jan 19 16:04:51 2021 by root via cibadmin on kvm1-cr
  * 2 nodes configured
  * 3 resource instances configured

Node List:
  * Online: [ kvm1-cr kvm2-cr ]

Full List of Resources:
  * fence-kvm1    (stonith:fence_xvm):    Started kvm2-cr
  * fence-kvm2    (stonith:fence_xvm):    Started kvm1-cr
  * Resource Group: HA-LVM:
    * my-vg    (ocf::heartbeat:LVM-activate):    Started kvm1-cr

Daemon Status:
  corosync: active/enabled
  pacemaker: active/enabled
  pcsd: active/enabled
```

从输出可以看出：资源组 HA-LVM 创建成功并启动。

扫描逻辑卷的信息，示例命令如下：

```
[root@kvm1 ~]# lvscan
  ACTIVE  '/dev/cl/swap' [3.96 GiB] inherit
  ACTIVE  '/dev/cl/home' [24.61 GiB] inherit
  ACTIVE  '/dev/cl/root' [<50.42 GiB] inherit
  ACTIVE  '/dev/vgkvm1/lvkvm1' [19.98 GiB] inherit
```

有了 LVM 资源，就可以在其上创建文件系统类型的资源了。创建时可将逻辑卷指定为/dev/vgkvm1/lvkvm1，文件系统类型是 XFS，挂载路径是/lvmcluster，新资源也属于资源组 HA-LVM，示例命令如下：

```
[root@kvm1 ~]# pcs resource list | grep -i filesystem
ocf:heartbeat:Filesystem - Manages filesystem mounts


[root@kvm1 ~]# pcs resource create my-fs \
  ocf:heartbeat:Filesystem \
  device=/dev/vgkvm1/lvkvm1 \
  directory=/lvmcluster \
  fstype=xfs --group HA-LVM
```

查看群集状态，会在资源组 HA-LVM 看到有两个资源，即 my-vg 和 my-fs，示例命令如下：

```
[root@kvm1 ~]# pcs status
Cluster name: cluster1
Cluster Summary:
  * Stack: corosync
  * Current DC: kvm1-cr (version 2.0.4-6.el8_3.1-2deceaa3ae) - partition with quorum
  * Last updated: Tue Jan 19 16:09:27 2021
  * Last change:  Tue Jan 19 16:08:53 2021 by root via cibadmin on kvm1-cr
  * 2 nodes configured
  * 4 resource instances configured

Node List:
  * Online: [ kvm1-cr kvm2-cr ]

Full List of Resources:
  * fence-kvm1    (stonith:fence_xvm):     Started kvm2-cr
  * fence-kvm2    (stonith:fence_xvm):     Started kvm1-cr
  * Resource Group: HA-LVM:
    * my-vg    (ocf::heartbeat:LVM-activate):     Started kvm1-cr
    * my-fs    (ocf::heartbeat:Filesystem):     Started kvm1-cr

Daemon Status:
  corosync: active/enabled
  pacemaker: active/enabled
  pcsd: active/enabled
```

在 kvm1 上进行文件系统的读写测试，示例命令如下：

```
[root@kvm1 ~]# mount | grep lvmcluster
/dev/mapper/vgkvm1-lvkvm1 on /lvmcluster type xfs (rw,relatime,seclabel,attr2,inode64,
logbufs=8,logbsize=32k,noquota)
```

```
[root@kvm1 ~]# echo "kvm1" >> /lvmcluster/1.txt

[root@kvm1 ~]# ls -l /lvmcluster/
total 4
-rw-r--r--. 1 root root 5 Jan 19 16:29 1.txt
```

将 kvm1 设置为待机状态，对群集组资源进行迁移测试，示例命令如下：

```
[root@kvm1 ~]# pcs node standby kvm1-cr

[root@kvm1 ~]# pcs status
Cluster name: cluster1
Cluster Summary:
  * Stack: corosync
  * Current DC: kvm1-cr (version 2.0.4-6.el8_3.1-2deceaa3ae) - partition with quorum
  * Last updated: Tue Jan 19 16:32:34 2021
  * Last change:  Tue Jan 19 16:32:31 2021 by root via cibadmin on kvm1-cr
  * 2 nodes configured
  * 4 resource instances configured

Node List:
  * Node kvm1-cr: standby
  * Online: [ kvm2-cr ]

Full List of Resources:
  * fence-kvm1    (stonith:fence_xvm):     Started kvm2-cr
  * fence-kvm2    (stonith:fence_xvm):     Started kvm2-cr
  * Resource Group: HA-LVM:
    * my-vg    (ocf::heartbeat:LVM-activate):     Started kvm2-cr
    * my-fs    (ocf::heartbeat:Filesystem):       Started kvm2-cr

Daemon Status:
  corosync: active/enabled
  pacemaker: active/enabled
  pcsd: active/enabled
```

资源组 HA-LVM 已经迁移到 kvm2，所以在 kvm1 上无法再访问那个 LVM 资源了，示例命令如下：

```
[root@kvm1 ~]# vgs
  VG #PV #LV #SN Attr   VSize   VFree
  cl   1   3   0 wz--n- <79.00g    0

[root@kvm1 ~]# mount | grep lvmcluster
…无输出…

[root@kvm1 ~]# ls -l /lvmcluster/
total 0
```

而在 kvm2 上可以进行文件系统的读写，示例命令如下：

```
[root@kvm2 ~]# vgs -o+systemid
  VG     #PV #LV #SN Attr   VSize   VFree System ID
  cl       1   3   0 wz--n- <79.00g    0
  vgkvm1   1   1   0 wz--n-  19.98g    0  kvm2

[root@kvm2 ~]# df -h -T /lvmcluster/
Filesystem                Type  Size  Used Avail Use% Mounted on
/dev/mapper/vgkvm1-lvkvm1 xfs    20G  175M   20G   1% /lvmcluster

[root@kvm2 ~]# hostname >> /lvmcluster/2.txt

[root@kvm2 ~]# ls -l /lvmcluster/
total 8
-rw-r--r--. 1 root root 5 Jan 19 16:29 1.txt
-rw-r--r--. 1 root root 5 Jan 19 16:38 2.txt
```

下面将群集恢复为原有的状态，示例命令如下：

```
[root@kvm2 ~]# pcs node unstandby kvm1-cr

[root@kvm2 ~]# pcs status
Cluster name: cluster1
Cluster Summary:
  * Stack: corosync
  * Current DC: kvm1-cr (version 2.0.4-6.el8_3.1-2deceaa3ae) - partition with quorum
  * Last updated: Mon Jan 18 22:23:11 2021
  * Last change:  Mon Jan 18 22:23:09 2021 by root via cibadmin on kvm2-cr
  * 2 nodes configured
  * 4 resource instances configured

Node List:
  * Online: [ kvm1-cr kvm2-cr ]

Full List of Resources:
  * fence-kvm1    (stonith:fence_xvm):     Started kvm2-cr
  * fence-kvm2    (stonith:fence_xvm):     Started kvm1-cr
  * Resource Group: HA-LVM:
    * my-vg    (ocf::heartbeat:LVM-activate):     Started kvm2-cr
    * my-fs    (ocf::heartbeat:Filesystem):       Started kvm2-cr

Daemon Status:
  corosync: active/enabled
  pacemaker: active/enabled
  pcsd: active/enabled
```

将资源组迁回 kvm1 上，示例命令如下：

```
[root@kvm2 ~]# pcs resource move HA-LVM
Warning: Creating location constraint 'cli-ban-HA-LVM-on-kvm2-cr' with a score of -INFINITY for resource HA-LVM on kvm2-cr.
        This will prevent HA-LVM from running on kvm2-cr until the constraint is removed
        This will be the case even if kvm2-cr is the last node in the cluster

[root@kvm2 ~]# pcs status
Cluster name: cluster1
Cluster Summary:
  * Stack: corosync
  * Current DC: kvm1-cr (version 2.0.4-6.el8_3.1-2deceaa3ae) - partition with quorum
  * Last updated: Tue Jan 19 16:46:14 2021
  * Last change:  Tue Jan 19 16:46:06 2021 by root via crm_resource on kvm2-cr
  * 2 nodes configured
  * 4 resource instances configured

Node List:
  * Node kvm1-cr: standby
  * Online: [ kvm2-cr ]

Full List of Resources:
  * fence-kvm1    (stonith:fence_xvm):    Started kvm2-cr
  * fence-kvm2    (stonith:fence_xvm):    Started kvm2-cr
  * Resource Group: HA-LVM:
    * my-vg    (ocf::heartbeat:LVM-activate):    Stopped
    * my-fs    (ocf::heartbeat:Filesystem):      Stopped

Daemon Status:
  corosync: active/enabled
  pacemaker: active/ena
```

检查目前文件系统的状态，示例命令如下：

```
[root@kvm1 ~]# vgs -o+systemid
  VG     #PV #LV #SN Attr   VSize   VFree System ID
  cl       1   3   0 wz--n- <79.00g    0
  vgkvm1   1   1   0 wz--n-  19.98g    0  kvm1

[root@kvm1 ~]# ls -l /lvmcluster/
total 8
-rw-r--r--. 1 root root 5 Jan 19 16:29 1.txt
-rw-r--r--. 1 root root 5 Jan 19 16:38 2.txt
```

清除在移动资源组时生成的约束，示例命令如下：

```
[root@kvm1 ~]# pcs constraint --full
Location Constraints:
  Resource: HA-LVM
    Disabled on:
      Node: kvm2-cr (score:-INFINITY) (role:Started) (id:cli-ban-HA-LVM-on-kvm2-cr)
Ordering Constraints:
Colocation Constraints:
Ticket Constraints:

[root@kvm1 ~]# pcs constraint delete cli-ban-HA-LVM-on-kvm2-cr

[root@kvm1 ~]# pcs constraint
Location Constraints:
Ordering Constraints:
Colocation Constraints:
Ticket Constraints:
```

### 2.6.4 配置 SELinux

虚拟机的配置文件和磁盘映像放置在 iSCSI 的 LUN 中，而不是默认的/etc/libvirt/qemu 和/var/lib/libvirt/images。在启用 SELinux 的情况下，必须为虚拟机的配置文件和磁盘映像文件所在的目录设置正确的 SELinux 上下文。

首先查看默认目录的上下文，示例命令如下：

```
[root@kvm1 ~]# getenforce
Enforcing

[root@kvm1 ~]# ls -d -Z /etc/libvirt/qemu
system_u:object_r:virt_etc_rw_t:s0 /etc/libvirt/qemu

[root@kvm1 ~]# ls -d -Z /var/lib/libvirt/images
system_u:object_r:virt_image_t:s0 /var/lib/libvirt/images
```

创建群集中虚拟机所使用的目录结构，示例命令如下：

```
[root@kvm1 ~]# mkdir /lvmcluster/qemu_config

[root@kvm1 ~]# mkdir /lvmcluster/imgs
```

参照默认配置文件和磁盘映像原有目录的上下文，在所有的节点上修改目录的上下文，示例命令如下：

```
[AllNodes ~]# semanage fcontext -a -e /etc/libvirt/qemu \
    /lvmcluster/qemu_config

[AllNodes ~]# semanage fcontext -a -e /var/lib/libvirt/images \
    /lvmcluster/imgs

[AllNodes ~]# restorecon -R -v /lvmcluster/
```

## 2.6.5 创建虚拟机资源

准备好虚拟机的配置文件和磁盘映像，并放置到 iSCSI 的 LUN 中，示例命令如下：

```
[root@kvm1 ~]# ls -Z /lvmcluster/qemu_config/
unconfined_u:object_r:virt_etc_rw_t:s0 centos6.10a.xml

[root@kvm1 ~]# ls -Z /lvmcluster/imgs/
system_u:object_r:svirt_image_t:s0:c363,c384 centos6.10a.qcow2
```

然后编辑虚拟机配置文件中的映像文件的 source file，示例命令如下：

```
[root@kvm1 ~]# vi /lvmcluster/qemu_config/centos6.10a.xml
...
<disk type='file' device='disk'>
<driver name='qemu' type='qcow2'/>
<source file='/lvmcluster/imgs/centos6.10a.qcow2'/>
<target dev='vda' bus='virtio'/>
<address type='pci' domain='0x0000' bus='0x00' slot='0x07' function='0x0'/>
</disk>
...
```

创建虚拟机资源命令与基于 NFS 群集类似，其中要注意新资源 centos6.10a 属于资源组 HA-LVM，示例命令如下：

```
[root@kvm1 ~]# pcs resource create centos6.10a_res VirtualDomain  \
  hypervisor="qemu:///system"  \
  config="/lvmcluster/qemu_config/centos6.10a.xml"  \
  migration_transport=ssh  \
  op start timeout="120s"  \
  op stop timeout="120s"  \
  op monitor timeout="30"  \
    interval="10"  meta allow-migrate="true" priority="100"  \
  op migrate_from interval="0" timeout="120s"  \
  op migrate_to interval="0" timeout="120" \
  --group HA-LVM

Assumed agent name 'ocf:heartbeat:VirtualDomain' (deduced from 'VirtualDomain')
```

查看群集状态，最终会查到 centos6.10a_res 的状态为 Started，示例命令如下：

```
[root@kvm1 ~]# pcs status
Cluster name: cluster1
Cluster Summary:
  * Stack: corosync
  * Current DC: kvm1-cr (version 2.0.4-6.el8_3.1-2deceaa3ae) - partition with quorum
  * Last updated: Wed Jan 20 12:00:12 2021
  * Last change:  Wed Jan 20 11:59:36 2021 by root via crm_resource on kvm1-cr
  * 2 nodes configured
  * 5 resource instances configured

Node List:
  * Online: [ kvm1-cr kvm2-cr ]

Full List of Resources:
  * fence-kvm1    (stonith:fence_xvm):    Started kvm2-cr
  * fence-kvm2    (stonith:fence_xvm):    Started kvm1-cr
  * Resource Group: HA-LVM:
    * my-vg    (ocf::heartbeat:LVM-activate):    Started kvm1-cr
    * my-fs    (ocf::heartbeat:Filesystem):      Started kvm1-cr
    * centos6.10a_res    (ocf::heartbeat:VirtualDomain):    Started kvm1-cr

Daemon Status:
  corosync: active/enabled
  pacemaker: active/enabled
  pcsd: active/enabled
```

可以在 libvirt 中查到新的虚拟机。这种虚拟机没有 XML 配置文件，所以是临时型的虚拟机，示例命令如下：

```
[root@kvm1 ~]# virsh list --all
Id   Name          State
------------------------------
8    centos6.10a   running
```

## 2.6.6 群集测试

与前面测试的方法类似，登录到用于测试的虚拟机 centos6.10a 的控制台，在其中使用 vi 编辑器创建一个新文件，输入一些内容但不保存。

将节点 kvm1 设置为待机模式，这会触发资源组 HA-LVM 的迁移，示例命令如下：

```
[root@kvm1 ~]# pcs node standby kvm1-cr
```

马上查看群集的状态。会发现相对于资源 my-vg 和 myfs 来讲，centos6.10a_res 迁移得最慢，它显示的中间状态有以下两种。

(1) centos6.10a_res(ocf::heartbeat:VirtualDomain): Stopping kvm1-cr

(2) centos6.10a_res(ocf::heartbeat:VirtualDomain): Starting kvm2-cr

群集最终的状态如下：

```
[root@kvm1 ~]# pcs status
Cluster name: cluster1
Cluster Summary:
  * Stack: corosync
  * Current DC: kvm1-cr (version 2.0.4-6.el8_3.1-2deceaa3ae) - partition with quorum
  * Last updated: Wed Jan 20 12:04:21 2021
  * Last change:  Wed Jan 20 12:04:08 2021 by root via cibadmin on kvm1-cr
  * 2 nodes configured
  * 5 resource instances configured

Node List:
  * Node kvm1-cr: standby
  * Online: [ kvm2-cr ]

Full List of Resources:
  * fence-kvm1    (stonith:fence_xvm):     Started kvm2-cr
  * fence-kvm2    (stonith:fence_xvm):     Started kvm2-cr
  * Resource Group: HA-LVM:
    * my-vg    (ocf::heartbeat:LVM-activate):     Started kvm2-cr
    * my-fs    (ocf::heartbeat:Filesystem):       Started kvm2-cr
    * centos6.10a_res    (ocf::heartbeat:VirtualDomain):     Started kvm2-cr

Daemon Status:
  corosync: active/enabled
  pacemaker: active/enabled
  pcsd: active/enabled
```

在节点 kvm2 查看虚拟机列表，示例命令如下：

```
[root@kvm2 ~]# virsh list
Id   Name           State
------------------------------------
6    centos6.10a   running
```

虽然虚拟机运行正常，但是由于不是实时迁移，所以原有内存中的数据会丢失，虚拟机中的业务会发生中断，这不是最佳的虚拟机群集的架构。

## 2.6.7 删除群集资源

下面将删除资源为后续的实验清理环境，示例命令如下：

```
[root@kvm1 ~]# pcs node unstandby kvm1-cr

[root@kvm1 ~]# pcs resource delete centos6.10a_res
Attempting to stop: centos6.10a_res... Stopped

[root@kvm1 ~]# pcs resource clear HA-LVM
Removing constraint: cli-ban-HA-LVM-on-kvm2-cr

[root@kvm1 ~]# pcs status
Cluster name: cluster1
Cluster Summary:
  * Stack: corosync
  * Current DC: kvm1-cr (version 2.0.4-6.el8_3.1-2deceaa3ae) - partition with quorum
  * Last updated: Wed Jan 20 12:19:07 2021
  * Last change:  Wed Jan 20 12:19:02 2021 by root via crm_resource on kvm1-cr
  * 2 nodes configured
  * 4 resource instances configured

Node List:
  * Online: [ kvm1-cr kvm2-cr ]

Full List of Resources:
  * fence-kvm1     (stonith:fence_xvm):     Started kvm2-cr
  * fence-kvm2     (stonith:fence_xvm):     Started kvm1-cr
  * Resource Group: HA-LVM:
    * my-vg    (ocf::heartbeat:LVM-activate):     Started kvm1-cr
    * my-fs    (ocf::heartbeat:Filesystem):       Started kvm1-cr

Daemon Status:
  corosync: active/enabled
  pacemaker: active/enabled
  pcsd: active/enabled
```

由于后续的实验将使用 lvm2-lockd 和 gfs2，所以将资源组 HA-LVM 也删除，只留下一个基本的群集环境，示例命令如下：

```
[root@kvm1 ~]# pcs resource delete  HA-LVM
Removing group: HA-LVM (and all resources within group)
Stopping all resources in group: HA-LVM...
Deleting Resource - my-vg
Deleting Resource (and group) - my-fs
```

```
[root@kvm1 ~]# pcs status
Cluster name: cluster1
Cluster Summary:
  * Stack: corosync
  * Current DC: kvm1-cr (version 2.0.4-6.el8_3.1-2deceaa3ae) - partition with quorum
  * Last updated: Wed Jan 20 12:31:34 2021
  * Last change:  Wed Jan 20 12:31:29 2021 by root via cibadmin on kvm1-cr
  * 2 nodes configured
  * 2 resource instances configured

Node List:
  * Online: [ kvm1-cr kvm2-cr ]

Full List of Resources:
  * fence-kvm1    (stonith:fence_xvm):    Started kvm2-cr
  * fence-kvm2    (stonith:fence_xvm):    Started kvm1-cr

Daemon Status:
  corosync: active/enabled
  pacemaker: active/enabled
  pcsd: active/enabled
```

在 kvm1 上删除 LV、VG 和 PV,示例命令如下:

```
[root@kvm1 ~]# lvremove /dev/vgkvm1/lvkvm1

[root@kvm1 ~]# vgremove vgkvm1

[root@kvm1 ~]# pvremove /dev/sda1
```

在所有节点上执行以下操作以便恢复原有的 LVM 配置,示例命令如下:

```
[AllNodes ~]# vi /etc/lvm/lvm.conf
#将 system_id_source 修改为原来的值
        system_id_source = "none"

[AllNodes ~]# cp /boot/initramfs-$(uname -r).img /boot/initramfs-$(uname -r).img.
$(date +%m-%d-%H%M%S).bak

[AllNodes ~]# dracut -f -v

[AllNodes ~]# reboot
```

## 2.7 基于 iSCSI 的 KVM 群集 2

在 RHEL/CentOS 8 中，主动/主动模式 LVM 卷需要有 lvmlockd 守护程序的支持，这样可以避免多个节点进行重叠更改，从而防止 LVM 元数据及其逻辑卷的损坏。

本实验将在共享的 LVM 上创建 GFS2(Global File System 2)文件系统，它允许群集中的所有节点并发访问同一共享块存储。

本实验还是利用上一实验的环境：基本的群集环境已经安装完毕，iSCSI 目标(服务器)与启动器(客户端)也已经配置完成。

**提示**：在 RHEL/CentOS 7 及更早的版本中，lvmlockd 的功能是由 clvmd(Cluster LVM daemon)实现的。

### 2.7.1 安装软件包

在群集的所有节点，安装 lvm2-lockd、gfs2-utils 和 dlm 包。

CentOS 8 软件仓库中包括 lvm2-lockd 和 gfs2-utils 软件包，可以直接在所有节点上进行安装，示例命令如下：

```
[AllNodes ~]# dnf -y install lvm2-lockd gfs2-utils
```

当前的 CentOS 8 软件仓库不提供 dlm，但是可以先安装其所依赖的库 dlm-lib，示例命令如下：

```
[root@kvm1 ~]# dnf repolist
repo id         repo name
appstream       CentOS Linux 8 - AppStream
baseos          CentOS Linux 8 - BaseOS
epel            Extra Packages for Enterprise Linux 8 - x86_64
epel-modular    Extra Packages for Enterprise Linux Modular 8 - x86_64
extras          CentOS Linux 8 - Extras
ha              CentOS Linux 8 - HighAvailability

[root@tomkvm1 ~]# dnf list | grep dlm
dlm-lib.i686      4.0.9-3.el8   baseos
dlm-lib.x86_64    4.0.9-3.el8   baseos

[root@kvm1 ~]# dnf -y install dlm-lib
```

dlm 包含在 RHEL 8 Resilient Storage 中，这是一个需要订阅的服务。CentOS 8 并没有提供它的编译包，所以我们需要通过源 RPM 包进行编译。

首先下载源 RPM 包，示例命令如下：

```
[root@kvm1 ~]# cat /etc/redhat-release
centos Linux release 8.3.2011

[root@kvm1 ~]# uname -a
Linux kvm1 4.18.0-240.1.1.el8_3.x86_64 #1 SMP Thu Nov 19 17:20:08 UTC 2020 x86_64 x86_64
x86_64 GNU/Linux

[root@kvm1 ~]# wget http://vault.centos.org/8.3.2011/BaseOS/Source/SPackages/dlm-4.0.9
-3.el8.src.rpm
```

在一个节点上编译。编译源 RPM 需要其他软件的支持，所以先安装它们，示例命令如下：

```
[root@kvm1 ~]# yum -y install make gcc rpm-build corosynclib-devel glibc-kernheaders
libxml2-devel pacemaker-libs-devel systemd-devel

[root@kvm1 ~]# rpmbuild --rebuild dlm-4.0.9-3.el8.src.rpm

[root@kvm1 ~]# tree rpmbuild/
rpmbuild/
├── BUILD
├── BUILDROOT
├── RPMS
│   └── x86_64
│       ├── dlm-4.0.9-3.el8.x86_64.rpm
│       ├── dlm-debuginfo-4.0.9-3.el8.x86_64.rpm
│       ├── dlm-debugsource-4.0.9-3.el8.x86_64.rpm
│       ├── dlm-devel-4.0.9-3.el8.x86_64.rpm
│       ├── dlm-lib-4.0.9-3.el8.x86_64.rpm
│       └── dlm-lib-debuginfo-4.0.9-3.el8.x86_64.rpm
├── SOURCES
├── SPECS
└── SRPMS
```

在所有节点上安装 dlm 软件包，示例命令如下：

```
[AllNodes ~]# rpm -ivh rpmbuild/RPMS/x86_64/dlm-4.0.9-3.el8.x86_64.rpm
```

dlm 软件包中的 3 个命令是构建群集所必需的，示例命令如下：

```
[root@kvm1 ~]# rpm -ql dlm | grep sbin
/usr/sbin/dlm_controld
/usr/sbin/dlm_stonith
/usr/sbin/dlm_tool
```

## 2.7.2 在群集中创建 LVM 卷组及文件系统资源

通过以下命令查看群集当前的非默认属性：

```
[root@kvm1 ~]#pcs property
Cluster Properties:
cluster-infrastructure: corosync
cluster-name: cluster1
dc-version: 2.0.4-6.el8_3.1-2deceaa3ae
have-watchdog: false
stonith-enabled: true
```

属性 no-quorum-policy 的默认值是 stop，表示一旦丢失仲裁(不到法定人数)，剩余分区上的所有资源将立即停止。通常，这个默认值是最安全和最佳的选项，但不适合 GFS2 资源，需要将其设置为 freeze，示例命令如下：

```
[root@kvm1 ~]#pcs property set no-quorum-policy=freeze

[root@kvm1 ~]#pcs property
Cluster Properties:
cluster-infrastructure: corosync
cluster-name: cluster1
dc-version: 2.0.4-6.el8_3.1-2deceaa3ae
have-watchdog: false
no-quorum-policy: freeze
stonith-enabled: true
```

文件系统 GFS2 依赖 LVM 卷，LVM 卷又依赖 dlm 资源提供的锁定服务，所以将它们组织到资源组中以方便统一管理。下面将创建一个名为 dlm 的资源，它属于资源组 locking，示例命令如下：

```
[root@kvm1 ~]#pcs resource create dlm --group locking \
    ocf:pacemaker:controld op monitor interval=30s on-fail=fence

[root@kvm1 ~]#pcs status
Cluster name: cluster1
Cluster Summary:
  * Stack: corosync
  * Current DC: kvm2-cr (version 2.0.4-6.el8_3.1-2deceaa3ae) - partition with quorum
  * Last updated: Wed Jan 20 15:51:00 2021
  * Last change:  Wed Jan 20 15:50:56 2021 by root via cibadmin on kvm1-cr
  * 2 nodes configured
```

```
  * 3 resource instances configured

Node List:
  * Online: [ kvm1-cr kvm2-cr ]

Full List of Resources:
  * fence-kvm1    (stonith:fence_xvm):    Started kvm2-cr
  * fence-kvm2    (stonith:fence_xvm):    Started kvm1-cr
  * Resource Group: locking:
    * dlm    (ocf::pacemaker:controld):    Started kvm1-cr

Daemon Status:
  corosync: active/enabled
  pacemaker: active/enabled
  pcsd: active/enabled
```

每个节点上都需要有 dlm 资源，这可以通过克隆 locking 资源组实现，示例命令如下：

```
[root@kvm1 ~]# pcs resource clone locking interval=true

[root@kvm1 ~]# pcs status --full
Cluster name: cluster1
Cluster Summary:
  * Stack: corosync
  * Current DC: kvm2-cr (2) (version 2.0.4-6.el8_3.1-2deceaa3ae) - partition with quorum
  * Last updated: Wed Jan 20 15:53:27 2021
  * Last change:  Wed Jan 20 15:52:16 2021 by root via cibadmin on kvm1-cr
  * 2 nodes configured
  * 4 resource instances configured

Node List:
  * Online: [ kvm1-cr (1) kvm2-cr (2) ]

Full List of Resources:
  * fence-kvm1    (stonith:fence_xvm):    Started kvm2-cr
  * fence-kvm2    (stonith:fence_xvm):    Started kvm1-cr
  * Clone Set: locking-clone [locking]:
    * Resource Group: locking:0:
      * dlm    (ocf::pacemaker:controld):    Started kvm1-cr
    * Resource Group: locking:1:
      * dlm    (ocf::pacemaker:controld):    Started kvm2-cr

Migration Summary:

Tickets:

PCSD Status:
```

```
  kvm1 - cr: Online
  kvm2 - cr: Online

Daemon Status:
  corosync: active/enabled
  pacemaker: active/enabled
  pcsd: active/enabled
```

使用--full 选项可以查看更详细的状态，从中可以看到 locking 资源组在群集的每个节点上都处于启动状态。

有了 dlm，就可以创建 lvmlockd 资源了，示例命令如下：

```
[root@kvm1 ~]# pcs resource create lvmlockd --group locking \
    ocf:heartbeat:lvmlockd op monitor interval=30s on-fail=fence

[root@kvm1 ~]# pcs status --full
Cluster name: cluster1
Cluster Summary:
  * Stack: corosync
  * Current DC: kvm2-cr (2) (version 2.0.4-6.el8_3.1-2deceaa3ae) - partition with quorum
  * Last updated: Wed Jan 20 15:55:16 2021
  * Last change:  Wed Jan 20 15:54:48 2021 by root via cibadmin on kvm1-cr
  * 2 nodes configured
  * 6 resource instances configured

Node List:
  * Online: [ kvm1-cr (1) kvm2-cr (2) ]

Full List of Resources:
  * fence-kvm1     (stonith:fence_xvm):      Started kvm2-cr
  * fence-kvm2     (stonith:fence_xvm):      Started kvm1-cr
  * Clone Set: locking-clone [locking]:
    * Resource Group: locking:0:
      * dlm        (ocf::pacemaker:controld):     Started kvm1-cr
      * lvmlockd   (ocf::heartbeat:lvmlockd):     Started kvm1-cr
    * Resource Group: locking:1:
      * dlm        (ocf::pacemaker:controld):     Started kvm2-cr
      * lvmlockd   (ocf::heartbeat:lvmlockd):     Started kvm2-cr

Migration Summary:

Tickets:

PCSD Status:
  kvm1-cr: Online
```

```
  kvm2-cr: Online

Daemon Status:
  corosync: active/enabled
  pacemaker: active/enabled
  pcsd: active/enabled
```

创建成功后，再验证 lvmlockd 守护程序是否正在集群的每个节点上运行，示例命令如下：

```
[root@kvm1 ~]# ps -ef | grep lvmlockd
root     15970     1    0 15:54 ?       00:00:00 lvmlockd -p /run/lvmlockd.pid -A 1 -g dlm
root     16181     1955 0 15:56 pts/0   00:00:00 grep --color=auto lvmlockd

[root@kvm1 ~]# ssh kvm2 "ps -ef | grep lvmlockd"
root     3600      1    0 15:54 ?       00:00:00 lvmlockd -p /run/lvmlockd.pid -A 1 -g dlm
root     3817      3816 0 15:56 ?       00:00:00 bash -c ps -ef | grep lvmlockd
root     3833      3817 0 15:56 ?       00:00:00 grep lvmlockd
```

在群集的一个节点上创建一个共享的 LVM 卷组。在创建卷组时，需要提供--shared 选项，示例命令如下：

```
[root@kvm1 ~]# pvcreate /dev/sda1
  Physical volume "/dev/sda1" successfully created.

[root@kvm1 ~]# pvscan
  PV /dev/vda2   VG cl            lvm2 [<79.00 GiB / 0    free]
  PV /dev/sda1                    lvm2 [19.99 GiB]
  Total: 2 [<98.99 GiB] / in use: 1 [<79.00 GiB] / in no VG: 1 [19.99 GiB]

[root@kvm1 ~]# vgcreate --shared shared-vg1 /dev/sda1
  Volume group "shared-vg1" successfully created
  VG shared-vg1 starting dlm lockspace
  Starting locking.  Waiting until locks are ready...

[root@kvm1 ~]# vgscan
  Found volume group "cl" using metadata type lvm2
  Found volume group "shared-vg1" using metadata type lvm2

[root@kvm1 ~]# vgs
  VG         #PV #LV #SN Attr   VSize   VFree
  cl           1   3   0 wz--n- <79.00g      0
  shared-vg1   1   0   0 wz--ns  19.98g 19.98g
```

由于是共享的卷组，所以在另外一个节点上也会看到相同的信息，示例命令如下：

```
[root@kvm2 ~]# pvscan
  PV /dev/vda2    VG cl                 lvm2 [<79.00 GiB / 0     free]
  Reading VG shared-vg1 without a lock.
  PV /dev/sda1    VG shared-vg1         lvm2 [19.98 GiB / 19.98 GiB free]
  Total: 2 [98.98 GiB] / in use: 2 [98.98 GiB] / in no VG: 0 [0   ]

[root@kvm2 ~]# vgscan
  Found volume group "cl" using metadata type lvm2
  Reading VG shared-vg1 without a lock.
  Found volume group "shared-vg1" using metadata type lvm2

[root@kvm2 ~]# vgs
  Reading VG shared-vg1 without a lock.
  VG         #PV #LV #SN Attr   VSize   VFree
  cl           1   3   0 wz--n- <79.00g     0
  shared-vg1   1   0   0 wz--ns  19.98g 19.98g
```

使用 vgchange 命令为共享卷组启动锁管理器，示例命令如下：

```
[root@kvm2 ~]# vgchange --lock-start shared-vg1
  VG shared-vg1 starting dlm lockspace
  Starting locking.  Waiting until locks are ready...
```

卷组就绪后，就可以创建共享逻辑卷了。新逻辑卷的名称为 shared-lv1，使用全部的可用空间，示例命令如下：

```
[root@kvm1 ~]# lvcreate --activate sy -l 100%FREE -n shared-lv1 shared-vg1
WARNING: xfs signature detected on /dev/shared-vg1/shared-lv1 at offset 0. Wipe it? [y/n]: y
  Wiping xfs signature on /dev/shared-vg1/shared-lv1.
  Logical volume "shared-lv1" created.
```

在每个节点上验证是否创建成功，示例命令如下：

```
[root@kvm1 ~]# lvscan
  ACTIVE            '/dev/cl/swap' [3.96 GiB] inherit
  ACTIVE            '/dev/cl/home' [24.61 GiB] inherit
  ACTIVE            '/dev/cl/root' [<50.42 GiB] inherit
  ACTIVE            '/dev/shared-vg1/shared-lv1' [19.98 GiB] inherit

[root@kvm1 ~]# ssh kvm2 lvscan
  ACTIVE            '/dev/cl/swap' [3.96 GiB] inherit
  ACTIVE            '/dev/cl/home' [24.61 GiB] inherit
  ACTIVE            '/dev/cl/root' [<50.42 GiB] inherit
  inactive          '/dev/shared-vg1/shared-lv1' [19.98 GiB] inherit
```

使用 GFS2 文件系统格式化逻辑卷，示例命令如下：

```
[root@kvm1 ~]# mkfs.gfs2 -j2 -p lock_dlm -t cluster1:gfs2-test1 /dev/shared-vg1/shared-lv1
/dev/shared-vg1/shared-lv1 is a symbolic link to /dev/dm-3
This will destroy any data on /dev/dm-3
Are you sure you want to proceed? [y/n] y
Discarding device contents (may take a while on large devices): Done
Adding journals: Done
Building resource groups: Done
Creating quota file: Done
Writing superblock and syncing: Done
Device:                    /dev/shared-vg1/shared-lv1
Block size:                4096
Device size:               19.98 GB (5238784 blocks)
Filesystem size:           19.98 GB (5238782 blocks)
Journals:                  2
Journal size:              64MB
Resource groups:           82
Locking protocol:          "lock_dlm"
Lock table:                "cluster1:gfs2-test1"
UUID:                      47804020-bb3e-4cc5-ad94-5f23b7b3f218
```

下面需要为逻辑卷再创建一个 LVM-activate 类型资源，它将自动激活逻辑卷。将这个新资源放置到一个新的资源组 shared-vg1，示例命令如下：

```
[root@kvm1 ~]# pcs resource create sharedlv1 \
  --group shared-vg1 ocf:heartbeat:LVM-activate \
  lvname=shared-lv1 \
  vgname=shared-vg1 \
  activation_mode=shared \
  vg_access_mode=lvmlockd
```

克隆资源组 shared-vg1，保证在所有节点上启动该资源，示例命令如下：

```
[root@kvm1 ~]# pcs resource clone shared-vg1 interleave=true
```

查看群集的当前状态，示例命令如下：

```
[root@kvm1 ~]# pcs status --full
Cluster name: cluster1
Cluster Summary:
  * Stack: corosync
  * Current DC: kvm2-cr (2) (version 2.0.4-6.el8_3.1-2deceaa3ae) - partition with quorum
  * Last updated: Wed Jan 20 16:38:06 2021
```

```
  * Last change:   Wed Jan 20 16:37:56 2021 by root via cibadmin on kvm1-cr
  * 2 nodes configured
  * 8 resource instances configured

Node List:
  * Online: [ kvm1-cr (1) kvm2-cr (2) ]

Full List of Resources:
  * fence-kvm1      (stonith:fence_xvm):             Started kvm2-cr
  * fence-kvm2      (stonith:fence_xvm):             Started kvm1-cr
  * Clone Set: locking-clone [locking]:
    * Resource Group: locking:0:
      * dlm         (ocf::pacemaker:controld):       Started kvm1-cr
      * lvmlockd    (ocf::heartbeat:lvmlockd):       Started kvm1-cr
    * Resource Group: locking:1:
      * dlm         (ocf::pacemaker:controld):       Started kvm2-cr
      * lvmlockd    (ocf::heartbeat:lvmlockd):       Started kvm2-cr
  * Clone Set: shared-vg1-clone [shared-vg1]:
    * Resource Group: shared-vg1:0:
      * sharedlv1   (ocf::heartbeat:LVM-activate):   Started kvm1-cr
    * Resource Group: shared-vg1:1:
      * sharedlv1   (ocf::heartbeat:LVM-activate):   Started kvm2-cr

Migration Summary:

Tickets:

PCSD Status:
  kvm1-cr: Online
  kvm2-cr: Online

Daemon Status:
  corosync: active/enabled
  pacemaker: active/enabled
  pcsd: active/enabled
```

从输出可以看出群集有 dlm(controld)、lvmlockd(lvmlockd)、sharedlv1(LVM-activate)3 个资源,dlm 与 lvmlockd 组成了 1 个名为 locking 的资源组,sharedlv1 在 1 个名为 shared-vg1 的资源组中,并且通过克隆的方式保证每个节点上都运行这些资源。

由于资源组有依赖关系,所以要创建次序约束,以确保先启动包含 dlm 和 lvmlockd 资源的资源组 locking,再启动资源组 shared-vg1,示例命令如下:

```
[root@kvm1 ~]# pcs constraint order start locking-clone then shared-vg1-clone
Adding locking-clone shared-vg1-clone (kind: Mandatory) (Options: first-action=start
then-action=start)

[root@kvm1 ~]# pcs constraint
Location Constraints:
Ordering Constraints:
  start locking-clone then start shared-vg1-clone (kind:Mandatory)
Colocation Constraints:
Ticket Constraints:
```

需要创建位置约束，以确保资源组 shared-vg1 与资源组 locking 在同一节点上，示例命令如下：

```
[root@kvm1 ~]# pcs constraint colocation add shared-vg1-clone with locking-clone

[root@kvm1 ~]# pcs constraint
Location Constraints:
Ordering Constraints:
  start locking-clone then start shared-vg1-clone (kind:Mandatory)
Colocation Constraints:
  shared-vg1-clone with locking-clone (score:INFINITY)
Ticket Constraints:
```

在集群中的每个节点上，验证逻辑卷是否处于活动状态(有可能有延迟钟)，示例命令如下：

```
[root@kvm1 ~]# lvs
  LV        VG          Attr        LSize   Pool Origin Data%  Meta%  Move Log Cpy%
Sync Convert
  home      cl          -wi-ao----  24.61g
  root      cl          -wi-ao---- <50.42g
  swap      cl          -wi-ao----   3.96g
  shared-lv1 shared-vg1 -wi-a-----  19.98g

[root@kvm1 ~]# ssh kvm2 lvs
  LV        VG          Attr        LSize   Pool Origin Data%  Meta%  Move Log Cpy%
Sync Convert
  home      cl          -wi-ao----  24.61g
  root      cl          -wi-ao---- <50.42g
  swap      cl          -wi-ao----   3.96g
  shared-lv1 shared-vg1 -wi-a-----  19.98g
```

在所有节点上创建目录，然后将创建的文件系统挂载到此目录，示例命令如下：

```
[root@kvm1 ~]# mkdir /mnt/gfs1

[root@kvm1 ~]# ssh kvm2 "mkdir /mnt/gfs1"
```

创建文件系统资源，这将自动在所有节点上挂载GFS2文件系统。新资源属于组shared-vg1，将逻辑卷/dev/shared-vg1/shared-lv1挂载到目录/mnt/gfs1中，示例命令如下：

```
[root@kvm1 ~]# pcs resource describe Filesystem
Assumed agent name 'ocf:heartbeat:Filesystem' (deduced from 'Filesystem')
ocf:heartbeat:Filesystem - Manages filesystem mounts

Resource script for Filesystem. It manages a Filesystem on a
shared storage medium.
...

[root@kvm1 ~]# pcs resource create sharedfs1 \
  --group shared-vg1 ocf:heartbeat:Filesystem \
  device="/dev/shared-vg1/shared-lv1" \
  directory="/mnt/gfs1" \
  fstype="gfs2" \
  options=noatime \
  op monitor interval=10s on-fail=fence
```

查看群集状态，每个节点均有sharedfs1资源，示例命令如下：

```
[root@kvm1 ~]# pcs status  --full
Cluster name: cluster1
Cluster Summary:
  * Stack: corosync
  * Current DC: kvm2-cr (2) (version 2.0.4-6.el8_3.1-2deceaa3ae) - partition with quorum
  * Last updated: Wed Jan 20 16:51:07 2021
  * Last change:  Wed Jan 20 16:50:48 2021 by root via cibadmin on kvm1-cr
  * 2 nodes configured
  * 10 resource instances configured

Node List:
  * Online: [ kvm1-cr (1) kvm2-cr (2) ]

Full List of Resources:
  * fence-kvm1      (stonith:fence_xvm):          Started kvm2-cr
  * fence-kvm2      (stonith:fence_xvm):          Started kvm1-cr
  * Clone Set: locking-clone [locking]:
    * Resource Group: locking:0:
```

```
      * dlm         (ocf::pacemaker:controld):      Started kvm1-cr
      * lvmlockd    (ocf::heartbeat:lvmlockd):      Started kvm1-cr
    * Resource Group: locking:1:
      * dlm         (ocf::pacemaker:controld):      Started kvm2-cr
      * lvmlockd    (ocf::heartbeat:lvmlockd):      Started kvm2-cr
  * Clone Set: shared-vg1-clone [shared-vg1]:
    * Resource Group: shared-vg1:0:
      * sharedlv1   (ocf::heartbeat:LVM-activate):  Started kvm1-cr
      * sharedfs1   (ocf::heartbeat:Filesystem):    Started kvm1-cr
    * Resource Group: shared-vg1:1:
      * sharedlv1   (ocf::heartbeat:LVM-activate):  Started kvm2-cr
      * sharedfs1   (ocf::heartbeat:Filesystem):    Started kvm2-cr

Migration Summary:

Tickets:

PCSD Status:
  kvm1-cr: Online
  kvm2-cr: Online

Daemon Status:
  corosync: active/enabled
  pacemaker: active/enabled
  pcsd: active/enabled
```

检查每个节点上的挂载情况，示例命令如下：

```
[root@kvm1 ~]# mount | grep gfs2
/dev/mapper/shared--vg1-shared--lv1 on /mnt/gfs1 type gfs2 (rw,noatime,seclabel)

[root@kvm1 ~]# ssh kvm2 "mount | grep gfs2"
/dev/mapper/shared--vg1-shared--lv1 on /mnt/gfs1 type gfs2 (rw,noatime,seclabel)
```

测试每个节点是否可以读写，示例命令如下：

```
[root@kvm1 ~]# echo kvm1 > /mnt/gfs1/1.txt

[root@kvm1 ~]# ssh kvm2 "echo kvm2 > /mnt/gfs1/2.txt"

[root@kvm1 ~]# ls -l /mnt/gfs1/
total 16
-rw-r--r--. 1 root root 5 Jan 20 16:52 1.txt
-rw-r--r--. 1 root root 5 Jan 20 16:53 2.txt

[root@kvm1 ~]# cat /mnt/gfs1/*.txt
kvm1
kvm2
```

**提示**：不应将群集中的文件系统添加到/etc/fstab文件中，因为它将作为Pacemaker群集资源进行管理。

### 2.7.3 禁用SELinux

根据RHEL的官方文档*Support Policies for RHEL Resilient Storage-gfs2 with SELinux*（https://access.redhat.com/articles/3244861），RHEL8支持GFS2文件系统，但是如果在挂载时不指定context选项，则会影响性能。

但是在当前版本的实验中，如果在创建文件系统的命令的options中添加SELinux的上下文，则在启动虚拟机时会出错，示例命令如下：

```
[root@kvm1 ~]# pcs resource create sharedfs1 \
  --group shared-vg1 ocf:heartbeat:Filesystem \
  device="/dev/shared-vg1/shared-lv1" \
  directory="/mnt/gfs1" \
  fstype="gfs2" \
  options='noatime,context=system_u:object_r:virt_image_t:s0'\
  op monitor interval=10s on-fail=fence
```

在启动虚拟机时的出错信息如下：

```
[root@kvm1 ~]# virsh start centos6.10a
error: Failed to start domain centos6.10a
error: internal error: process exited while connecting to monitor: 2021-01-21T06:43:09.
934405Z qemu-kvm: -blockdev {"node-name":"libvirt-1-format","read-only":false,
"driver":"qcow2","file":"libvirt-1-storage","backing":null}: Could not reopen file:
Permission denied
```

根据日志出错信息，即使修改文件的SELinux上下文也无法排除错误。通过综合考虑，下面将禁用所有群集节点的SELinux，示例命令如下：

```
[AllNodes ~]# vi /etc/sysconfig/seLinux
修改为 SELINUX=disabled

[AllNodes ~]# reboot

[AllNodes ~]# getenforce
Disabled
```

如果启动虚拟机时出现以下错误，则需要将虚拟机配置文件中与SELinux相关的属性删除，示例命令如下：

```
[root@kvm1 ~]# virsh start centos6.10a
error: Failed to start domain centos6.10a
error: unsupported configuration: Security driver model 'seLinux' is not available

[root@kvm1 ~]# virsh edit centos6.10a
删除<seclabel type='dynamic' model='seLinux' relabel='yes'/>

[root@kvm1 ~]# virsh start centos6.10a
Domain centos6.10a started
```

### 2.7.4 准备测试用的虚拟机并测试实时迁移

由于有共享存储，所以可以进行实时迁移。在创建群集资源之前，要保证此虚拟机可以在群集的节点进行实时迁移。

将虚拟机从kvm1迁移到kvm2，示例命令如下：

```
[root@kvm1 ~]# virsh migrate --persistent --undefinesource centos6.10a \
  qemu+ssh://kvm2/system

[root@kvm1 ~]# virsh list --all
Id   Name   State
--------------------

[root@kvm1 ~]# virsh --connect=qemu+ssh://kvm2/system list
Id   Name           State
------------------------------
1    centos6.10a   running
```

将虚拟机从kvm2迁回kvm1，示例命令如下：

```
[root@kvm1 ~]# virsh --connect=qemu+ssh://kvm2/system migrate --persistent \
    --undefinesource centos6.10a qemu+ssh://kvm1/system

[root@kvm1 ~]# virsh list
Id   Name           State
------------------------------
3    centos6.10a   running
```

与实时迁移相关的知识，可参见本书的“第1章　实现虚拟机迁移”。

### 2.7.5 在群集中创建虚拟机资源

准备好虚拟机的配置文件和磁盘映像，并放置到GFS2卷中，示例命令如下：

```
[root@kvm1 ~]# virsh shutdown centos6.10a

[root@kvm1 ~]# mkdir /mnt/gfs1/config/

[root@kvm1 ~]# mkdir /mnt/gfs1/imgs/

[root@kvm1 ~]# cp /etc/libvirt/qemu/centos6.10a.xml /mnt/gfs1/config/

[root@kvm1 ~]# cp /var/lib/libvirt/images/centos6.10a.qcow2 /mnt/gfs1/imgs/

[root@kvm1 ~]# virsh undefine centos6.10a
```

然后编辑虚拟机配置文件中的映像文件的 source file,示例命令如下：

```
[root@kvm1 ~]# vi /mnt/gfs1/config/centos6.10a.xml
...
<disk type='file' device='disk'>
<driver name='qemu' type='qcow2'/>
<source file='/mnt/gfs1/imgs/centos6.10a.qcow2'/>
<backingStore/>
<target dev='vda' bus='virtio'/>
<address type='pci' domain='0x0000' bus='0x00' slot='0x07' function='0x0'/>
</disk>
...
```

创建虚拟机资源命令与前面类似,本次仅使用最少的选项参数,其他的参数使用默认值,示例命令如下：

```
[root@kvm1 ~]# pcs resource create centos6.10a_res VirtualDomain \
  hypervisor="qemu:///system" \
  config="/mnt/gfs1/config/centos6.10a.xml" \
  migration_transport=ssh \
  meta allow-migrate="true"

Assumed agent name 'ocf:heartbeat:VirtualDomain' (deduced from 'VirtualDomain')

[root@kvm1 ~]# virsh list
Id   Name          State
------------------------------
3    centos6.10a   running
```

查看群集状态,示例命令如下：

```
[root@kvm1 ~]# pcs status --full
Cluster name: cluster1
```

```
Cluster Summary:
  * Stack: corosync
  * Current DC: kvm1-cr (1) (version 2.0.4-6.el8_3.1-2deceaa3ae) - partition with quorum
  * Last updated: Thu Jan 21 16:24:56 2021
  * Last change:  Thu Jan 21 16:24:46 2021 by root via crm_resource on kvm1-cr
  * 2 nodes configured
  * 11 resource instances configured

Node List:
  * Online: [ kvm1-cr (1) kvm2-cr (2) ]

Full List of Resources:
  * fence-kvm1      (stonith:fence_xvm):                 Started kvm1-cr
  * fence-kvm2      (stonith:fence_xvm):                 Started kvm2-cr
  * Clone Set: locking-clone [locking]:
    * Resource Group: locking:0:
      * dlm         (ocf::pacemaker:controld):           Started kvm1-cr
      * lvmlockd    (ocf::heartbeat:lvmlockd):           Started kvm1-cr
    * Resource Group: locking:1:
      * dlm         (ocf::pacemaker:controld):           Started kvm2-cr
      * lvmlockd    (ocf::heartbeat:lvmlockd):           Started kvm2-cr
  * Clone Set: shared-vg1-clone [shared-vg1]:
    * Resource Group: shared-vg1:0:
      * sharedlv1   (ocf::heartbeat:LVM-activate):       Started kvm1-cr
      * sharedfs1   (ocf::heartbeat:Filesystem):         Started kvm1-cr
    * Resource Group: shared-vg1:1:
      * sharedlv1   (ocf::heartbeat:LVM-activate):       Started kvm2-cr
      * sharedfs1   (ocf::heartbeat:Filesystem):         Started kvm2-cr
  * centos6.10a_res (ocf::heartbeat:VirtualDomain):      Started kvm1-cr

Migration Summary:

Tickets:

PCSD Status:
  kvm1-cr: Online
  kvm2-cr: Online

Daemon Status:
  corosync: active/enabled
  pacemaker: active/enabled
  pcsd: active/enabled
```

添加次序约束，以确保虚拟机资源在资源组 shared-vg1 之后启动，示例命令如下：

```
[root@kvm1 ~]#pcs constraint order start shared-vg1-clone then centos6.10a_res
Adding shared-vg1-clone centos6.10a_res (kind: Mandatory) (Options: first-action=start
then-action=start)

[root@kvm1 ~]#pcs constraint
Location Constraints:
Ordering Constraints:
  start locking-clone then start shared-vg1-clone (kind:Mandatory)
  start shared-vg1-clone then start centos6.10a_res (kind:Mandatory)
Colocation Constraints:
  shared-vg1-clone with locking-clone (score:INFINITY)
Ticket Constraints:
```

## 2.7.6 群集测试

由于是共享存储，此群集的功能与基于 NFS 的 KVM 群集功能相同，所以可以很好地实现以下 4 种测试：

(1) pcs resource move。

(2) pcs node standby。

(3) pcs cluster stop。

(4) 强制关闭一个节点。

测试的结果与本书“2.5.5 群集测试”相同，这里不再赘述了。

如果测试没有通过，则需要检查群集中各资源的状态。本实验环境中群集资源的相互关系如图 2-10 所示。

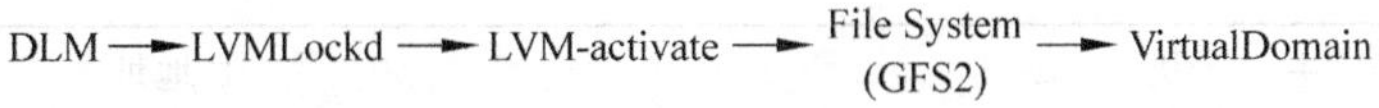

DLM → LVMLockd → LVM-activate → File System (GFS2) → VirtualDomain

图 2-10 基于 iSCSI 的主动/主动群集资源的相互关系

## 2.7.7 删除群集资源

删除与虚拟机、LVM 卷组有关的群集资源，保留 dlm、lvmlockd 即资源组 locking，因为它们还会在下一个实验用到，示例命令如下：

```
[root@kvm1 ~]#pcs resource delete centos6.10a_res

[root@kvm1 ~]#pcs resource delete shared-vg1
```

删除节点与 iSCSI 存储之间的会话与配置，示例命令如下：

```
[AllNodes ~]# iscsiadm -m node -u
[AllNodes ~]# iscsiadm -m node -o delete
[AllNodes ~]# systemctl stop iscsid
[AllNodes ~]# systemctl disable iscsid
```

## 2.8 基于DRBD的KVM群集构建

DRBD是Distributed Replicated Block Device的缩写，是一个用软件实现的、服务器之间镜像块设备内容的存储复制解决方案，可以认为是基于网络的RAID1。

对于小型KVM虚拟化应用，为了降低硬件成本，可以使用两台物理宿主机构建高可用群集，如图2-11所示。在每台宿主机上安装DRBD组件，将不同宿主机上的块设备进行复制，构建出一个无单点故障的共享的虚拟存储。虚拟机的映像文件保存在其中，从而构建高可用群集。

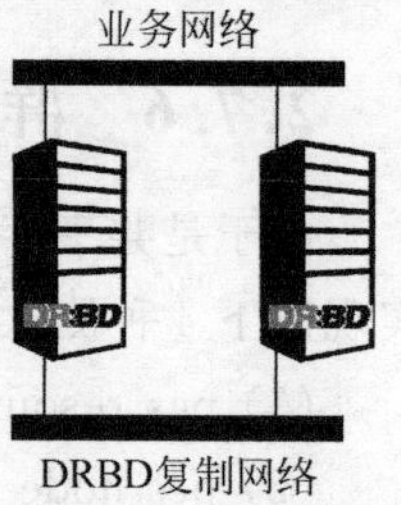

图2-11 DRBD架构

**提示**：由于DRBD的实时复制对网络要求比较高，所以通常使用与业务网络分开的专用网络。

下面的实验将构建一个基于DRBD的主动/主动群集，它们之间的复制使用独立的存储复制网络，IP地址规划如表2-6所示。

**表2-6 基于DRBD的KVM群集实验的IP地址规划**

| 主　机 | IP地址 |
|---|---|
| kvm1 | 10.0.1.231 |
| kvm2 | 10.0.1.232 |

群集部分的配置与上一实验类似，不过将由iSCSI服务器提供存储改由DRBD来提供。群集中的资源及相互关系如图2-12所示。

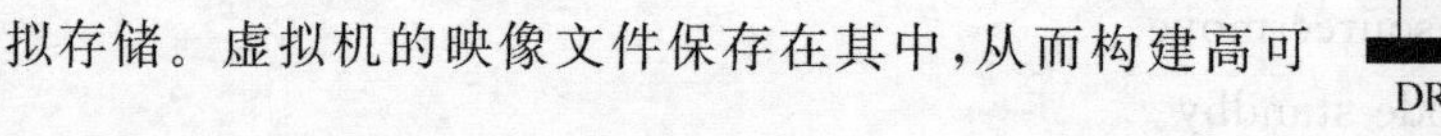

图2-12 基于DRBD的主动/主动群集资源的相互关系

还是利用上一实验的环境。由于基本的群集环境已经安装完毕，所以重点将放到DRBD安装与配置、在群集中创建DRBD资源。

## 2.8.1 DRBD基本原理

DRBD的核心功能通过Linux的内核实现。它位于文件系统之下，比文件系统更加靠近操作系统内核及IO栈。DRBD对文件系统来讲是透明的。下面通过4个核心概念来讲解DRBD的基本原理。

**1. 资源**

在DRBD中，资源是特指某复制的存储设备的所有方面。包括资源名称、DRBD设备(/dev/drbd*m*，这里*m*是数字)、磁盘配置(使本地数据可以为DRBD所用)、网络配置(与对方通信)。

本实验会将一个LVM逻辑卷用于构建DRBD的资源。

**2. 角色**

每个资源被分为主要角色(Primary)或辅助角色(Secondary)。

(1) 主要角色的DRBD设备：可以不受限制地读和写。例如：创建和映射文件系统，对于块设备的访问。

(2) 辅助角色的DRBD设备：仅接受来自对方的所有修改，但是不能被应用程序读写，甚至只读也不行。

(3) 角色可以改变。

**3. 工作模式**

(1) 单主模式：传统的高可靠性集群方案。

(2) 多主模式：需要采用共享群集文件系统，如GFS/GFS2和OCFS2，适用于需要多个节点并发访问数据的场合。本实验将采用此工作模式。

**4. 复制(镜像)方法**

(1) 协议A：异步复制协议。成功写入本地后立即返回，数据放在发送缓存中，随后再进行发送。这种方式速度快，但有可能会丢失数据。

(2) 协议B：内存同步(半同步)复制协议。成功写入本地并将数据发送到对方后立即返回。如果双机掉电，则数据可能丢失。

(3) 协议C：同步复制协议。成功写入本地并等待对方也成功写入后再返回，比协议A和B的可行性高。当然，如果双机掉电或磁盘同时损坏，则会有数据丢失的风险。本实验将采用此种复制方法。

## 2.8.2 安装DRBD软件

DRBD由两部分组成：内核模块和用户空间的管理工具。

CentOS 8的EPEL(Extra Packages for Enterprise Linux 8)仓库中仅提供了DRBD的用户空间的管理工具。查看CentOS 8软件仓库及DRBD组件的信息，示例命令如下：

```
[root@kvm1 ~]# dnf repolist
repo id        repo name
appstream      CentOS Linux 8 - AppStream
baseos         CentOS Linux 8 - BaseOS
epel           Extra Packages for Enterprise Linux 8 - x86_64
epel-modular Extra Packages for Enterprise Linux Modular 8 - x86_64
extras         CentOS Linux 8 - Extras
ha             CentOS Linux 8 - HighAvailability

[root@kvm1 ~]# yum list | grep drbd
collectd-drbd.x86_64              5.9.0-5.el8        epel
drbd.x86_64                       9.13.1-1.el8       epel
drbd-bash-completion.x86_64       9.13.1-1.el8       epel
drbd-pacemaker.x86_64             9.13.1-1.el8       epel
drbd-rgmanager.x86_64             9.13.1-1.el8       epel
drbd-udev.x86_64                  9.13.1-1.el8       epel
drbd-utils.x86_64                 9.13.1-1.el8       epel
drbd-xen.x86_64                   9.13.1-1.el8       epel
drbdlinks.noarch                  1.29-1.el8         epel
```

CentOS 8 还需要 DRBD 内核模式的驱动程序，既可以从 DRBD 项目的官方网站（https://www.linbit.com/）下载源码后编译并安装，也可以从 ELRepo 软件仓库中下载并安装。

**提示**：不要将 elrepo.org 的 ELRepo 与 CentOS epel(Extra Packages for Enterprise Linux)软件仓库混淆，前者是独立的项目，它支持 Red Hat Enterprise Linux(RHEL)及其衍生版本(Scientific Linux，如 CentOS 等)。

本实验将使用 ELRepo 软件仓库中的内核模块和用户空间的管理工具。

在所有节点上安装 ELRepo 仓库的配置文件，示例命令如下：

```
[AllNodes ~]# dnf -y install https://www.elrepo.org/elrepo-release-8.el8.elrepo.noarch.rpm

[root@kvm1 ~]# dnf repolist
repo id          repo name
appstream        CentOS Linux 8 - AppStream
baseos           CentOS Linux 8 - BaseOS
elrepo           ELRepo.org Community Enterprise Linux Repository - el8
epel             Extra Packages for Enterprise Linux 8 - x86_64
epel-modular     Extra Packages for Enterprise Linux Modular 8 - x86_64
extras           CentOS Linux 8 - Extras
ha               CentOS Linux 8 - HighAvailability
```

在每个节点上都会增加一个新的软件仓库 elrepo。在其中搜索与 DRBD 相关的软件包，示例命令如下：

```
[root@kvm1 ~]# dnf list --repo elrepo | grep drbd
drbd90-utils.x86_64                9.13.1-1.el8.elrepo      @elrepo
kmod-drbd90.x86_64                 9.0.25-2.el8_3.elrepo    @elrepo
drbd90-utils-sysvinit.x86_64       9.13.1-1.el8.elrepo      elrepo
```

其中 kmod-drbd90.x86_64 是 DRBD 内核模块的驱动程序，drbd90-utils 是 DRBD 的用户空间的管理工具（与 CentOS 8 epel 中的 drbd-utils 类似）。

安装 ELRepo 软件仓库的这两个软件包，示例命令如下：

```
[AllNodes ~]# dnf -y install kmod-drbd90 drbd90-utils
```

## 2.8.3 准备用于复制的块设备

在配置 DRBD 之前，需要为其准备用于复制的块设备。

在实验环境中的两个节点上添加一块 80GB 的磁盘。在其上创建一个使用全部空间的基本分区，分区的名称为/dev/vdb1。查看每个节点的块设备信息，示例命令如下：

```
[root@kvm1 ~]# lsblk
NAME          MAJ:MIN RM  SIZE RO TYPE MOUNTPOINT
sr0            11:0    1  1024M  0 rom
vda           252:0    0   80G   0 disk
├─vda1        252:1    0   1G    0 part /boot
└─vda2        252:2    0   79G   0 part
├─cl-root     253:0    0   50.4G 0 lvm  /
├─cl-swap     253:1    0   4G    0 lvm  [SWAP]
└─cl-home     253:2    0   24.6G 0 lvm  /home
vdb           252:16   0   80G   0 disk
└─vdb1        252:17   0   80G   0 part


[root@kvm1 ~]# ssh kvm2 lsblk
NAME          MAJ:MIN RM  SIZE RO TYPE MOUNTPOINT
sr0            11:0    1  1024M  0 rom
vda           252:0    0   80G   0 disk
├─vda1        252:1    0   1G    0 part /boot
└─vda2        252:2    0   79G   0 part
├─cl-root     253:0    0   50.4G 0 lvm  /
├─cl-swap     253:1    0   4G    0 lvm  [SWAP]
└─cl-home     253:2    0   24.6G 0 lvm  /home
vdb           252:16   0   80G   0 disk
└─vdb1        252:17   0   80G   0 part
```

在所有的节点上依次创建物理卷、卷组和逻辑卷。逻辑卷使用卷组的全部空间，逻辑卷的名称为/dev/vgdrbd1/lvdrbd1，它们将用于DBRD的复制。示例命令如下：

```
[AllNodes ~]# pvcreate /dev/vdb1

[AllNodes ~]# vgcreate vgdrbd1 /dev/vdb1

[AllNodes ~]# lvcreate -n lvdrbd1 -l 100%FREE vgdrbd1

[AllNodes ~]# lvscan
  ACTIVE            '/dev/vgdrbd1/lvdrbd1' [<80.00 GiB] inherit
  ACTIVE            '/dev/cl/swap' [3.96 GiB] inherit
  ACTIVE            '/dev/cl/home' [24.61 GiB] inherit
  ACTIVE            '/dev/cl/root' [<50.42 GiB] inherit
```

## 2.8.4 DRBD配置

DRBD的所有方面由配置文件/etc/drbd.conf进行控制，它是一个主配置文件，通常只是一个“大框架”。查看配置文件，示例命令如下：

```
[root@kvm1 ~]# cat /etc/drbd.conf
# You can find an example in  /usr/share/doc/drbd.../drbd.conf.example

include "drbd.d/global_common.conf";
include "drbd.d/*.res";
```

它通过include指令将外部文件包含到配置文件中。不建议对/etc/drbd.conf进行修改，而是在/etc/drbd.d/global_common.conf中修改全局共用的属性，在/etc/drbd.d/目录创建以.res结尾的文件用于配置DRBD的资源。

修改全局共用属性，示例命令如下：

```
[root@kvm1 ~]# vi /etc/drbd.d/global_common.conf
将 usage-count yes;     修改为     usage-count no;
```

本实验仅修改global_common.conf中的usage-count属性。DRBD研发团队使用这个属性来统计DRBD软件在全球的安装情况。

创建资源的配置文件，示例命令如下：

```
[root@kvm1 ~]# vi /etc/drbd.d/r0.res
resource r0 {
    protocol C;
    meta-disk internal;
```

```
        device /dev/drbd0;
        disk /dev/vgdrbd1/lvdrbd1;
        syncer {
            verify-alg sha1;
        }
        on kvm1 {
            address  10.0.1.231:7789;
        }
        on kvm2 {
            address  10.0.1.232:7789;
        }
        net {
            allow-two-primaries;
        }
    }
```

创建新的以.res结尾的文件用于复制DRBD的资源，包括以下属性。

(1) resource r0：资源名称为r0。

(2) protocol C：使用完全同步来复制DRBD资源，这意味着只有在确认了本地和远程磁盘写入后，才认为写入操作成功。

(3) meta-disk internal：在磁盘中包含元数据。

(4) device /dev/drbd0：生成DRBD设备的名称。

(5) disk /dev/vgdrbd1/lvdrbd1：用于复制的磁盘，也就是前面所创建的逻辑卷。

(6) syncer：同步器的属性。

(7) on：指定参与复制节点的属性，包括主机名、IP地址、端口等。

(8) net：对DRBD属性进行更细粒度的调整。allow-two-primaries表示允许多主复制。

将这两个配置文件复制到所有群集节点，示例命令如下：

```
[root@kvm1 ~]# scp /etc/drbd.d/global_common.conf kvm2:/etc/drbd.d

[root@kvm1 ~]# scp /etc/drbd.d/r0.res  kvm2:/etc/drbd.d
```

必须确保防火墙允许DRBD复制的出入站。在前面的配置中，使用的是默认端口号7789。在本实验环境中，将存储专用网络设置为可信任的网络，这是一种简单的配置方法。示例命令如下：

```
[AllNodes ~]# firewall-cmd --permanent --zone=trusted \
--add-source=10.0.1.0/24

[AllNodes ~]# firewall-cmd --reload
```

在激活资源 r0 之前，先查看一下进程跟踪文件和内核模块，示例命令如下：

```
[root@kvm1 ~]# cat /proc/drbd
cat: /proc/drbd: No such file or directory

[root@kvm1 ~]# lsmod | grep drbd
无输出
```

在 kvm1 上激活资源 r0，再次考察进程跟踪文件和内核模块，会看到有相应的变化。示例命令如下：

```
[root@kvm1 ~]# drbdadm create - md r0
initializing activity log
initializing bitmap (2560 KB) to all zero
Writing meta data...
New drbd meta data block successfully created.

[root@kvm1 ~]# cat /proc/drbd
version: 9.0.25 - 2 (api:2/proto:86 - 117)
GIT - hash: 0c392d83236898cce84cf8829235551173d5c1b9 build by akemi@Build64R8, 2020 - 11 -
08 02:16:35
Transports (api:16):

[root@kvm1 ~]# lsmod | grep drbd
drbd                   643072  0
libcrc32c              16384   4 nf_conntrack,nf_nat,xfs,drbd
```

在 kvm2 上也激活资源 r0，示例命令如下：

```
[root@kvm2 ~]# drbdadm create - md r0
```

DRBD 的管理主要通过 drbdadm 命令实现。通过它的 status 子命令来查看节点与资源的状态，示例命令如下：

```
[root@kvm1 ~]# drbdadm status
r0 role:Primary
  disk:UpToDate
  kvm2 role:Secondary
    replication:SyncSource peer - disk:Inconsistent done:18.20
```

输出的信息含义如下。

（1）r0 role：Primary：当前节点（kvm1）是资源 r0 的主要角色。

（2）disk：UpToDate：它的磁盘资源处于最新的状态。

（3）kvm2 role：Secondary：kvm2 是资源 r0 的辅助角色。

(4) replication：SyncSource peer-disk：Inconsistent done：18.20：正在复制，从源进行同步，磁盘处于不一致的状态，复制已经完成18.20%。

等复制结束再查看状态。如果是单主复制，则其状态如下：

```
[root@kvm1 ~]# drbdadm status
r0 role:Primary
  disk:UpToDate
  kvm2 role:Secondary
    peer-disk:UpToDate
```

如果是多主复制，则其状态如下：

```
[root@kvm1 ~]# drbdadm status
r0 role:Primary
  disk:UpToDate
  kvm2 role:Primary
    peer-disk:UpToDate
```

**提示**：在复制时，可以使用ifstat或sar -n DEV 1 4命令来获得网络使用情况。

## 2.8.5 创建DLM及LVMLockd资源

查看群集当前的状态，示例命令如下：

```
[root@kvm1 ~]# pcs status --full
Cluster name: cluster1
Cluster Summary:
  * Stack: corosync
  * Current DC: kvm1-cr (1) (version 2.0.4-6.el8_3.1-2deceaa3ae) - partition with quorum
  * Last updated: Sat Jan 23 04:57:24 2021
  * Last change:  Sat Jan 23 04:56:24 2021 by root via cibadmin on kvm1-cr
  * 2 nodes configured
  * 2 resource instances configured

Node List:
  * Online: [ kvm1-cr (1) kvm2-cr (2) ]

Full List of Resources:
  * fence-kvm1    (stonith:fence_xvm):    Started kvm1-cr
  * fence-kvm2    (stonith:fence_xvm):    Started kvm2-cr

Migration Summary:

Tickets:
```

```
PCSD Status:
  kvm1 - cr: Online
  kvm2 - cr: Online

Daemon Status:
  corosync: active/enabled
  pacemaker: active/enabled
  pcsd: active/enabled
```

在当前的实验环境中已经安装群集软件并且配置了隔离设备。

查看群集资源的约束，示例命令如下：

```
[root@kvm1 ~]# pcs constraint
Location Constraints:
Ordering Constraints:
Colocation Constraints:
Ticket Constraints:
```

查看群集的属性，示例命令如下：

```
[root@kvm1 ~]# pcs property
Cluster Properties:
cluster - infrastructure: corosync
cluster - name: cluster1
dc - version: 2.0.4 - 6.el8_3.1 - 2deceaa3ae
have - watchdog: false
last - lrm - refresh: 1611218827
no - quorum - policy: freeze
stonith - enabled: true
```

群集属性 no-quorum-policy 的值为 freeze。

下面将创建一个名为 DLM 的资源，它属于资源组 locking，通过克隆将此资源在群集的每个节点上都置于活动状态，示例命令如下：

```
[root@kvm1 ~]# pcs resource create dlm -- group locking ocf:pacemaker:controld op monitor
interval = 30s on - fail = fence

[root@kvm1 ~]# pcs resource clone locking interval = true

[root@kvm1 ~]# pcs status -- full
Cluster name: cluster1
Cluster Summary:
   * Stack: corosync
   * Current DC: kvm1 - cr (1) (version 2.0.4 - 6.el8_3.1 - 2deceaa3ae)    partition with quorum
```

```
  * Last updated: Sat Jan 23 05:00:28 2021
  * Last change:  Sat Jan 23 05:00:20 2021 by root via cibadmin on kvm1-cr
  * 2 nodes configured
  * 4 resource instances configured

Node List:
  * Online: [ kvm1-cr (1) kvm2-cr (2) ]

Full List of Resources:
  * fence-kvm1    (stonith:fence_xvm):    Started kvm1-cr
  * fence-kvm2    (stonith:fence_xvm):    Started kvm2-cr
  * Clone Set: locking-clone [locking]:
    * Resource Group: locking:0:
      * dlm    (ocf::pacemaker:controld):    Started kvm1-cr
    * Resource Group: locking:1:
      * dlm    (ocf::pacemaker:controld):    Started kvm2-cr

Migration Summary:

Tickets:

PCSD Status:
  kvm1-cr: Online
  kvm2-cr: Online

Daemon Status:
  corosync: active/enabled
  pacemaker: active/enabled
  pcsd: active/enabled
```

在资源组 locking 中再添加 LVMLockd 资源，示例命令如下：

```
[root@kvm1 ~]# pcs resource create lvmlockd --group locking ocf:heartbeat:lvmlockd op
monitor interval=30s on-fail=fence

[root@kvm1 ~]# pcs status --full
Cluster name: cluster1
Cluster Summary:
  * Stack: corosync
  * Current DC: kvm1-cr (1) (version 2.0.4-6.el8_3.1-2deceaa3ae) - partition with quorum
  * Last updated: Sat Jan 23 05:02:00 2021
  * Last change:  Sat Jan 23 05:01:48 2021 by root via cibadmin on kvm1-cr
  * 2 nodes configured
  * 6 resource instances configured
```

```
Node List:
  * Online: [ kvm1-cr (1) kvm2-cr (2) ]

Full List of Resources:
  * fence-kvm1      (stonith:fence_xvm):          Started kvm1-cr
  * fence-kvm2      (stonith:fence_xvm):          Started kvm2-cr
  * Clone Set: locking-clone [locking]:
    * Resource Group: locking:0:
      * dlm         (ocf::pacemaker:controld):    Started kvm1-cr
      * lvmlockd    (ocf::heartbeat:lvmlockd):    Started kvm1-cr
    * Resource Group: locking:1:
      * dlm         (ocf::pacemaker:controld):    Started kvm2-cr
      * lvmlockd    (ocf::heartbeat:lvmlockd):    Started kvm2-cr

Migration Summary:

Tickets:

PCSD Status:
  kvm1-cr: Online
  kvm2-cr: Online

Daemon Status:
  corosync: active/enabled
  pacemaker: active/enabled
  pcsd: active/enabled
```

## 2.8.6 创建 DRBD 资源

创建 ocf：linbit：drbd 类型的资源 vm-drbd。将 master-max 设置为 2，会将每个节点的 DRBD 资源提升为 Primary，示例命令如下：

```
[root@kvm1 ~]# pcs resource create vm-drbd ocf:linbit:drbd \
  drbd_resource=r0 \
  promotable \
  promoted-max=2 \
  promoted-node-max=1 \
  clone-max=2 \
  clone-node-max=1 \
  notify=true
```

查看群集状态，示例命令如下：

```
[root@kvm1 ~]# pcs status --full
Cluster name: cluster1
Cluster Summary:
  * Stack: corosync
  * Current DC: kvm1-cr (1) (version 2.0.4-6.el8_3.1-2deceaa3ae) - partition with quorum
  * Last updated: Sat Jan 23 05:05:46 2021
  * Last change:  Sat Jan 23 05:03:55 2021 by root via cibadmin on kvm1-cr
  * 2 nodes configured
  * 8 resource instances configured

Node List:
  * Online: [ kvm1-cr (1) kvm2-cr (2) ]

Full List of Resources:
  * fence-kvm1      (stonith:fence_xvm):        Started kvm1-cr
  * fence-kvm2      (stonith:fence_xvm):        Started kvm2-cr
  * Clone Set: locking-clone [locking]:
    * Resource Group: locking:0:
      * dlm         (ocf::pacemaker:controld): Started kvm1-cr
      * lvmlockd    (ocf::heartbeat:lvmlockd): Started kvm1-cr
    * Resource Group: locking:1:
      * dlm         (ocf::pacemaker:controld): Started kvm2-cr
      * lvmlockd    (ocf::heartbeat:lvmlockd): Started kvm2-cr
  * Clone Set: vm-drbd-clone [vm-drbd] (promotable):
    * vm-drbd     (ocf::linbit:drbd):      Master kvm1-cr
    * vm-drbd     (ocf::linbit:drbd):      Master kvm2-cr

Node Attributes:
  * Node: kvm1-cr (1):
    * master-vm-drbd                        : 10000
  * Node: kvm2-cr (2):
    * master-vm-drbd                        : 10000

Migration Summary:

Tickets:

PCSD Status:
  kvm1-cr: Online
  kvm2-cr: Online

Daemon Status:
  corosync: active/enabled
  pacemaker: active/enabled
  pcsd: active/enabled
```

两个节点的 DRBD 资源 vm-drbd 均是主要角色，这是多主模式。

查看进程跟踪文件以了解 DRBD 的工作情况，示例命令如下：

```
[root@kvm1 ~]# cat /proc/drbd
version: 9.0.25-2 (api:2/proto:86-117)
GIT-hash: 0c392d83236898cce84cf8829235551173d5c1b9 build by akemi@Build64R8, 2020-11-
08 02:16:35
Transports (api:16): tcp (9.0.25-2)

[root@kvm1 ~]# ssh kvm2 "cat /proc/drbd"
version: 9.0.25-2 (api:2/proto:86-117)
GIT-hash: 0c392d83236898cce84cf8829235551173d5c1b9 build by akemi@Build64R8, 2020-11-
08 02:16:35
Transports (api:16): tcp (9.0.25-2)
[root@kvm1 ~]# drbdadm status
r0 role:Primary
  disk:UpToDate
  kvm2 role:Primary
    peer-disk:UpToDate
```

查看每个节点的块设备，示例命令如下：

```
[root@kvm1 ~]# lsblk
NAME                 MAJ:MIN RM  SIZE RO TYPE MOUNTPOINT
sr0                   11:0    1  1024M  0 rom
vda                  252:0    0  80G    0 disk
├─vda1               252:1    0  1G     0 part /boot
└─vda2               252:2    0  79G    0 part
├─cl-root            253:0    0  50.4G  0 lvm  /
├─cl-swap            253:1    0  4G     0 lvm  [SWAP]
└─cl-home            253:3    0  24.6G  0 lvm  /home
vdb                  252:16   0  80G    0 disk
└─vdb1               252:17   0  80G    0 part
└─vgdrbd1-lvdrbd1    253:2    0  80G    0 lvm
└─drbd0              147:0    0  80G    0 disk <-- DRBD 构建的设备

[root@kvm1 ~]# ssh kvm2 lsblk
NAME                 MAJ:MIN RM  SIZE RO TYPE MOUNTPOINT
sr0                   11:0    1  1024M  0 rom
vda                  252:0    0  80G    0 disk
├─vda1               252:1    0  1G     0 part /boot
└─vda2               252:2    0  79G    0 part
├─cl-root            253:0    0  50.4G  0 lvm  /
├─cl-swap            253:1    0  4G     0 lvm  [SWAP]
└─cl-home            253:3    0  24.6G  0 lvm  /home
vdb                  252:16   0  80G    0 disk
└─vdb1               252:17   0  80G    0 part
└─vgdrbd1-lvdrbd1    253:2    0  80G    0 lvm
└─drbd0              147:0    0  80G    0 disk <-- DRBD 构建的设备
```

每个节点都有一个由DRBD构建的块设备drbd0。在kvm1上将这个块设备创建成LVM的物理卷。由于DRBD的自动复制，所以在kvm2上也会看到这个新的物理卷。示例命令如下：

```
[root@kvm1 ~]# file /dev/drbd0
/dev/drbd0: block special (147/0)

[root@kvm1 ~]# pvcreate /dev/drbd0

[root@kvm1 ~]# ssh kvm2 "pvscan"
  PV /dev/vdb1     VG vgdrbd1         lvm2 [<80.00 GiB / 0    free]
  PV /dev/vda2     VG cl              lvm2 [<79.00 GiB / 0    free]
  PV /dev/drbd0                       lvm2 [79.99 GiB]
  Total: 3 [<238.99 GiB] / in use: 2 [158.99 GiB] / in no VG: 1 [79.99 GiB]
```

在kvm1创建新的共享式的卷组，示例命令如下：

```
[root@kvm1 ~]# vgcreate --shared shared-vg1 /dev/drbd0
  Volume group "shared-vg1" successfully created
  VG shared-vg1 starting dlm lockspace
  Starting locking.  Waiting until locks are ready...

[root@kvm1 ~]# vgs
  VG         #PV #LV #SN Attr   VSize   VFree
  cl           1   3   0 wz--n-  <79.00g     0
  shared-vg1   1   0   0 wz--ns  79.99g  79.99g
  vgdrbd1      1   1   0 wz--n-  <80.00g     0
```

在kvm2上扫描卷组，示例命令如下：

```
[root@kvm1 ~]# ssh kvm2 "vgscan"
  Found volume group "vgdrbd1" using metadata type lvm2
  Found volume group "cl" using metadata type lvm2
  Found volume group "shared-vg1" using metadata type lvm2
  Reading VG shared-vg1 without a lock.
```

对于收到的“Reading VG shared-vg1 without a lock.”提示，可以在kvm2上通过修改卷组的锁定状态来解决，示例命令如下：

```
[root@kvm1 ~]# ssh kvm2 "vgchange --lock-start shared-vg1"
  VG shared-vg1 starting dlm lockspace
  Starting locking.  Waiting until locks are ready...

[root@kvm1 ~]# ssh kvm2 "vgscan"
```

```
Found volume group "vgdrbd1" using metadata type lvm2
Found volume group "cl" using metadata type lvm2
Found volume group "shared-vg1" using metadata type lvm2
```

在新卷组 shared-vg1 上创建逻辑卷 shared-lv1，让其使用卷组的全部空间。创建完成后，在所有的节点都可以看这个新的逻辑卷。示例命令如下：

```
[root@kvm1 ~]# lvcreate -n shared-lv1 -l 100%FREE shared-vg1

[root@kvm1 ~]# lvscan
  ACTIVE            '/dev/vgdrbd1/lvdrbd1' [<80.00 GiB] inherit
  ACTIVE            '/dev/cl/swap' [3.96 GiB] inherit
  ACTIVE            '/dev/cl/home' [24.61 GiB] inherit
  ACTIVE            '/dev/cl/root' [<50.42 GiB] inherit
  ACTIVE            '/dev/shared-vg1/shared-lv1' [79.99 GiB] inherit

[root@kvm1 ~]# ssh kvm2 lvscan
  ACTIVE            '/dev/vgdrbd1/lvdrbd1' [<80.00 GiB] inherit
  ACTIVE            '/dev/cl/swap' [3.96 GiB] inherit
  ACTIVE            '/dev/cl/home' [24.61 GiB] inherit
  ACTIVE            '/dev/cl/root' [<50.42 GiB] inherit
  inactive          '/dev/shared-vg1/shared-lv1' [79.99 GiB] inherit
```

在逻辑卷上创建 GFS2 文件系统，示例命令如下：

```
[root@kvm1 ~]# mkfs.gfs2 -j2 -p lock_dlm -t cluster1:gfs2-test1 /dev/shared-vg1/
shared-lv1
/dev/shared-vg1/shared-lv1 is a symbolic link to /dev/dm-4
This will destroy any data on /dev/dm-4
Are you sure you want to proceed? [y/n] y
Discarding device contents (may take a while on large devices): Done
Adding journals: Done
Building resource groups: Done
Creating quota file: Done
Writing superblock and syncing: Done
Device:                    /dev/shared-vg1/shared-lv1
Block size:                4096
Device size:               79.99 GB (20969472 blocks)
Filesystem size:           79.99 GB (20969471 blocks)
Journals:                  2
Journal size:              128MB
Resource groups:           321
Locking protocol:          "lock_dlm"
Lock table:                "cluster1:gfs2-test1"
UUID:                      ee96d5f9-de6f-49a4-9d5a-6368e1874388
```

为逻辑卷创建一个 LVM-activate 类型资源 sharedlv1，并将其放置到一个新的资源组 shared-vg1 中。通过克隆这个资源组，让所有节点都有此资源，都可以激活该逻辑卷。示例命令如下：

```
[root@kvm1 ~]# pcs resource create sharedlv1 \
   --group shared-vg1 ocf:heartbeat:LVM-activate \
   lvname=shared-lv1 \
   vgname=shared-vg1 \
   activation_mode=shared \
   vg_access_mode=lvmlockd

[root@kvm1 ~]# pcs resource clone shared-vg1 interleave=true
```

**提示**：创建后，如果在群集状态中看到资源失败的信息：shared-vg1/shared-lv1：failed to activate，说明之前卷组已经被激活。可以将刚创建的资源组 shared-vg1 删除，然后重新创建。

查看群集状态，检查所有资源是否工作正常。示例命令如下：

```
[root@kvm1 ~]# pcs status --full
Cluster name: cluster1
Cluster Summary:
  * Stack: corosync
  * Current DC: kvm1-cr (1) (version 2.0.4-6.el8_3.1-2deceaa3ae) - partition with quorum
  * Last updated: Sat Jan 23 05:57:12 2021
  * Last change:  Sat Jan 23 05:52:42 2021 by root via cibadmin on kvm1-cr
  * 2 nodes configured
  * 10 resource instances configured

Node List:
  * Online: [ kvm1-cr (1) kvm2-cr (2) ]

Full List of Resources:
  * fence-kvm1      (stonith:fence_xvm):                Started kvm1-cr
  * fence-kvm2      (stonith:fence_xvm):                Started kvm2-cr
  * Clone Set: locking-clone [locking]:
    * Resource Group: locking:0:
      * dlm         (ocf::pacemaker:controld):          Started kvm1-cr
      * lvmlockd    (ocf::heartbeat:lvmlockd):          Started kvm1-cr
    * Resource Group: locking:1:
      * dlm         (ocf::pacemaker:controld):          Started kvm2-cr
      * lvmlockd    (ocf::heartbeat:lvmlockd):          Started kvm2-cr
  * Clone Set: vm-drbd-clone [vm-drbd] (promotable):
    * vm-drbd       (ocf::linbit:drbd):     Master kvm1-cr
    * vm-drbd       (ocf::linbit:drbd):     Master kvm2-cr
```

```
  * Clone Set: shared-vg1-clone [shared-vg1]:
    * Resource Group: shared-vg1:0:
      * sharedlv1    (ocf::heartbeat:LVM-activate):    Started kvm1-cr
    * Resource Group: shared-vg1:1:
      * sharedlv1    (ocf::heartbeat:LVM-activate):    Started kvm2-cr

Node Attributes:
  * Node: kvm1-cr (1):
    * master-vm-drbd                            : 10000
  * Node: kvm2-cr (2):
    * master-vm-drbd                            : 10000

Migration Summary:

Tickets:

PCSD Status:
  kvm1-cr: Online
  kvm2-cr: Online

Daemon Status:
  corosync: active/enabled
  pacemaker: active/enabled
  pcsd: active/enabled
```

与上一个实验类似，也需要创建群集的约束规则，包括以下两种约束。

(1) 依赖约束：先启动包含 dlm 和 lvmlockd 资源的资源组 locking，再启动 DRBD 资源组 vm-drbd-clone，最后启动资源组 shared-vg1。

(2) 位置约束：确保资源组 shared-vg1 与资源组 vm-drbd-clone、locking 在同一个节点上，示例命令如下：

```
[root@kvm1 ~]# pcs constraint order start locking-clone then vm-drbd-clone

[root@kvm1 ~]# pcs constraint order start vm-drbd-clone then \
    shared-vg1-clone

[root@kvm1 ~]# pcs constraint colocation add shared-vg1-clone \
    with vm-drbd-clone

[root@kvm1 ~]# pcs constraint colocation add shared-vg1-clone \
    with locking-clone

[root@kvm1 ~]# pcs constraint
Location Constraints:
```

```
Ordering Constraints:
  start locking-clone then start vm-drbd-clone (kind:Mandatory)
  start vm-drbd-clone then start shared-vg1-clone (kind:Mandatory)
Colocation Constraints:
  shared-vg1-clone with vm-drbd-clone (score:INFINITY)
  shared-vg1-clone with locking-clone (score:INFINITY)
```

## 2.8.7 创建 GFS2 文件系统资源

创建文件系统资源的目的是让所有节点自动挂载 GFS2 文件系统。

新资源属于 shared-vg1 资源组，将逻辑卷/dev/shared-vg1/shared-lv1 挂载到目录/mnt/gfs1 中，示例命令如下：

```
[root@kvm1 ~]# pcs resource create sharedfs1 \
   --group shared-vg1 ocf:heartbeat:Filesystem \
  device="/dev/shared-vg1/shared-lv1" \
  directory="/mnt/gfs1" \
  fstype="gfs2" \
  options=noatime \
  op monitor interval=10s on-fail=fence
```

查看群集状态，示例命令如下：

```
[root@kvm1 ~]# pcs status --full
Cluster name: cluster1
Cluster Summary:
  * Stack: corosync
  * Current DC: kvm1-cr (1) (version 2.0.4-6.el8_3.1-2deceaa3ae) - partition with quorum
  * Last updated: Sat Jan 23 06:13:46 2021
  * Last change:  Sat Jan 23 06:13:29 2021 by root via cibadmin on kvm1-cr
  * 2 nodes configured
  * 12 resource instances configured

Node List:
  * Online: [ kvm1-cr (1) kvm2-cr (2) ]

Full List of Resources:
  * fence-kvm1    (stonith:fence_xvm):              Started kvm1-cr
  * fence-kvm2    (stonith:fence_xvm):              Started kvm2-cr
  * Clone Set: locking-clone [locking]:
    * Resource Group: locking:0:
      * dlm       (ocf::pacemaker:controld):        Started kvm1-cr
      * lvmlockd  (ocf::heartbeat:lvmlockd):        Started kvm1-cr
    * Resource Group: locking:1:
```

```
        * dlm           (ocf::pacemaker:controld):       Started kvm2-cr
        * lvmlockd      (ocf::heartbeat:lvmlockd):       Started kvm2-cr
    * Clone Set: vm-drbd-clone [vm-drbd] (promotable):
      * vm-drbd         (ocf::linbit:drbd):              Master kvm1-cr
      * vm-drbd         (ocf::linbit:drbd):              Master kvm2-cr
    * Clone Set: shared-vg1-clone [shared-vg1]:
      * Resource Group: shared-vg1:0:
        * sharedlv1     (ocf::heartbeat:LVM-activate):   Started kvm1-cr
        * sharedfs1     (ocf::heartbeat:Filesystem):     Started kvm1-cr
      * Resource Group: shared-vg1:1:
        * sharedlv1     (ocf::heartbeat:LVM-activate):   Started kvm2-cr
        * sharedfs1     (ocf::heartbeat:Filesystem):     Started kvm2-cr

Node Attributes:
    * Node: kvm1-cr (1):
      * master-vm-drbd                        : 10000
    * Node: kvm2-cr (2):
      * master-vm-drbd                        : 10000

Migration Summary:

Tickets:

PCSD Status:
  kvm1-cr: Online
  kvm2-cr: Online

Daemon Status:
  corosync: active/enabled
  pacemaker: active/enabled
  pcsd: active/enabled
[root@kvm1 ~]#
```

检查每个节点的文件系统的挂载情况，示例命令如下：

```
[root@kvm1 ~]# mount | grep gfs2
/dev/mapper/shared--vg1-shared--lv1 on /mnt/gfs1 type gfs2 (rw,noatime)

[root@kvm1 ~]# ssh kvm2 "mount | grep gfs2"
/dev/mapper/shared--vg1-shared--lv1 on /mnt/gfs1 type gfs2 (rw,noatime)
```

重新启动所有节点的群集服务，再次验证配置是否正确，示例命令如下：

```
[root@kvm1 ~]# pcs cluster stop --all
kvm1-cr: Stopping Cluster (pacemaker)...
kvm2-cr: Stopping Cluster (pacemaker)...
kvm2-cr: Stopping Cluster (corosync)...
kvm1-cr: Stopping Cluster (corosync)...

[root@kvm1 ~]# pcs cluster start --all
kvm1-cr: Starting Cluster...
kvm2-cr: Starting Cluster...
```

检查群集中所有资源是否处于正常工作状态，示例命令如下：

```
[root@kvm1 ~]# pcs status --full
Cluster name: cluster1
Cluster Summary:
  * Stack: corosync
  * Current DC: kvm2-cr (2) (version 2.0.4-6.el8_3.1-2deceaa3ae) - partition with quorum
  * Last updated: Sat Jan 23 06:19:45 2021
  * Last change:  Sat Jan 23 06:13:29 2021 by root via cibadmin on kvm1-cr
  * 2 nodes configured
  * 12 resource instances configured

Node List:
  * Online: [ kvm1-cr (1) kvm2-cr (2) ]

Full List of Resources:
  * fence-kvm1     (stonith:fence_xvm):              Started kvm1-cr
  * fence-kvm2     (stonith:fence_xvm):              Started kvm2-cr
  * Clone Set: locking-clone [locking]:
    * Resource Group: locking:0:
      * dlm        (ocf::pacemaker:controld):        Started kvm2-cr
      * lvmlockd   (ocf::heartbeat:lvmlockd):        Started kvm2-cr
    * Resource Group: locking:1:
      * dlm        (ocf::pacemaker:controld):        Started kvm1-cr
      * lvmlockd   (ocf::heartbeat:lvmlockd):        Started kvm1-cr
  * Clone Set: vm-drbd-clone [vm-drbd] (promotable):
    * vm-drbd      (ocf::linbit:drbd):               Master kvm2-cr
    * vm-drbd      (ocf::linbit:drbd):               Master kvm1-cr
  * Clone Set: shared-vg1-clone [shared-vg1]:
    * Resource Group: shared-vg1:0:
      * sharedlv1  (ocf::heartbeat:LVM-activate):    Started kvm2-cr
      * sharedfs1  (ocf::heartbeat:Filesystem):      Started kvm2-cr
    * Resource Group: shared-vg1:1:
      * sharedlv1  (ocf::heartbeat:LVM-activate):    Started kvm1-cr
      * sharedfs1  (ocf::heartbeat:Filesystem):      Started kvm1-cr
```

```
Node Attributes:
  * Node: kvm1 - cr (1):
    * master - vm - drbd                                : 10000
  * Node: kvm2 - cr (2):
    * master - vm - drbd                                : 10000

Migration Summary:

Tickets:

PCSD Status:
  kvm1 - cr: Online
  kvm2 - cr: Online

Daemon Status:
  corosync: active/enabled
  pacemaker: active/enabled
  pcsd: active/enabled
[root@kvm1 ~]#
```

### 2.8.8 后续配置

本实验后续操作与“2.7 基于 iSCSI 的 KVM 群集 2”基本相同，主要包括以下步骤：

(1) 在所有节点上禁用 SELinux。

(2) 准备虚拟机，要保证虚拟机实时迁移成功。

(3) 在群集中创建虚拟机资源。

(4) 测试虚拟机。

详细的操作本书就不再赘述了。

## 2.9 本章小结

本章讲解如何通过构建高可用群集以实现虚拟机的高可用，在 3 种常见的存储上做了 4 个典型的群集实验。由于配置群集涉及的组件多，所以操作步骤比较繁杂，在生产中可以考虑通过脚本来自动化执行。

第 3 章将讲解如何实现嵌套虚拟化。

# 第3章

# 实现嵌套虚拟化

嵌套虚拟化是指在一个虚拟机之中再运行另外一个虚拟机。由于嵌套虚拟化有一些功能上的限制，所以通常不建议在生产环境中使用，它主要用于开发和测试。有多种嵌套虚拟化的方法，本章将介绍通过 KVM 实现嵌套虚拟化。

**本章要点**

- 嵌套虚拟化的原理。
- L1 级别宿主机的准备。
- L2 级别 KVM 宿主机的配置。
- L2 级别 VMware ESXi 宿主机的配置。
- L2 级别 Microsoft Hyper-V 宿主机的配置。

## 3.1 嵌套虚拟化的原理

为了保证虚拟化平台的安全和稳定，默认情况下 L1 级别（也称为第一层）的 Hypervisor 会阻止其他软件使用 CPU 的虚拟化功能，所以其上的虚拟机就无法再充当 Hypervisor 了。

如果在 L1 级别的 Hypervisor 上启动了对嵌套虚拟化的支持，它就会向其虚拟机公开硬件虚拟化扩展。这些虚拟机可以充当 L2 级别的 Hypervisor，安装并运行自己的虚拟机，如图 3-1 所示。图中箭头表示硬件虚拟化扩展。

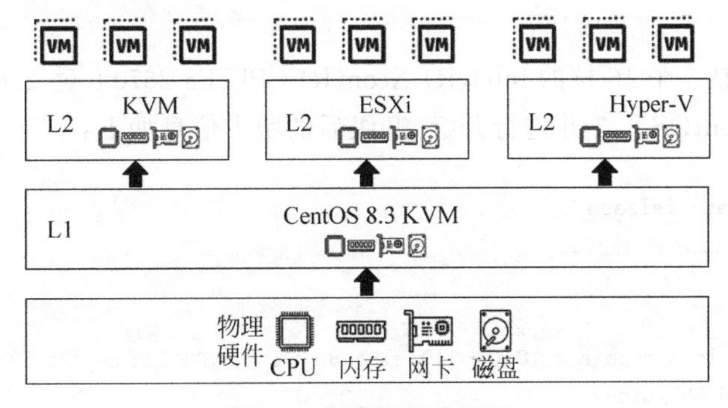


图 3-1 嵌套虚拟化原理

要实现嵌套虚拟化，需要 L1 和 L2 级别都支持 Hypervisor。RHEL/CentOS 8.2 及更高版本全面支持 Intel CPU 上的嵌套虚拟化功能。Red Hat 公司当前仅在 Intel CPU 嵌套虚拟化提供技术支持。在 AMD、IBM POWER9 和 IBM Z 系统上的嵌套虚拟化仅作为技术预览提供，因此 Red Hat 公司不提供支持服务。

早期的 Linux 发行版本例如 CentOS 7.2，无法实现 Microsoft Hyper-V 2012 和 2016 的嵌套虚拟化。随着技术的发展，RHEL/CentOS 8.3 可以很好地支持 Microsoft Hyper-V 2019 的嵌套虚拟化。

## 3.2 L1 级别宿主机的准备

嵌套虚拟化对运行 L1 级别 Hypervisor 宿主机的硬件要求比较高。下面的实验中使用一台 HP Z420 工作站，通过 lshw 命令查看其硬件信息，示例命令如下：

```
# lshw -short
H/W path    Class        Description
================================
            system       HP Z420 Workstation (LJ449AV)
/0          bus          1589
/0/0        memory       64KiB BIOS
/0/4        processor    Intel(R) Xeon(R) CPU E5-2670 0 @ 2.60GHz
/0/4/5      memory       512KiB L1 cache
/0/4/6      memory       2MiB L2 cache
/0/4/7      memory       20MiB L3 cache
/0/44       memory       32GiB System Memory
/0/44/0     memory       8GiB DIMM DDR3 Synchronous Registered (Buffered) 1333 MHz (0.8 ns)
/0/44/1     memory       DIMM Synchronous [empty]
/0/44/2     memory       8GiB DIMM DDR3 Synchronous Registered (Buffered) 1333 MHz (0.8 ns)
/0/44/3     memory       DIMM Synchronous [empty]
/0/44/4     memory       DIMM Synchronous [empty]
/0/44/5     memory       8GiB DIMM DDR3 Synchronous Registered (Buffered) 1333 MHz (0.8 ns)
/0/44/6     memory       DIMM Synchronous [empty]
/0/44/7     memory       8GiB DIMM DDR3 Synchronous Registered (Buffered) 1333 MHz (0.8 ns)
...
```

这台宿主机包括一个 16 核的 Intel(R) Xeon(R) CPU E5-2670 0 @ 2.60GHz，32GB 内存。在其上安装 CentOS 8.3 并进行升级，升级后的版本信息如下：

```
# cat /etc/redhat-release
CentOS Linux release 8.3.2011

# uname -a
Linux localhost.localdomain 4.18.0-240.el8.x86_64 #1 SMP Fri Sep 25 19:48:47 UTC 2020 x86_
64 x86_64 x86_64 GNU/Linux
```

HP Z420 工作站的 BIOS 默认启用了虚拟化支持，可以通过检查/proc/cpuinfo 是否包含 vmx 和 ept 标志来确认，示例命令如下：

```
#cat /proc/cpuinfo | egrep '(vmx|ept)'
flags           : fpu vme de pse tsc msr pae mce cx8 apic sep mtrr pge mca cmov pat pse36 clflush
dts acpi mmx fxsr sse sse2 ss ht tm pbe syscall nx pdpe1gb rdtscp lm constant_tsc arch_perfmon
pebs bts rep_good nopl xtopology nonstop_tsc cpuid aperfmperf pni pclmulqdq dtes64 monitor ds_
cpl vmx smx est tm2 ssse3 cx16 xtpr pdcm pcid dca sse4_1 sse4_2 x2apic popcnt tsc_deadline_timer
aes xsave avx lahf_lm epb pti ssbd ibrs ibpb stibp tpr_shadow vnmi flexpriority ept vpid
xsaveopt dtherm ida arat pln pts md_clear flush_l1d
...
```

检查内核参数是否启用了嵌套虚拟化的支持，示例命令如下：

```
#cat /sys/module/kvm_intel/parameters/nested
0
```

如果命令返回 1 或 Y，则表示已经启用了该功能。如果命令返回 0 或 N，则表示未启用。

有两种方法启用对嵌套虚拟化的支持。

第 1 种方法是使用 modprobe 命令进行临时启用，示例命令如下：

```
#modprobe -r kvm_intel

#modprobe kvm_intel nested=1
```

先通过 modprobe 命令卸载 kvm_intel 模块，然后使用选项 nested=1 启用嵌套功能。

第 2 种方法是修改配置文件/etc/modprobe.d/kvm.conf 中的选项，这样可以在下次重新启动宿主机之后一直生效，示例命令如下：

```
#vi /etc/modprobe.d/kvm.conf
#Setting modprobe kvm_intel/kvm_amd nested = 1
#only enables Nested Virtualization until the next reboot or
#module reload. Uncomment the option applicable
#to your system below to enable the feature permanently.
#
#User changes in this file are preserved across upgrades.
#
#For Intel
#options kvm_intel nested=1
options kvm_intel nested=1
#
#For AMD
#options kvm_amd nested=1
```

如果是 AMD 的 CPU，则其内核模块名称为 kvm_amd，选项名为 kvm_amd nested，所以会有一些不同，示例命令如下：

```
#modprobe -r kvm_amd

#modprobe kvm_amd nested=1

#vi /etc/modprobe.d/kvm.conf
添加
options kvm_amd nested=1
```

## 3.3 L2 级别 KVM 宿主机的配置

实验中 L2 级别的 KVM 宿主机使用 CentOS 8.3，安装方法与普通的 KVM 虚拟化宿主机相同。

### 3.3.1 虚拟机配置（Intel）

安装完成后，通过 virsh edit 命令修改虚拟机的配置文件，主要编辑 CPU 的设置。示例命令如下：

```
#virsh edit centos8.3
将原有 CPU 参数
  <cpu mode='custom' match='exact' check='full'>
    <model fallback='forbid'>SandyBridge-IBRS</model>
    <vendor>Intel</vendor>
    <feature policy='require' name='vme'/>
    <feature policy='require' name='ss'/>
    <feature policy='require' name='pcid'/>
    <feature policy='require' name='hypervisor'/>
    <feature policy='require' name='arat'/>
    <feature policy='require' name='tsc_adjust'/>
    <feature policy='require' name='umip'/>
    <feature policy='require' name='md-clear'/>
    <feature policy='require' name='stibp'/>
    <feature policy='require' name='arch-capabilities'/>
    <feature policy='require' name='ssbd'/>
    <feature policy='require' name='xsaveopt'/>
    <feature policy='require' name='pdpe1gb'/>
    <feature policy='require' name='ibpb'/>
    <feature policy='require' name='amd-ssbd'/>
    <feature policy='require' name='skip-l1dfl-vmentry'/>
    <feature policy='require' name='pschange-mc-no'/>
```

```
</cpu>
修改为
<cpu mode='host-passthrough'/>
```

如果 L1 级别宿主机采用的是 Intel 的 CPU，通过设置<cpu mode='host-passthrough'/>，则可以使 L2 宿主机像 L1 宿主机一样使用 CPU 的虚拟化特性。

也可以使用 virt-manager 设置虚拟机 CPU 的模式，如图 3-2 所示。

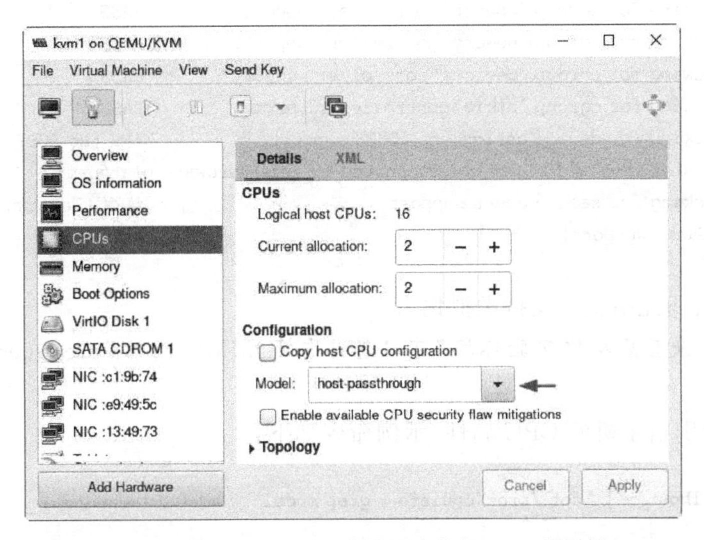


图 3-2　在 virt-manager 中设置虚拟机 CPU 的属性

目前 Cockpit 还不支持修改虚拟机的 CPU 的模式。

启动 L2 级别的虚拟机 CentOS 8.3，通过 SSH 登录，示例命令如下：

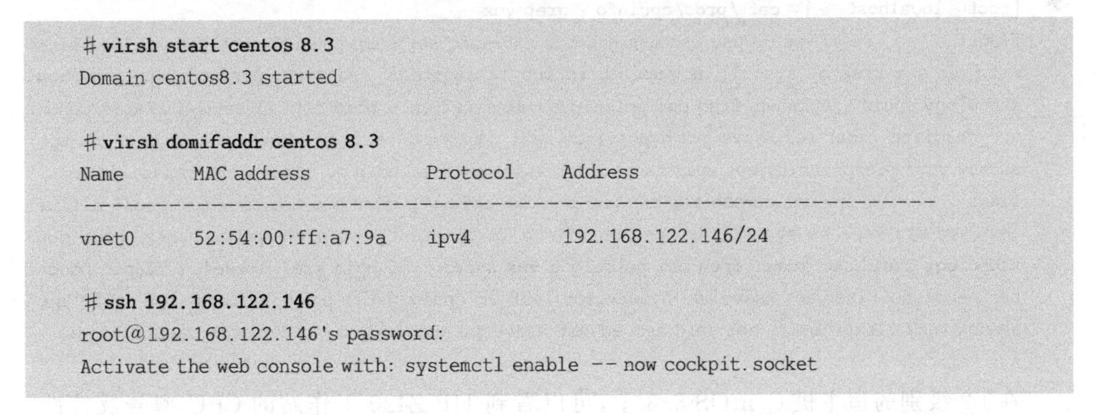

```
#virsh start centos 8.3
Domain centos8.3 started

#virsh domifaddr centos 8.3
Name     MAC address          Protocol    Address
-------------------------------------------------------------------------------
vnet0    52:54:00:ff:a7:9a    ipv4        192.168.122.146/24

#ssh 192.168.122.146
root@192.168.122.146's password:
Activate the web console with: systemctl enable --now cockpit.socket
```

L2 级别宿主机的 IP 地址是 192.168.122.146，使用 SSH 登录。通过虚拟化验证工具 virt-host-validate 进行检查，示例命令如下：

```
[root@localhost ~]# virt-host-validate
  QEMU: Checking for hardware virtualization                    : PASS
  QEMU: Checking if device /dev/kvm exists                      : PASS
  QEMU: Checking if device /dev/kvm is accessible               : PASS
  QEMU: Checking if device /dev/vhost-net exists                : PASS
  QEMU: Checking if device /dev/net/tun exists                  : PASS
  QEMU: Checking for cgroup 'cpu' controller support            : PASS
  QEMU: Checking for cgroup 'cpuacct' controller support        : PASS
  QEMU: Checking for cgroup 'cpuset' controller support         : PASS
  QEMU: Checking for cgroup 'memory' controller support         : PASS
  QEMU: Checking for cgroup 'devices' controller support        : PASS
  QEMU: Checking for cgroup 'blkio' controller support          : PASS
  QEMU: Checking for device assignment IOMMU support           : WARN (No ACPI DMAR table found,
IOMMU either disabled in BIOS or not supported by this hardware platform)
  QEMU: Checking for secure guest support                       : WARN (Unknown if this platform
has Secure Guest support)
```

输出结果说明此宿主机支持虚拟化。

**提示**：对于嵌套虚拟化实验环境，可以忽略与 IOMMU 和 Secure Guest support 有关的警告。

查看 L2 级别宿主机的 CPU 属性，示例命令如下：

```
[root@localhost ~]# cat /proc/cpuinfo | grep model
model           : 45
model name      : Intel(R) Xeon(R) CPU E5-2670 0 @ 2.60GHz
model           : 45
model name      : Intel(R) Xeon(R) CPU E5-2670 0 @ 2.60GHz

[root@localhost ~]# cat /proc/cpuinfo | grep vmx
flags           : fpu vme de pse tsc msr pae mce cx8 apic sep mtrr pge mca cmov pat pse36 clflush
mmx fxsr sse sse2 ss syscall nx pdpe1gb rdtscp lm constant_tsc arch_perfmon rep_good nopl
xtopology cpuid tsc_known_freq pni pclmulqdq vmx ssse3 cx16 pcid sse4_1 sse4_2 x2apic popcnt
tsc_deadline_timer aes xsave avx hypervisor lahf_lm cpuid_fault pti ssbd ibrs ibpb stibp tpr_
shadow vnmi flexpriority ept vpid tsc_adjust xsaveopt arat umip md_clear arch_capabilities
flags           : fpu vme de pse tsc msr pae mce cx8 apic sep mtrr pge mca cmov pat pse36 clflush
mmx fxsr sse sse2 ss syscall nx pdpe1gb rdtscp lm constant_tsc arch_perfmon rep_good nopl
xtopology cpuid tsc_known_freq pni pclmulqdq vmx ssse3 cx16 pcid sse4_1 sse4_2 x2apic popcnt
tsc_deadline_timer aes xsave avx hypervisor lahf_lm cpuid_fault pti ssbd ibrs ibpb stibp tpr_
shadow vnmi flexpriority ept vpid tsc_adjust xsaveopt arat umip md_clear arch_capabilities
```

在 L2 级别的宿主机 CentOS 8.3 上，可以看到 HP Z420 工作站的 CPU 型号及特性。有了这些特性，就可以在其上再创建虚拟机了。创建的过程与直接在 L1 宿主机上创建虚拟机类似，这里不再赘述。

### 3.3.2　虚拟机配置(AMD)

如果 L1 级别宿主机采用的是 AMD 的 CPU，则配置 L2 级别宿主机的方法与配置 Intel 的 CPU 类似，也是将虚拟机 CPU 配置为使用 host-passthrough 模式，示例命令如下：

```
#virsh edit centos 8.3
将 CPU 属性修改为
<cpu mode = 'host - passthrough'/>
```

如果要求 L2 级别宿主机使用特定的 CPU 而不是 host-passthrough，则需在 CPU 配置中添加< feature policy='require' name='vmx'/>，示例命令如下：

```
#virsh edit centos 8.3
将 CPU 属性修改为
<cpu mode  = 'custom' match  = 'exact' check = 'partial'>
  <model fallback = 'allow'> Haswell - noTSX </model>
  <feature policy = 'require' name = 'vmx'/>
</cpu>
```

## 3.4　L2 级别 VMware ESXi 宿主机的配置

### 3.4.1　VMware ESXi 下载与安装

VMware vSphere Hypervisor ESXi 是一个商业产品，但是 VMware 公司提供了 60 天试用版。下面的实验使用的版本是 VMware vSphere Hypervisor(ESXi) 6.7，它的 ISO 文件的下载网址为 https://my.vmware.com/en/web/vmware/evalcenter?p=free-esxi6。

在下载页面中会显示文件的 MD5 及 SHA 类型的摘要信息，如图 3-3 所示。

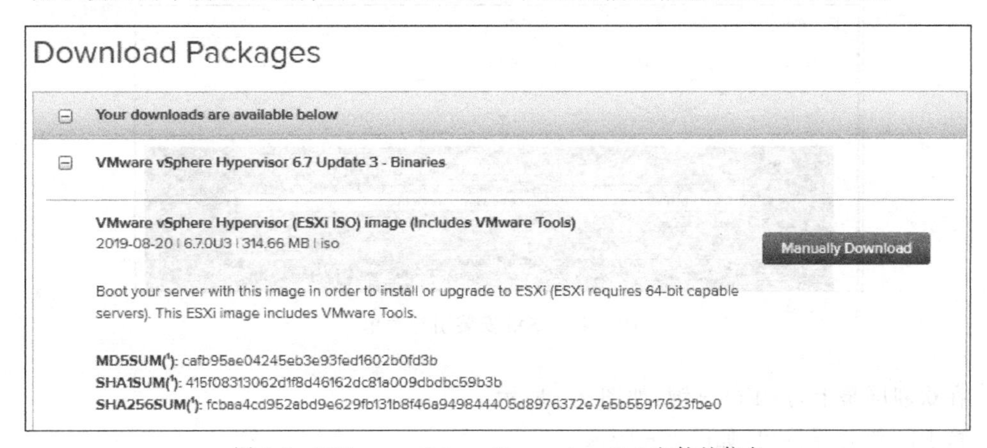


图 3-3　VMware vSphere Hypervisor 6.7 文件的信息

在安装前，建议通过这些信息来检查所下载文件的完整性，示例命令如下：

```
# md5sum /iso/VMware - VMvisor - Installer - 6.7.0.update03 - 14320388.x86_64.iso
cafb95ae04245eb3e93fed1602b0fd3b  /iso/VMware - VMvisor - Installer - 6. 7. 0. update03 -
14320388.x86_64.iso
```

由于 Cockpit 和 virt-manager 没有适合 ESXi 的操作系统参数，所以需要通过 virt-install 命令来创建能够运行 ESXi 的虚拟机，示例命令如下：

```
# virt - install -- name = esxi6.7u3 \
-- cpu host - passthrough \
-- ram 4096 -- vcpus = 4 \
-- virt - type = kvm -- hvm \
-- cdrom /iso/VMware - VMvisor - Installer - 6.7.0.update03 - 14320388.x86_64.iso \
-- network bridge = virbr1,model = e1000 \
-- graphics spice,listen = 127.0.0.1 \
-- graphics vnc \
-- video qxl \
-- disk pool = vm,size = 80,sparse = true,bus = ide,format = qcow2 \
-- boot cdrom,hd -- noautoconsole -- force
```

需要注意以下选项参数。

(1) CPU：host-passthrough。

(2) 网卡类型：E1000。

(3) 磁盘接口类型：IDE。

(4) 显卡类型：QXL。

**提示**：ESXi 的硬件兼容列表很短，这些是经过验证的虚拟硬件类型的组合。

从 ISO 文件启动虚拟机，会出现安装的 ESXi 引导菜单，如图 3-4 所示。

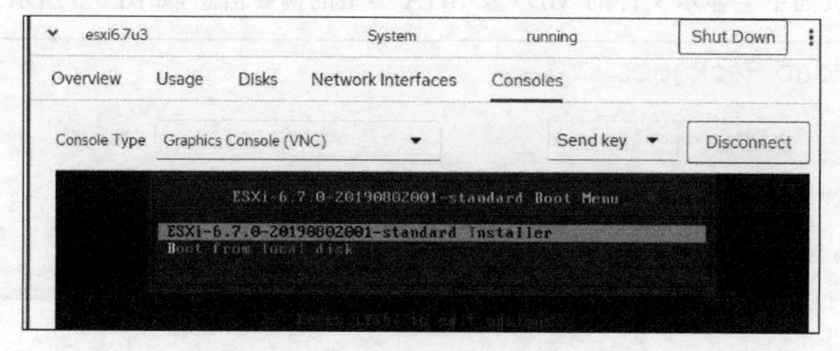


图 3-4　ESXi 安装引导菜单

在欢迎屏幕上，按 Enter 键，如图 3-5 所示。

按 F11 键接受许可，如图 3-6 所示。

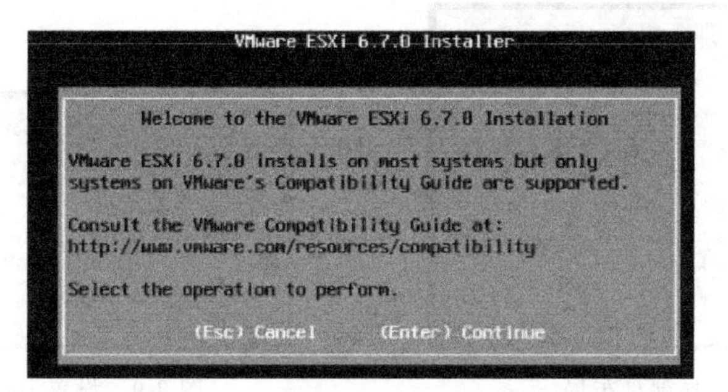


图 3-5 安装 ESXi 的欢迎信息

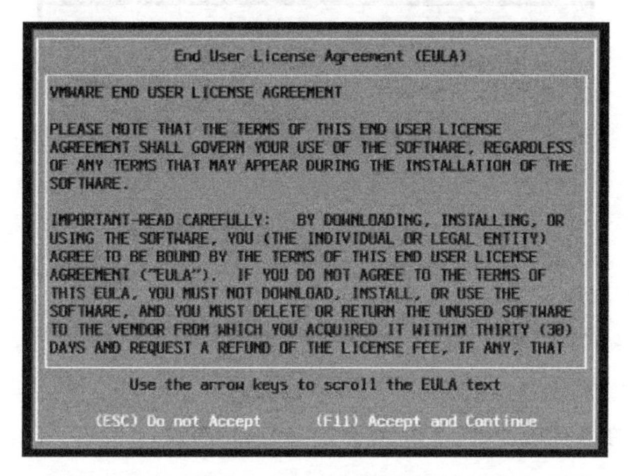


图 3-6 ESXi 的许可协议

按 Enter 键选择单个磁盘作为默认安装驱动器，如图 3-7 所示。

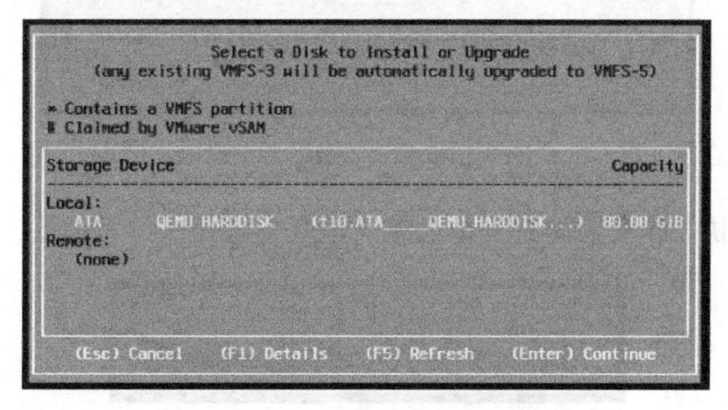


图 3-7 选择要安装或升级的磁盘

将键盘布局选为 US Default，按 Enter 键，如图 3-8 所示。

输入 root 的初始密码，如图 3-9 所示。

如果收到 CPU 或其他设备的兼容性警告，则可按 Enter 键继续，如图 3-10 所示。

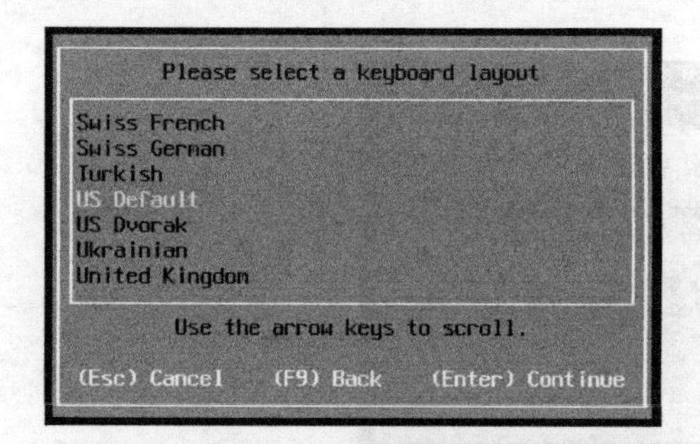


图 3-8 选择键盘布局

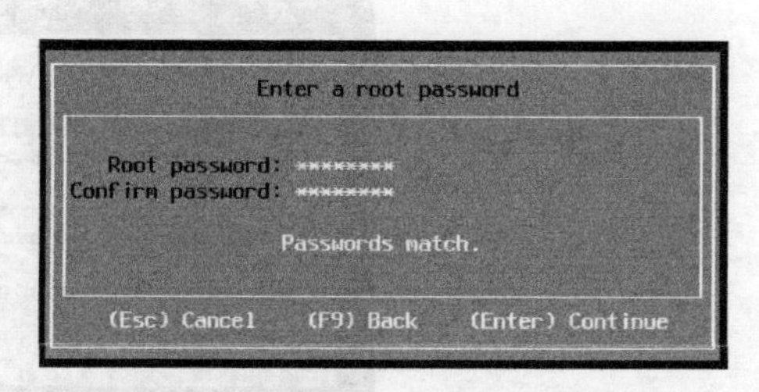


图 3-9 设置 root 用户的密码

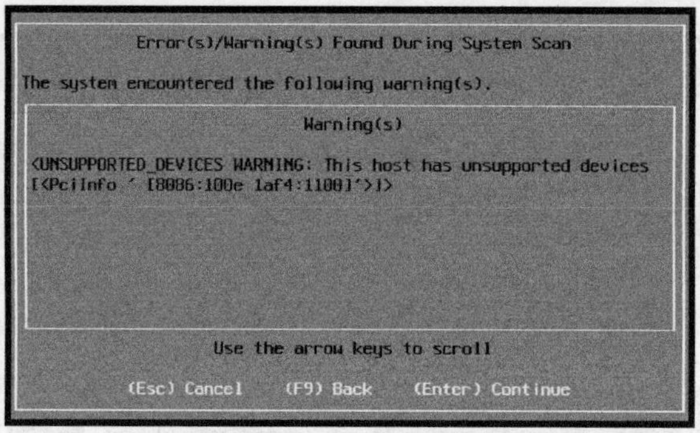

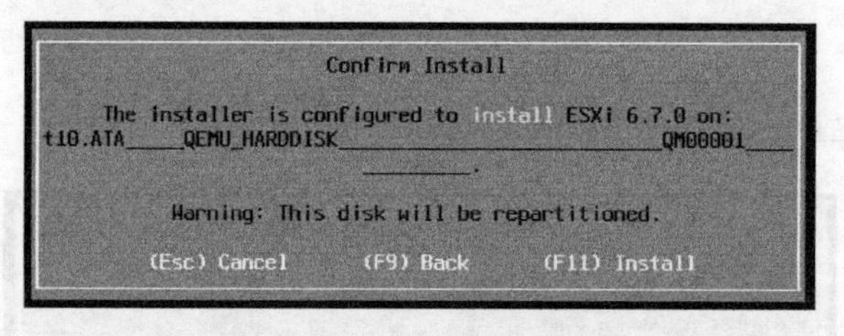


图 3-10 硬件兼容性警告

确认将会安装在目标磁盘,按 F11 键开始安装,如图 3-11 所示。

Confirm Install

The installer is configured to install ESXi 6.7.0 on:
t10.ATA_____QEMU_HARDDISK_______________________________QM00001____
_________.

Warning: This disk will be repartitioned.

(Esc) Cancel (F9) Back (F11) Install

图 3-11 确认安装目标磁盘

安装的速度很快。在安装的过程中会显示进度条,如图 3-12 所示。

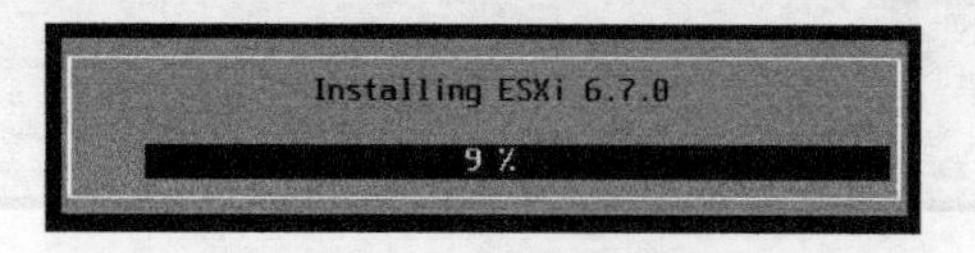


图 3-12 安装进度

安装结束后,提示断开安装介质,如图 3-13 所示。

重新启动后,会出现 ESXi 服务器的管理入口页面,如图 3-14 所示。可以使用浏览器访问屏幕上显示的地址,这是一个功能丰富的基于 Web 的管理工具。

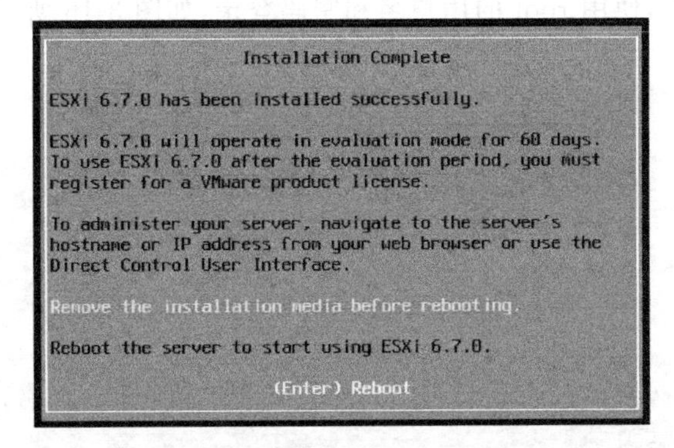


图 3-13　安装完成

VMware ESXi 6.7.0 (VMKernel Release Build 14320388)

Red Hat KVM

4 x Intel(R) Xeon(R) CPU E5-2670 0 @ 2.60GHz
4 GiB Memory

To manage this host go to:
http://192.168.1.210/ (DHCP)
http://[fe80::5054:ff:fe3f:c372]/ (STATIC)

<F2> Customize System/View Logs　　<F12> Shut Down/Restart

图 3-14　ESXi 服务器的管理入口页面

也可以按 F2 键，输入在安装过程中设置的 root 密码，这样就会进入系统设置界面。可以在其中完成一些最基本的设置，如图 3-15 所示。

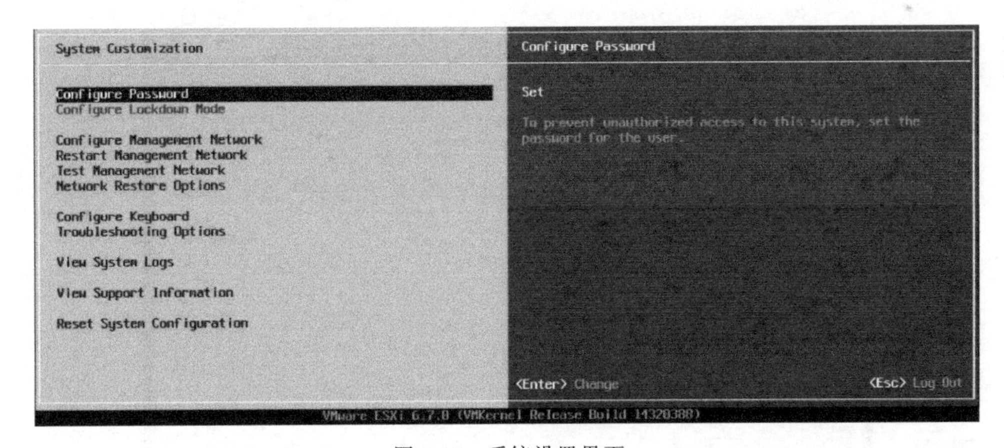


图 3-15　系统设置界面

## 3.4.2　VMware ESXi 管理

通过 Web 浏览器（建议使用对 HTML5 支持比较好的浏览器，如 Chrome、Firefox 等）

打开 ESXi 的管理页面，使用 root 的用户名和密码登录，如图 3-16 所示。

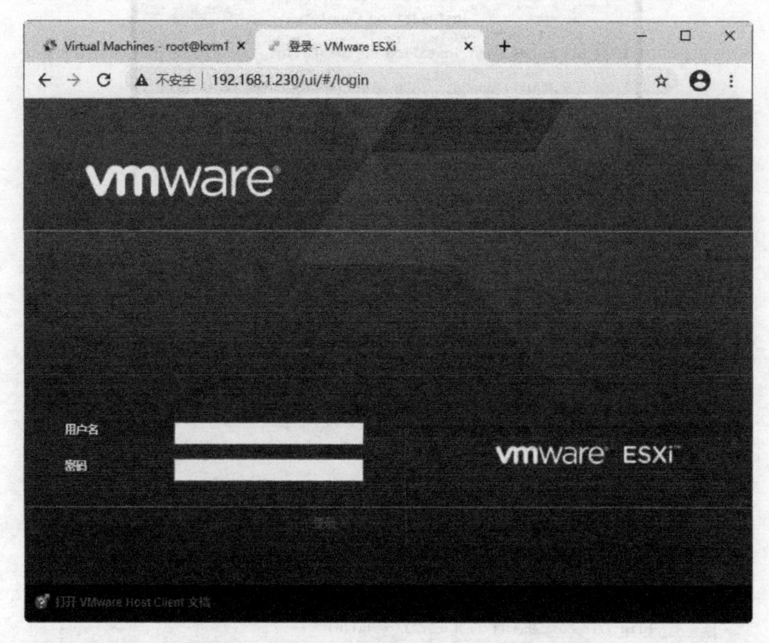


图 3-16 VMware Host Client 登录界面

这是一个 HTML/JavaScript 的应用程序，是由 ESXi 主机直接提供的轻量级管理界面，如图 3-17 所示。

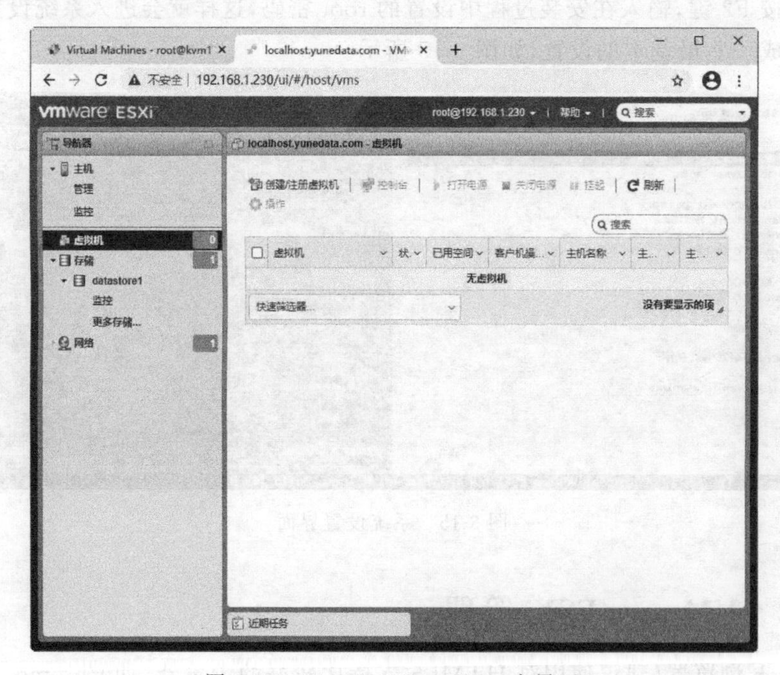


图 3-17 VMware Host Client 主界面

下面在 ESXi 中再创建一个虚拟机。首先单击“创建/注册虚拟机”按钮，就会出现创建新虚拟机的向导，如图 3-18 所示。

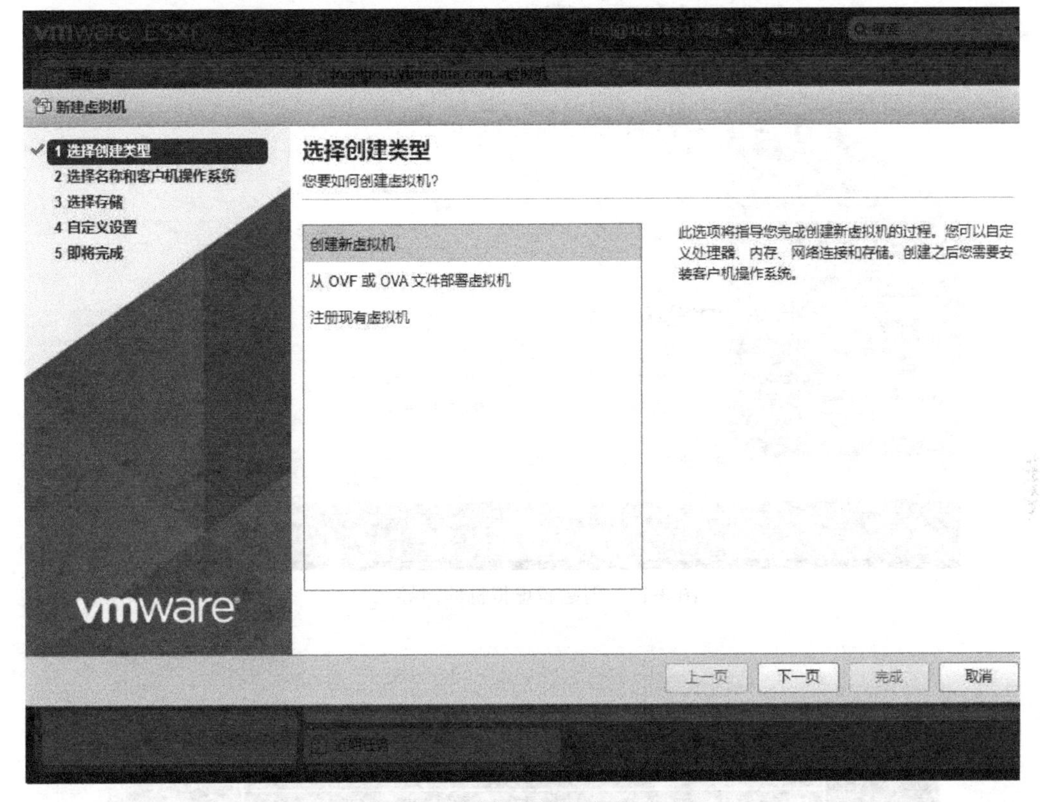


图 3-18　创建新虚拟机的向导(一)

向导共分为 5 个步骤，分别是：

(1) 选择创建类型。

(2) 选择名称和客户机操作系统。

(3) 选择存储。

(4) 自定义设置。

(5) 即将完成。

首先选择“创建新的虚拟机”，然后单击“下一步”按钮。向导的提示信息友好、简洁，此处不再赘述详细操作过程。向导最后的界面如图 3-19 所示。

默认情况下，新虚拟的电源是关闭的。启动虚拟机之后，应该看到预览屏幕的更改。单击“控制台”，然后单击“打开浏览器控制台”，这样就可以获得虚拟机的控制台了，如图 3-20 所示。

图 3-19 创建新虚拟机的向导(二)

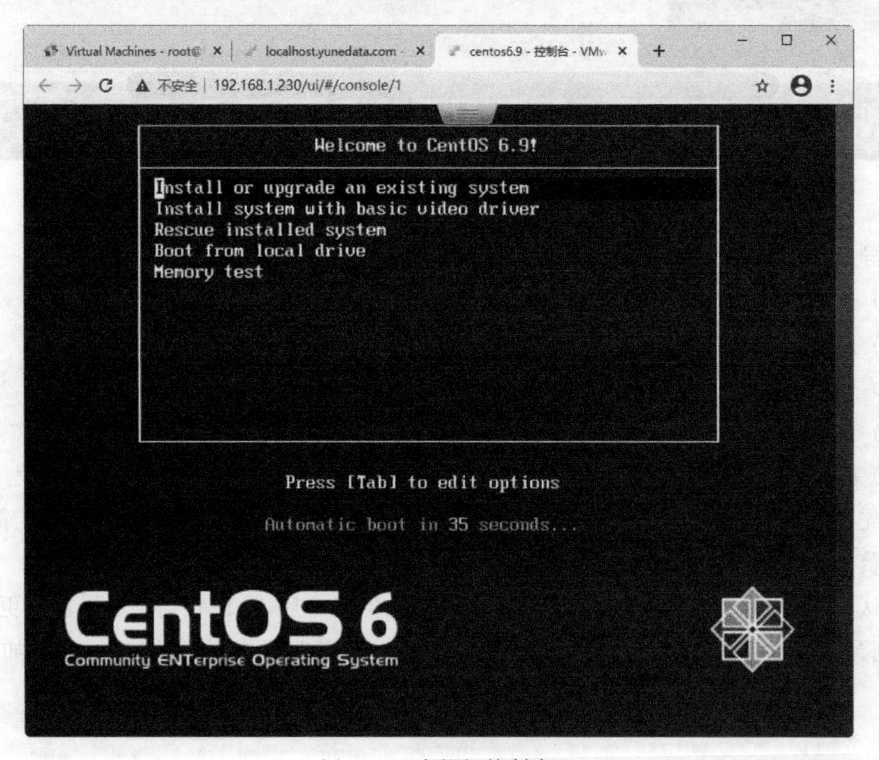


图 3-20 虚拟机控制台

可以很顺利地完成操作系统的安装,后续操作不再赘述。虚拟机的详细信息如图 3-21 所示。

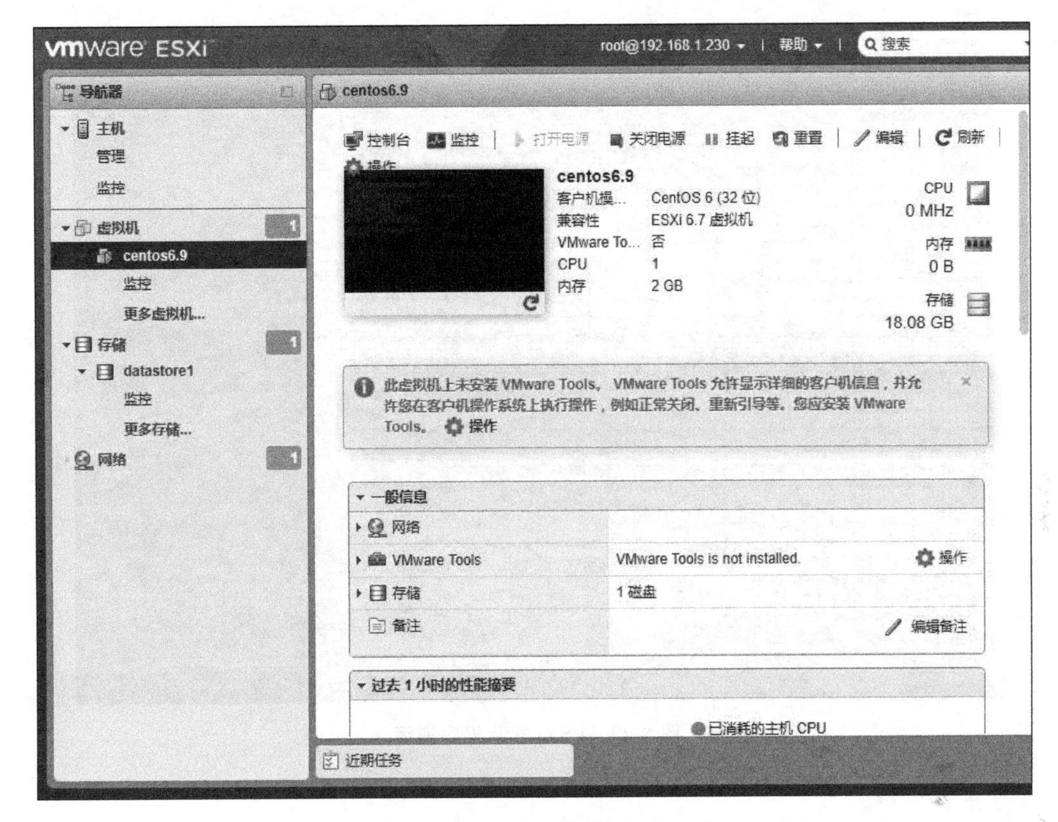


图 3-21　虚拟机的详细信息

## 3.4.3　实验中遇到的问题

如前所述,ESXi 的硬件兼容列表很短,所以在做嵌套实验时会遇到不少问题。以下是可以重复再现但无法解决的问题。

(1) CentOS 8.3+ESXi 6.7 组合：只能选择 e1000 网卡。如果选择的是 e1000e 网卡,则会出现紫色的系统崩溃屏幕,如图 3-22 所示。

(2) CentOS 8.3+ESXi7(7.0U1-16850804)组合：由于 ESXi7 已不再支持 e1000 网卡,所以只能选择 e1000e 网卡。安装 ESXi7 没有问题,但是无法出现 VMware Host Client 的登录界面。

除了上述实验所使用的 CentOS 8.3+ESXi 6.7(6.7.0.update03-14320388)组合之外,还在 CentOS 7.2+ESXi6.0(6.0.0.update02-3620759)组合下通过了实验。

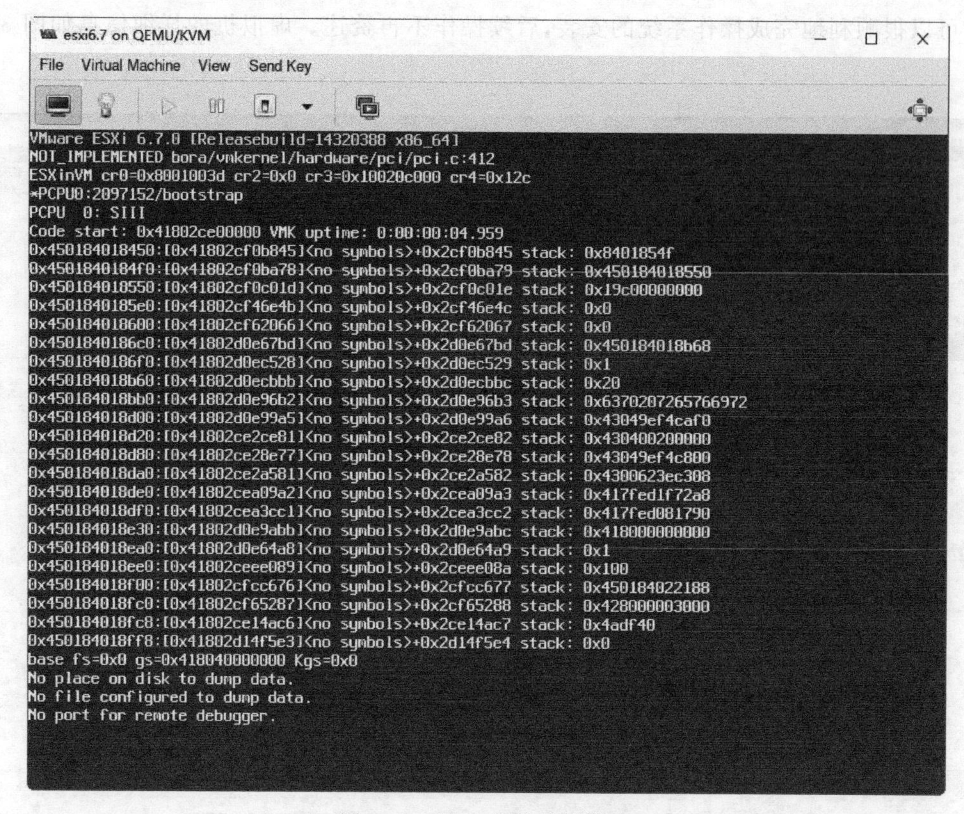


图 3-22 ESXi 安装程序崩溃

## 3.5 L2 级别 Microsoft Hyper-V 宿主机的配置

本次实验将采用 CentOS 8.3＋Microsoft Windows Server 2019 简体中文服务器版(17763.973)的组合。

由于本次实验环境采用的是 Intel 的 CPU,所以将 L2 级别虚拟机的 CPU 设置为< cpu mode＝'host-passthrough'/>。

安装 Windows Server 2019 后,可以在其中通过 SecurAble 工具来检测是否支持虚拟化。如果支持,则会显示绿色的 Yes,如图 3-23 所示。

**提示**：SecurAble 是一个免费的系统硬件检测工具,其下载网址为 https://securable.en.softonic.com/。

启动管理工具的服务器管理,在其中单击“添加角色和功能”,如图 3-24 所示。

在向导中单击“下一步”按钮。根据需求选择“安装类型”。由于是本地安装,所以选择“基于角色或基于功能的安装”,单击“下一步”按钮,如图 3-25 所示。

选择服务器,单击“下一步”按钮,如图 3-26 所示。

图 3-23　检查 CPU 的特性

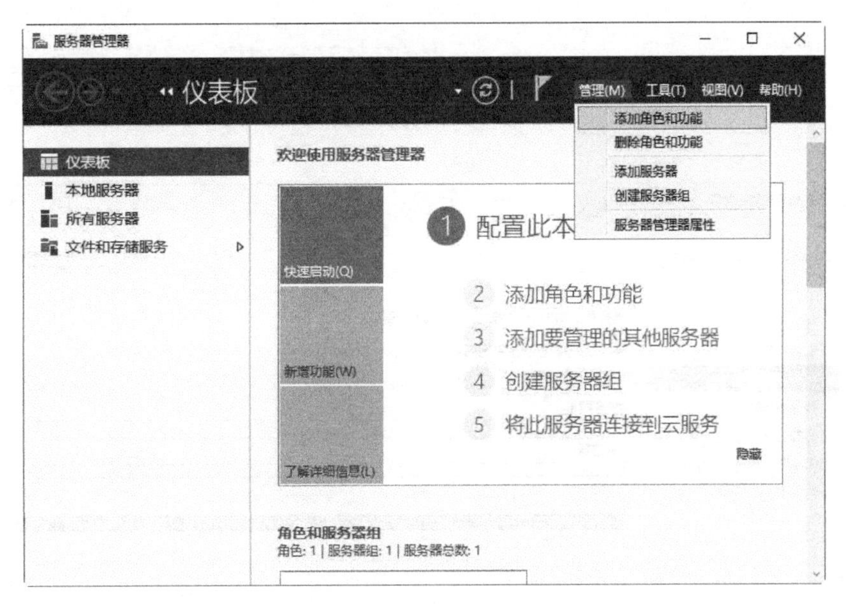


图 3-24　服务管理器

选中 Hyper-V 功能，在弹出的窗口中单击“添加功能”按钮，如图 3-27～图 3-29 所示。根据向导提示还需要完成以下步骤：

(1) 是否需要添加其他角色？

(2) 是否需要添加其他功能？

(3) 选择网卡，创建虚拟交换机。

(4) 根据实际需求选择“允许发送和接收虚拟机的实时迁移”。

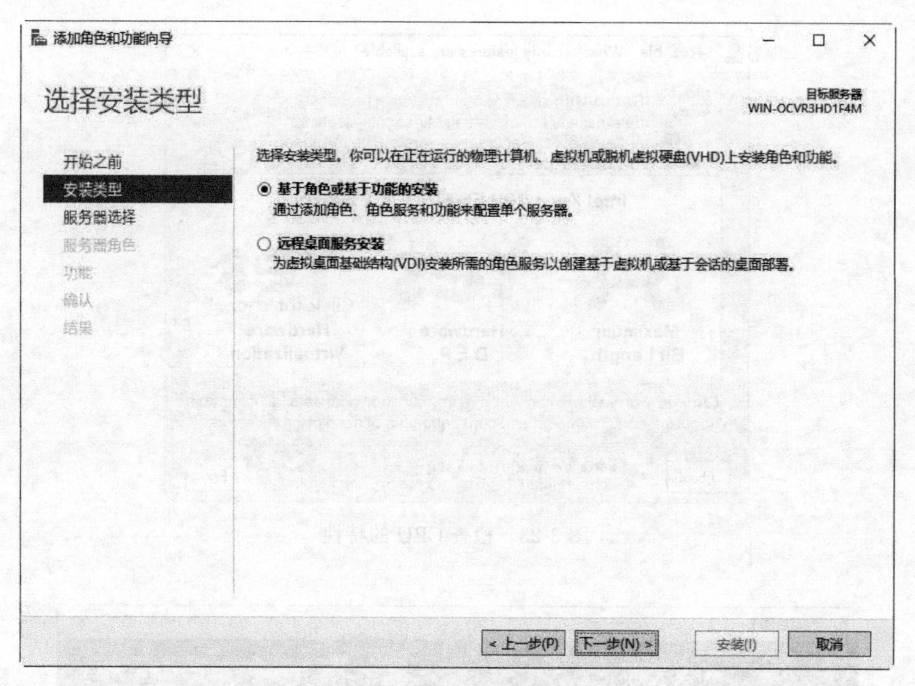


图 3-25　安装类型

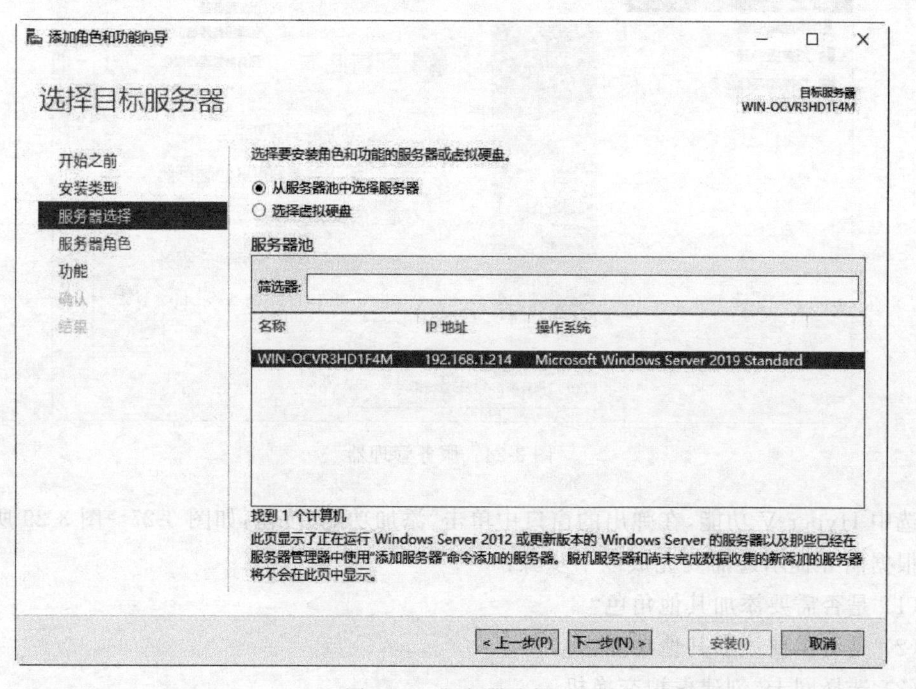


图 3-26　服务器选择

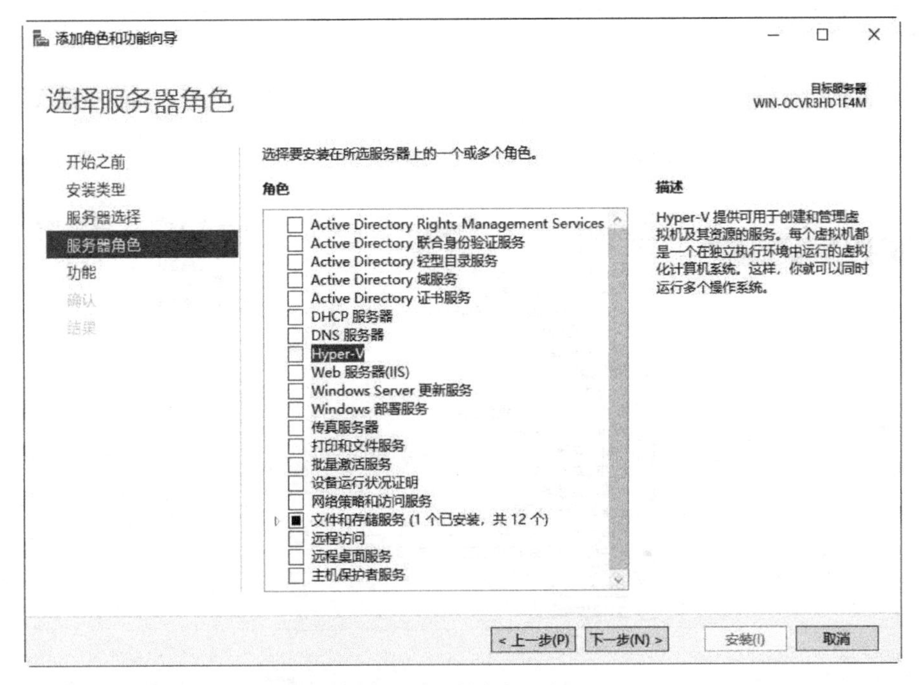


图 3-27　服务器角色选择(一)

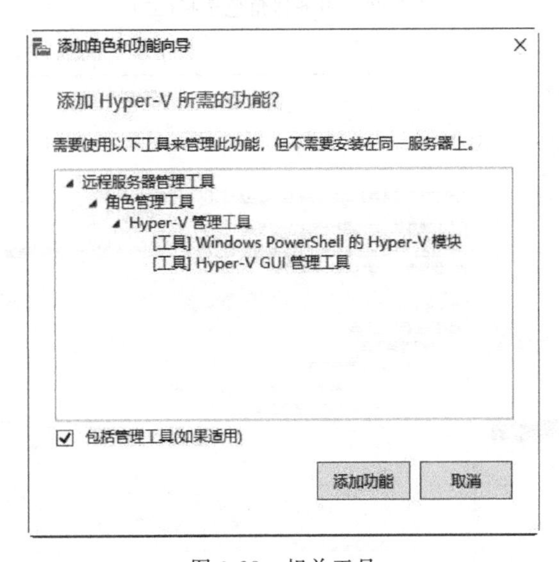


图 3-28　相关工具

(5) 选择虚拟硬盘和虚拟机配置文件的默认存储位置。

在向导的最后的确认页面中单击“安装”按钮进行安装，如图 3-30 所示。安装完成后单击“关闭”按钮，然后重启服务器。

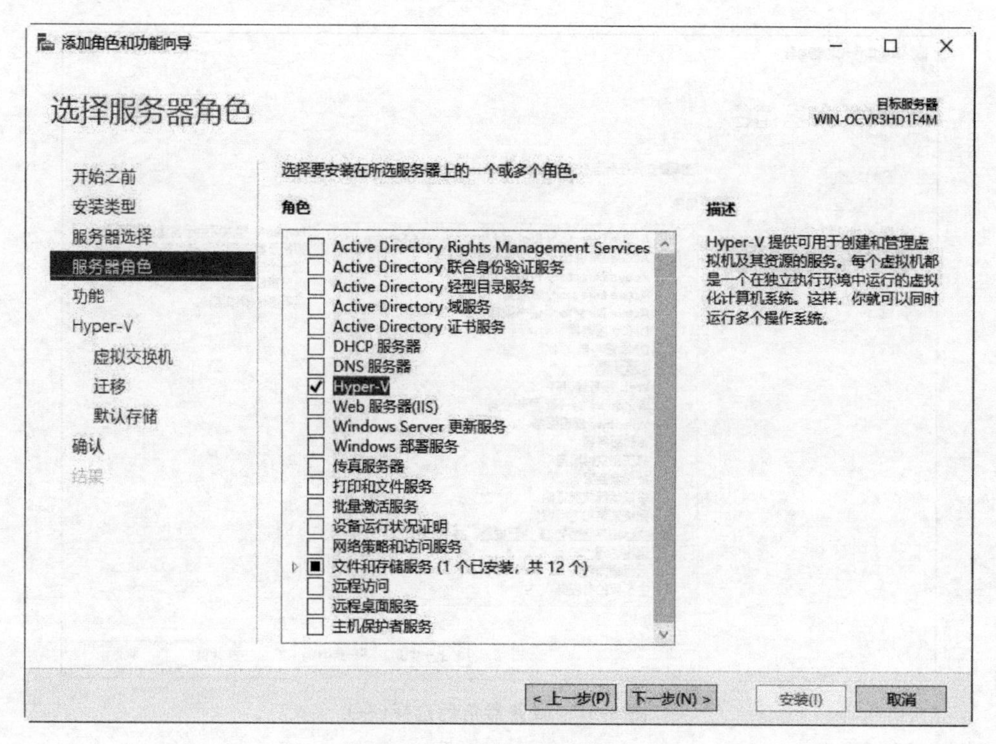


图 3-29 服务器角色选择(二)

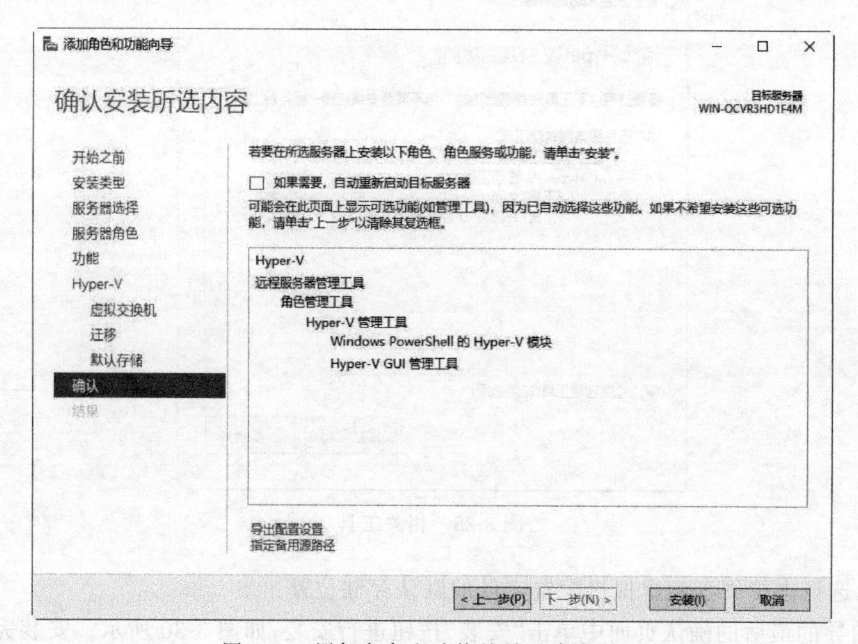


图 3-30 添加角色和功能的最后确认页面

Hyper-V 的功能到此已经安装完成,接下来需要创建虚拟机、安装操作系统、连接网络并使用虚拟机,如图 3-31 和图 3-32 所示。此处不再赘述详细操作过程。

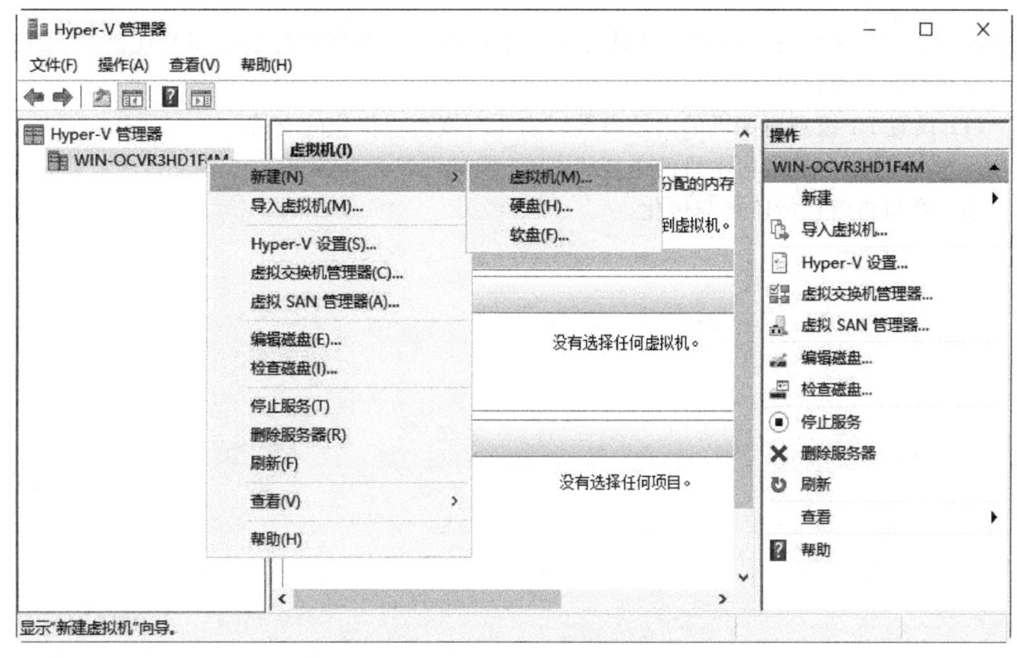


图 3-31 通过 Hyper-V 管理器创建虚拟机

图 3-32 虚拟机管理

## 3.6 本章小结

嵌套虚拟化主要用于开发和测试环境。本章讲解了如何构建基于 KVM 的嵌套虚拟化。有两个关键点：

(1) 设置 L1 级别宿主机的内核参数以启用对嵌套虚拟化的支持。

(2) 将充当 L2 级别宿主机的虚拟机 CPU 模式设置为 host-passthrough。

第 4 章将讲解性能监视与优化。

# 第4章 性能监视与优化

对虚拟化平台进行性能监视与优化是一个挑战，除了要监控宿主机之外还要监控虚拟机，例如：运行数据库的虚拟机与运行 Web 服务器虚拟机对资源的访问方式是不同的。本章先介绍性能监视与优化的思路和工具，然后讲解针对 CPU、内存、存储和网络的优化技术。

**本章要点**

- ❑ 性能监视与优化概述。
- ❑ 性能监控工具。
- ❑ 使用 Tuned 进行性能优化。
- ❑ VirtIO 驱动程序。
- ❑ CPU 优化技术。
- ❑ 内存优化技术。
- ❑ 网络优化技术。
- ❑ 存储优化技术。

## 4.1 性能监视与优化概述

监视与优化这两个动作密不可分、循环往复。通过监视系统性能，我们可以了解工作负荷及对服务器资源的需求，掌握性能的变化趋势，分析数据以便确定系统瓶颈，诊断系统问题并确定优化措施，执行优化措施并且监视变化，确认优化措施是否有效……

在此过程中，建立性能基线尤为重要。性能基线是在一段时间内在典型的工作负荷和用户连接数量的情况下收集的服务器性能数据，例如：将 5 分钟内 CPU 平均利用率不超过 70%作为基线。在确定性能基线时，应当了解服务器所执行的任务的特点及时间，例如：每周日凌晨 3 点需要生成周报表，CPU 利用率达到 90%，大约持续 30 分钟。及早建立性能基线有助于快速发现和解决性能瓶颈问题，建议在业务部署阶段就建立性能基线，然后和实际性能进行对比。

在进行性能监视时，建议通过日志保存监视的数据，同时要尽量减少性能监视本身对服务器所造成的影响。创建性能警报，当达到阈值时向管理员发送通知、自动进行相应操作。

造成性能瓶颈的原因,要么是错误或非最优的配置,要么是业务负载重。对于前者,调整配置就可以迎刃而解,而对于后者,就需要扩容,扩容又分为以下两种扩展方式。

(1) 纵向扩展(Scale-up):使用更高主频的 CPU、更大内存、更大网络带宽……

(2) 横向扩展(Scale-out):使用更多的节点、组件来分担负荷……

对于虚拟化平台而言,影响性能的因素有很多,但最重要是的 CPU、内存、存储和网络这 4 个资源子系统。

## 4.2 Linux 性能监控及调优工具

Linux 性能监控及调优工具有几十种。每个工具针对的资源及工作范围不同,详细介绍它们的使用方法超出了本书的范围。在这里,推荐通过 Brendan D. Gregg 绘制的 5 张图来学习它们:

(1) Linux 观测工具。

(2) Linux 静态性能分析工具。

(3) Linux 基准测试工具。

(4) Linux 调优工具。

(5) Linux sar 命令。

Brendan D. Gregg 是 Netflix 的资深工程师,在他的个人网站 http://www.brendangregg.com 上提供了这 5 张图。Linux 观测工具如图 4-1 所示。

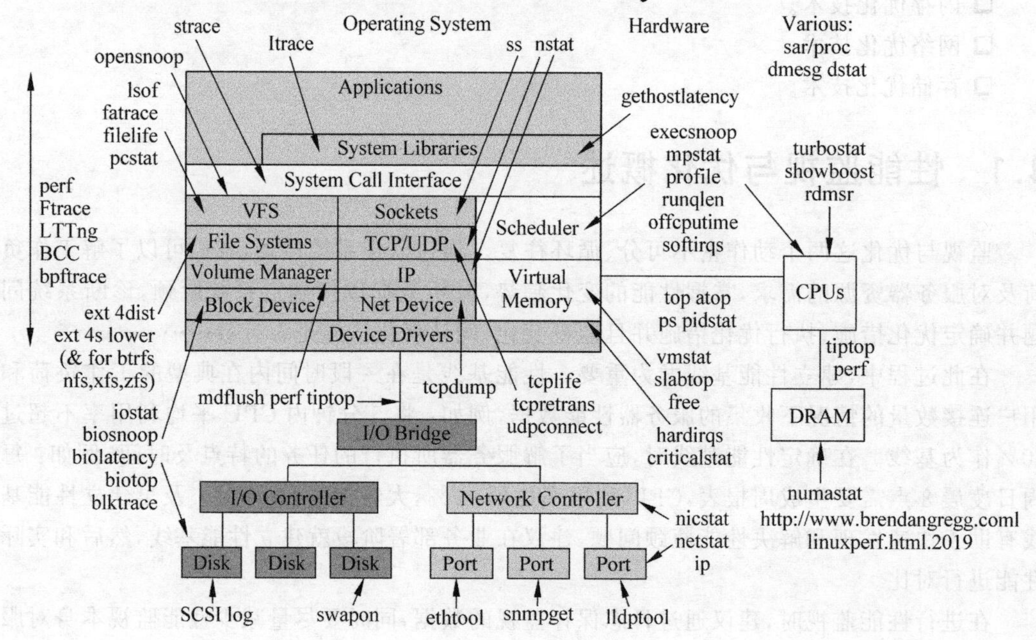


图 4-1 Brendan D. Gregg 绘制的 Linux 观测工具

至少需要了解图 4-1 中的这些常用命令。

(1) top：显示 Linux 整体运行状态及进程信息。

(2) ps：显示进程状态。

(3) uptime：显示系统运行了多长时间。

(4) netstat：显示网络连接、路由表、接口统计信息……

(5) free：显示系统中的可用和已用内存。

(6) vmstat：显示虚拟内存统计信息。

(7) sysctl：在运行时配置内核参数。

(8) tcpdump：包分析工具。

(9) /proc 目录下的文件：系统信息。

(10) pgrep：根据名称和其他属性查找进程。

(11) ls* 命令是获得硬件信息的命令，例如：lscpu、lsblk、lsmem、lsscsi、lstopo 等。

在后续的章节中，还会用到一些与虚拟机相关的监视及优化工具。

## 4.3 使用 Tuned 优化宿主机和 Linux 虚拟机的性能

Tuned 是 RHEL/CentOS 中最有价值的服务之一。管理员使用它可以针对不同的业务需求快速地调整多个系统参数，也可以让它先监视系统一段时间，然后提出有针对性的优化建议并执行。

不同的业务系统优化的需求是有差异的，有的希望实现高吞吐，有的希望达到低延迟，而有些可能更关心节能。同一个业务系统在不同时间段的需求也有可能不一样，例如视频转码服务器希望在转码工作时高速运行，而在没有任务时则进入低功耗状态。针对不同的 CPU 架构，调整的措施也有所不同，这些对管理员是一个巨大的挑战。

Tuned 可以很好地应对这些挑战。它的核心是配置文件(Profile)，可以针对不同的用例(Use Cases)调整系统，可以很方便地切换配置文件以应对需求的改变，也可以将多个配置文件组合在一起使用。

针对虚拟化环境，有两个预安装的配置文件可以直接使用。

(1) virtual-host：适用于宿主机。它是一种基于 throughput-performance 的配置文件，针对虚拟化宿主机进行了调整：减少了虚拟内存的交换性，增加了磁盘的预读值，并提供了更积极的脏页回写(Dirty Pages Writeback)。

(2) virtual-guest：适用于运行在 KVM 上的 RHEL/CentOS 虚拟机。它也是在 throughput-performance 配置文件上修改而来的，减少了虚拟内存的交换性并增加了磁盘预读值，不会禁用 disk barriers。

下面在实验环境中的宿主机上来做一下实验。

(1) 通过软件仓库来安装 Tuned，示例命令如下：

```
# cat /etc/redhat-release
CentOS Linux release 8.3.2011

# dnf -y install tuned
```

（2）启动 tuned 服务，示例命令如下：

```
# systemctl enable --now tuned

# systemctl status tuned
● tuned.service - Dynamic System Tuning Daemon
   Loaded: loaded (/usr/lib/systemd/system/tuned.service; enabled; vendor preset: enabled)
   Active: active (running) since Sun 2021-01-31 15:30:45 CST; 2h 10min ago
     Docs: man:tuned(8)
           man:tuned.conf(5)
           man:tuned-adm(8)
Main PID: 1297 (tuned)
    Tasks: 4 (limit: 203267)
   Memory: 24.6M
   CGroup: /system.slice/tuned.service
           └─1297 /usr/libexec/platform-python -Es /usr/sbin/tuned -l -P

Jan 31 15:30:44 tomkvm1 systemd[1]: Starting Dynamic System Tuning Daemon...
Jan 31 15:30:45 tomkvm1 systemd[1]: Started Dynamic System Tuning Daemon.
```

（3）验证 Tuned 配置文件是否处于激活状态，示例命令如下：

```
# tuned-adm active
Current active profile: throughput-performance

# tuned-adm verify
Verfication succeeded, current system settings match the preset profile.
See tuned log file ('/var/log/tuned/tuned.log') for details.
```

默认使用的配置文件是 throughput-performance。

（4）列出系统上所有可用的 Tuned 配置文件，示例命令如下：

```
# tuned-adm list
Available profiles:
- accelerator-performance    - Throughput performance based tuning with disabled higher
latency STOP states
- balanced                   - General non-specialized tuned profile
- desktop                    - Optimize for the desktop use-case
- hpc-compute                - Optimize for HPC compute workloads
- intel-sst                  - Configure for Intel Speed Select Base Frequency
```

```
- latency-performance         - Optimize for deterministic performance at the cost of
increased power consumption
- network-latency             - Optimize for deterministic performance at the cost of
increased power consumption, focused on low latency network performance
- network-throughput          - Optimize for streaming network throughput, generally
only necessary on older CPUs or 40G+ networks
- optimize-serial-console     - Optimize for serial console use.
- powersave                   - Optimize for low power consumption
- throughput-performance      - Broadly applicable tuning that provides excellent
performance across a variety of common server workloads
- virtual-guest               - Optimize for running inside a virtual guest
- virtual-host                - Optimize for running KVM guests
Current active profile: throughput-performance
```

这些预安装的配置文件针对不同的用例,可以直接使用。

- balanced:常规的配置文件。
- desktop:针对桌面应用进行了优化。
- latency-performance:以增加功耗为代价进行了优化。
- network-latency:以增加功耗为代价进行了优化,重点是降低网络的延迟。
- network-throughput:优化了网络吞吐,通常应用于较旧的CPU或40Gb/s以上网络。
- powersave:为节能进行了优化。
- throughput-performance:为在常见的服务器上的工作负荷进行了优化。
- virtual-guest:针对虚拟机进行了优化。
- virtual-host:针对虚拟化宿主机进行了优化。

**提示**:Tuned的配置文件存储在/usr/lib/tuned/目录中,每个配置文件都有其自己的子目录。Tuned配置文件由配置文件tuned.conf和可选的其他文件(例如:帮助程序、脚本)组成。有些第三方软件也会提供自定义的Tuned配置文件,例如Oracle数据库、PostgreSQL数据库等。

(5) 激活配置文件virtual-host,示例命令如下:

```
#tuned-adm profile virtual-host

#tuned-adm active
Current active profile: virtual-host
```

(6) 建议重新启动系统,示例命令如下:

```
#reboot
```

(7) 检查当前使用的配置文件，示例命令如下：

```
# tuned - adm active
Current active profile: virtual - host
```

如果是 KVM 中的 RHEL/CentOS 的虚拟机，则可以使用 virtual-guest 配置文件。示例命令如下：

```
# tuned - adm profile virtual - guest
```

## 4.4 VirtIO 驱动程序

在虚拟化世界中，总是将虚拟化与裸机系统进行对比。半虚拟化的驱动程序可以提高虚拟机的性能，甚至接近裸机系统的性能。建议在虚拟机中尽量使用半虚拟化驱动程序，尤其是当虚拟机运行具有大量 I/O 的任务和应用程序时。

VirtIO 是一种半虚拟化框架。当我们使用半虚拟化驱动程序时，虚拟机操作系统会意识到它正在 Hypervisor 上运行，并且包含充当前端的驱动程序。虚拟机的请求不经过 Hypervisor，而是直接发送给由 QEMU 提供的半虚拟化设备，通过宿主机上的驱动程序发送给真实的物理硬件设备。

常见的 VirtIO 设备驱动程序有以下几种。

(1) virtio-net：虚拟以太网卡驱动程序。

(2) virtio-blk：虚拟块设备(磁盘)驱动程序。

(3) virtio-balloon：用于管理虚拟机内存的气球设备。

(4) virtio-scsi：替代和改进的 virtio-blk。具有更好的扩展性，使用标准 SCSI 命令集，可以与 SCSI 设备直通。

(5) virtio-console：用于虚拟机和宿主机用户空间应用程序进行数据交换的设备。

(6) virtio-rng：提供了高质量随机数。

## 4.5 CPU 优化技术

从宿主机角度来看，每个虚拟机就是一个常规的进程。下面通过实验来验证一下。

首先查看宿主机上正在运行的虚拟机，示例命令如下：

```
# virsh list
Id   Name      State
----------------------------
1    win2019   running
```

有 1 个正在运行的虚拟机。通过 pgrep 命令可以看出这就是 1 个 Linux 进程，示例命令如下：

```
# pgrep -lfa qemu-kvm
5412 /usr/libexec/qemu-kvm -name guest=win2019,debug-threads=on -S -object secret,
id=masterKey0,format=raw,file=/var/lib/libvirt/qemu/domain-1-win2019/master-key.aes
-machine pc-q35-rhel8.2.0,accel=kvm,usb=off,vmport=off,dump-guest-core=off -cpu
host,hv-time,hv-relaxed,hv-vapic,hv-spinlocks=0x1fff -m 8192 -overcommit mem-lock
=off -smp 4,sockets=4,cores=1,threads=1 -uuid 2b8f0015-b42c-4813-a117
-2092f424ea50 -no-user-config -nodefaults -chardev socket,id=charmonitor,fd=38,
server,nowait -mon chardev=charmonitor,id=monitor,mode=control -rtc base=localtime,
driftfix=slew -global kvm-pit.lost_tick_policy=delay -no-hpet -no-shutdown -global
ICH9-LPC.disable_s3=1 -global ICH9-LPC.disable_s4=1 -boot strict=on -device pcie-
root-port,port=0x10,chassis=1,id=pci.1,bus=pcie.0,multifunction=on,addr=0x2
...
```

libvirt 保存在虚拟机 XML 配置文件中的属性都是传递给 qemu-kvm 进程的参数。

分配给虚拟机的 vCPU 是 Linux 线程，它们由宿主机上的调度程序管理。虚拟机 win2019 有 4 颗 vCPU，所以 qemu-kvm 进程就有 4 个线程，示例命令如下：

```
# virsh dumpxml win2019 | grep vcpu
  <vcpu placement='static'>4</vcpu>

# ps -T -p 5412
    PID    SPID TTY          TIME CMD
   5412    5412 ?        00:03:46 qemu-kvm
   5412    5415 ?        00:00:00 qemu-kvm
   5412    5420 ?        00:00:00 IO mon_iothread
   5412    5421 ?        00:02:49 CPU 0/KVM
   5412    5422 ?        00:00:27 CPU 1/KVM
   5412    5423 ?        00:00:27 CPU 2/KVM
   5412    5424 ?        00:00:21 CPU 3/KVM
   5412    5426 ?        00:00:03 SPICE Worker
   5412    5427 ?        00:00:00 vnc_worker
   5412    5989 ?        00:00:00 worker
   5412    5990 ?        00:00:00 worker
```

可以根据需要为虚拟机分配 vCPU。为了获得最佳性能，建议根据虚拟机操作系统的预期负载来分配所需的 vCPU。虽然可以分配超出所需数量的 vCPU，但是将来可能会导致扩展问题。

宿主机逻辑 CPU 的数量是物理 CPU 的插槽(socket)、内核(core)和线程(thread)的乘积。以这台宿主机为例，查看 CPU 的信息的示例命令如下：

```
# lscpu
Architecture:          x86_64
CPU op-mode(s):        32-bit, 64-bit
Byte Order:            Little Endian
CPU(s):                32
On-line CPU(s) list: 0-15
Thread(s) per core:    2
Core(s) per socket:    8
Socket(s):             2
NUMA node(s):          2
Vendor ID:             GenuineIntel
CPU family:            6
Model:                 45
Model name:            Intel(R) Xeon(R) CPU E5-2670 0 @ 2.60GHz
...
```

这台宿主机有两颗 Intel(R) Xeon(R) CPU E5-2670 的 CPU，这是 8 核的 CPU，每核有两个线程，所以这台宿主机共用 32 个逻辑 CPU。

有一个常见的错误说法，即所有虚拟机 vCPU 的数量之和应少于宿主系统中可用逻辑 CPU 的总数。例如：如果宿主机中可用的逻辑 CPU 总数为 32，错误的说法是如果有 4 个虚拟机，则每个虚拟机最多只能定义 8 个 vCPU。

KVM 支持超量分配 vCPU，但前提是虚拟机的负载应在允许的范围内，否则 vCPU 可能会被过量使用，如果 vCPU 负载增加到接近 100%，则可能会严重损害性能。

下面介绍一下 CPU 优化技术。调整和优化任何系统时，最重要的是在调整之前先建立性能基准，然后进行小的、单一的调整，观察这些更改对性能的影响。如此不断调整，直到达到所需的效果。

### 4.5.1 vCPU 的数量

考虑到 vCPU 过量使用对性能的影响，应确保为单个虚拟机分配的 vCPU 数量不超过宿主机中的逻辑 CPU 的总数。否则很有可能会导致性能显著下降。

virt-manager 在显示宿主机逻辑 CPU 数量的同时，会对超出数量的虚拟机发出警告，如图 4-2 所示。

而 Cockpit 和 virsh 则不会对 vCPU 的数量进行检查。如果宿主机有 16 个逻辑 CPU，在 Cockpit 中将 vCPU 设置为 32 个，则不会收到警告，如图 4-3 所示。

增加虚拟机 vCPU 的数量，并不一定会提高虚拟机的性能，这要视操作系统、软件而定。最佳策略是通过工具来测量并发现虚拟机性能指标与 vCPU 数量之间的规律后，再做决策。

KVM 支持 vCPU 的超量分配，即所有虚拟机的 vCPU 总和大于宿主机逻辑 CPU 数量（桌面虚拟化应用场景经常会出现超量分配），但是，在生产系统中实现此功能之前，一定要进行测试。

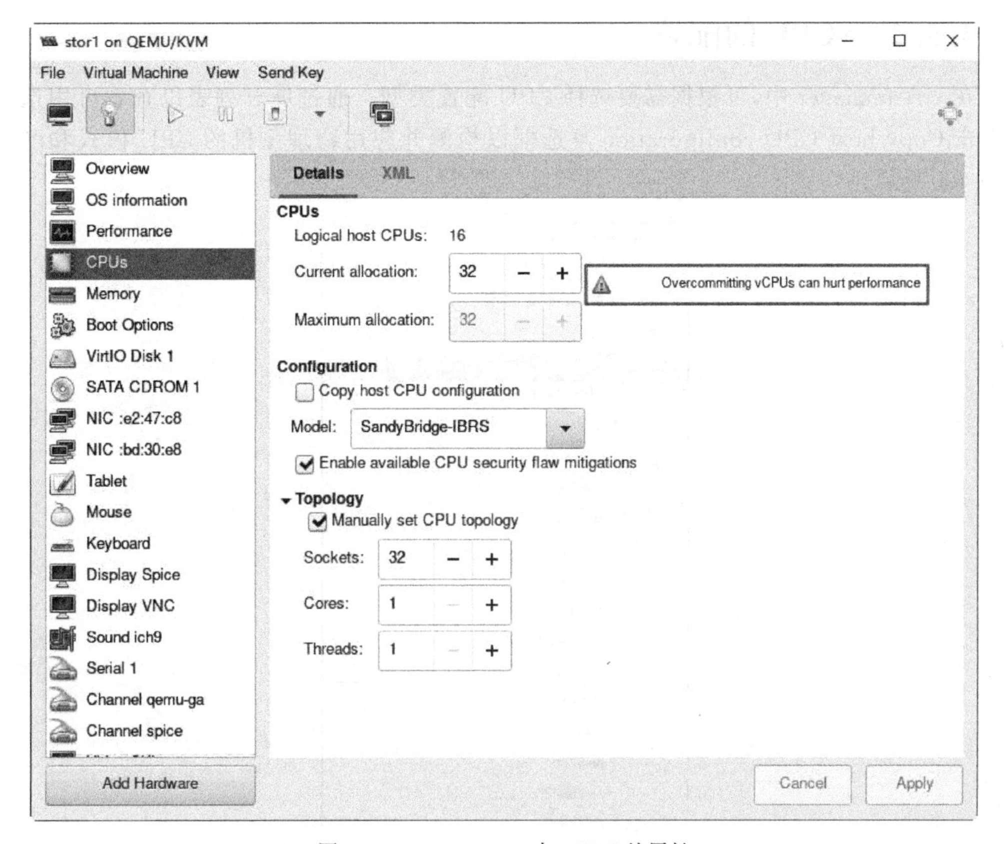


图 4-2　virt-manager 中 vCPU 的属性

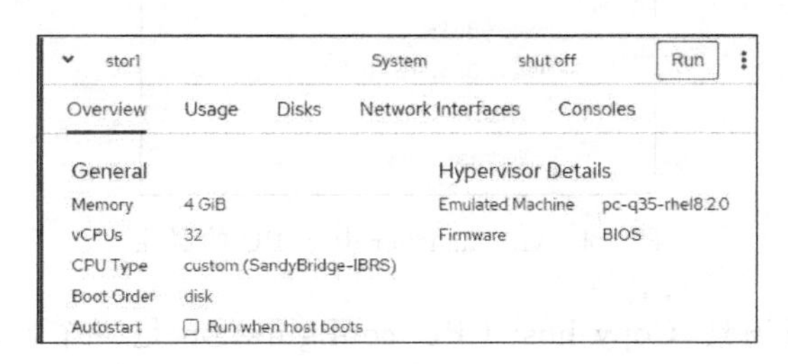


图 4-3　Cockpit 中虚拟机的属性

图 4-2 中所显示的虚拟机的 CPU 配置中，有 3 个主要部分：

(1) vCPU 的数量。

(2) vCPU 的配置。

(3) vCPU 的拓扑。

下面继续讲解第 2 部分，即 vCPU 配置。

**提示：** Cockpit 和 virt-manager 提供的配置选项比较少，而只有在命令行、配置文件中才能对物理 CPU 和 vCPU 进行细粒度配置。

### 4.5.2 vCPU 的配置

在 virt-manager 中，可根据需要选择 CPU 配置类型。通过展开列表以查看可用选项，或单击 Copy host CPU configuration 复选框以检测并应用物理主机的 CPU 模式和配置，如图 4-4 所示。

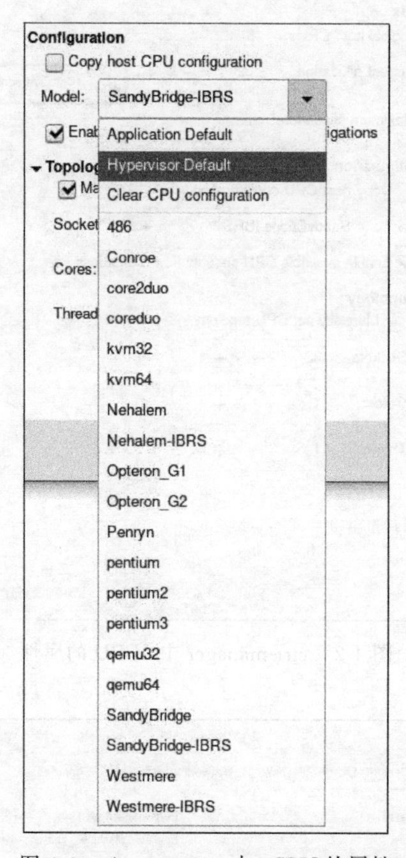


图 4-4 virt-manager 中 vCPU 的属性

对于大多数应用场景，Copy host CPU configuration 是一个比较好选择，它对应的 XML 配置如下：

```
<cpu mode = 'host - model' check = 'none'/>
```

在模式下拉列表框中有 3 个选项：

(1) Application Default：相当于选中 Copy host CPU configuration。

(2) Hypervisor Default：相当于将模式设置为 qemu64，对应的 XML 配置如下：

```
<cpu mode = 'custom' match = 'exact' check = 'none'>
  <model fallback = 'forbid'> qemu64 </model>
</cpu>
```

(3) Copy host CPU configuration：也相当于将模式设置为 qemu64。

qemu-kvm 为虚拟机提供的 qemu64 模式是一种通用的模式。通过这种模式可以让虚拟机在物理 CPU 有差异的宿主机之间进行实时迁移。

查询当前版本的 qemu-kvm 所支持的 CPU 模式，示例命令如下：

```
#/usr/libexec/qemu-kvm -cpu ?
Available CPUs:
x86 486                  (alias configured by machine type)
...
x86 SandyBridge          (alias configured by machine type)
x86 SandyBridge-IBRS     (alias of SandyBridge-v2)
x86 SandyBridge-v1       Intel Xeon E312xx (Sandy Bridge)
x86 SandyBridge-v2       Intel Xeon E312xx (Sandy Bridge, IBRS update)
...

Recognized CPUID flags:
  3dnow 3dnowext 3dnowprefetch abm ace2 ace2-en acpi adx aes amd-no-ssb
...

#
```

针对具体的场景，还可以通过 virsh edit 命令进行更细粒度的 vCPU 特性调整，示例命令如下：

```
#virsh edit win2019
...
  <cpu mode = 'custom' match = 'exact' check = 'full'>
    <model fallback = 'forbid'> SandyBridge-IBRS </model>
    <vendor> Intel </vendor>
    <feature policy = 'require' name = 'vme'/>
    <feature policy = 'require' name = 'ss'/>
    <feature policy = 'require' name = 'vmx'/>
    <feature policy = 'require' name = 'pcid'/>
    <feature policy = 'require' name = 'hypervisor'/>
    <feature policy = 'require' name = 'arat'/>
    <feature policy = 'require' name = 'tsc_adjust'/>
    <feature policy = 'require' name = 'umip'/>
    <feature policy = 'require' name = 'md-clear'/>
    <feature policy = 'require' name = 'stibp'/>
    <feature policy = 'require' name = 'arch-capabilities'/>
```

```
    <feature policy='require' name='ssbd'/>
    <feature policy='require' name='xsaveopt'/>
    <feature policy='require' name='pdpe1gb'/>
    <feature policy='require' name='ibpb'/>
    <feature policy='require' name='amd-ssbd'/>
    <feature policy='require' name='skip-l1dfl-vmentry'/>
    <feature policy='require' name='pschange-mc-no'/>
  </cpu>
...
```

## 4.5.3 vCPU 的拓扑

可以设置 vCPU 的插槽、核心和线程数，如图 4-5 和图 4-6 所示。

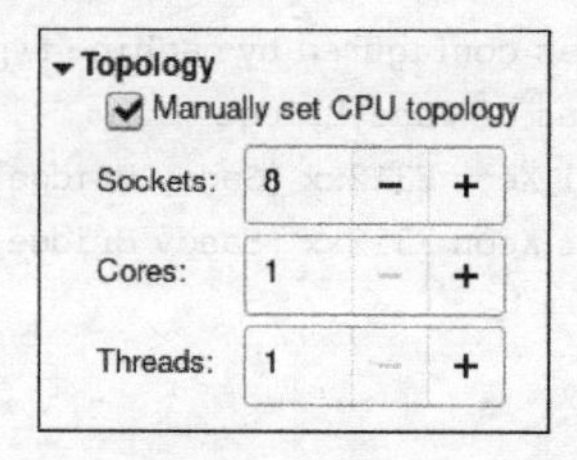


图 4-5 virt-manager 中 vCPU 的拓扑

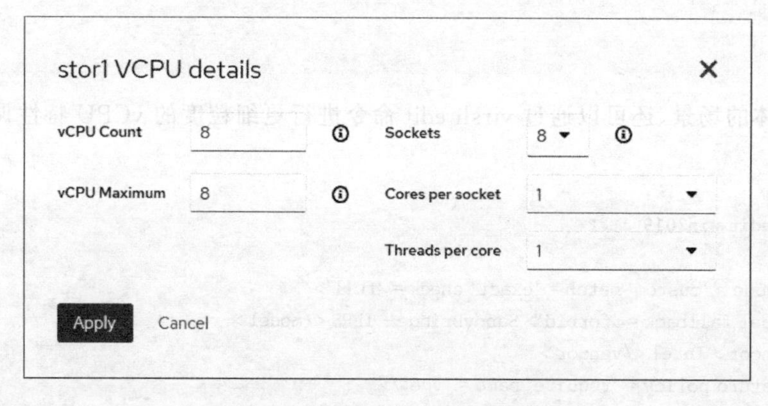


图 4-6 Cockpit 中 vCPU 的拓扑

CPU 拓扑最佳策略是根据需要确定所需插槽数量，只分配 1 个内核和 1 个线程，通常会提供最佳的性能结果。

## 4.5.4 非一致性内存访问(NUMA)的基本概念

早期版本的 RHEL/CentOS 中的 virt-manager 有一个称为固定(Pining)的 vCPU 选项，如图 4-7 所示。

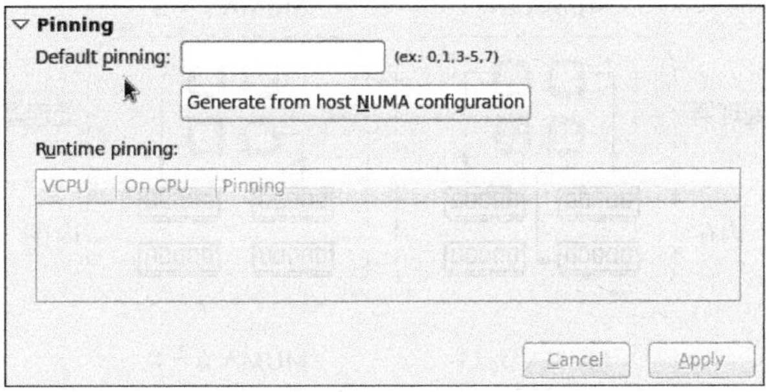

图 4-7　早期版本的 virt-manager 的 vCPU 固定选项

但是 RHEL/CentOS 8 中的 virt-manager 已经取消了这个选项。取消这个选项不是说它不重要，而是这个选项对性能影响很大，不正确的设置会造成很多问题。这个选项是一把"双刃剑"，针对特定的需求进行正确设置可以大大提高性能。

在学习这个选项之前，我们需要掌握 NUMA 的一些基本概念。

为了提高虚拟机的密度，我们总是希望宿主机拥有更高主频的 CPU、更多数量的 CPU、更多的 CPU 内核数、更大的内存，但早期的对称多处理器结构（Symmetric Multi-Processor，SMP）已经不能满足需求。这种架构中多个 CPU 对称工作，无主次或从属关系。各 CPU 共享相同的物理内存，每个 CPU 访问内存中的任何地址所需时间是相同的，因此 SMP 也被称为一致内存访问结构（Uniform Memory Access，UMA）。SMP 架构的主要问题是扩展能力有限，每个共享的环节都可能造成扩展时的瓶颈，而最受限制的则是内存。由于每个 CPU 必须通过相同的内存总线访问相同的内存资源，因此随着 CPU 数量的增加，内存访问冲突将迅速增加，最终会造成 CPU 资源的浪费，使 CPU 性能下降。实验证明，SMP 服务器 CPU 利用率最好的情况是 2～4 个 CPU。

非一致性内存访问（Non-Uniform Memory Access，NUMA）就是为了解决这个问题而出现的。利用 NUMA 技术，可以把几十个 CPU 组合在一个服务器内。其示意架构如图 4-8 所示。这种构架下，不同的物理内存和 CPU 核心属于不同的节点（Node），每个节点都有自己的集成内存控制器（Integrated Memory Controller，IMC）。在节点内部，架构类似于 SMP，使用 IMC 总线进行不同核心间的通信。在节点之间，通过快速路径互连（Quick Path Interconnect，QPI）进行通信。QPI 的延迟要高于 IMC，也就是说 CPU 访问内存有了远近（Remote/Local）之别，而且差别非常明显。由于这个特点，为了更好地发挥系统性能，应尽量减少不同 NUMA 节点之间的信息交互。图 4-8 的配置是最理想的：内存和 CPU 核心都在相同的 NUMA 节点内，在每个 NUMA 节点中都配置了内存，线程对内存的访问都发生在节点内。

跨节点（Cross-Node）的内存访问会慢很多，图 4-9 就是一个糟糕的 NUMA 布局。系统不均衡，多个线程对内存的访问会跨不同的 NUMA 节点，节点 2 的内存条是奇数，节点 3 没有足够内存，节点 4 无本地内存（最糟糕的情况）。

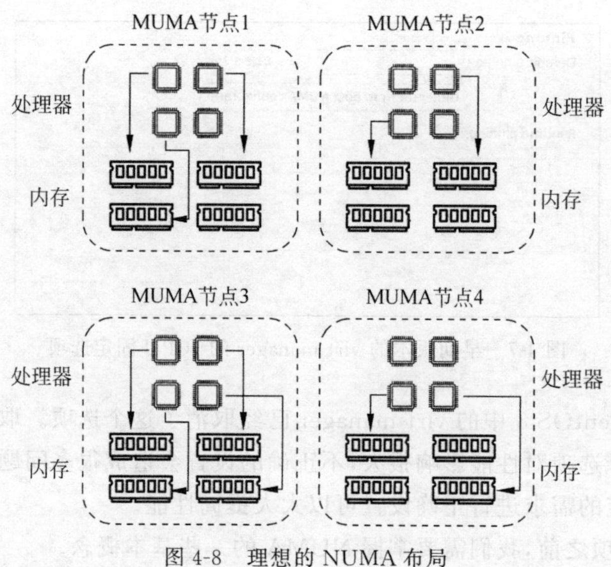


图 4-8 理想的 NUMA 布局

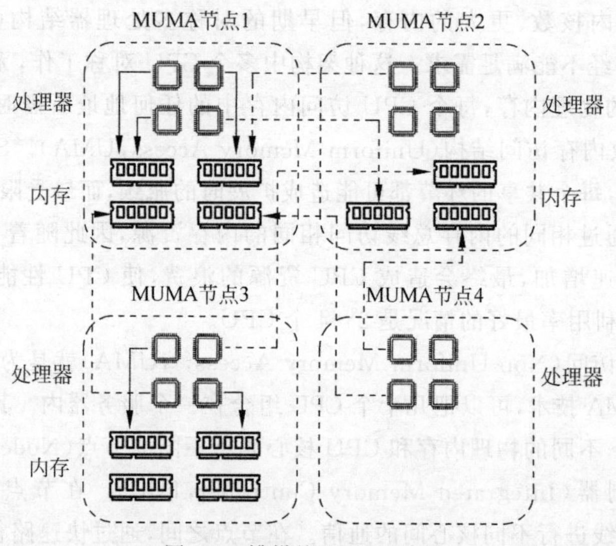


图 4-9 糟糕的 NUMA 布局

目前主流的 x86 服务器采用 NUMA 架构，在安装 CPU 和内存条时，需要严格按手册进行配置。图 4-10 就是 Dell PowerEdge R720 服务器手册中的内存配置准则。

Dell PowerEdge R720 服务器包含 24 个内存插槽，分为两组，每组 12 个插槽，每个处理器一组。将插槽 A1～A12 中的 DIMM 分配给处理器 1，并将插槽 B1～B12 中的 DIMM 分配给处理器 2。在具体安装时还应注意插槽上的颜色，白色、黑色和绿色代表不同的安插次序。

除了硬件之外，要考虑软件的配置，下面就在两台物理宿主机上做一下实验。

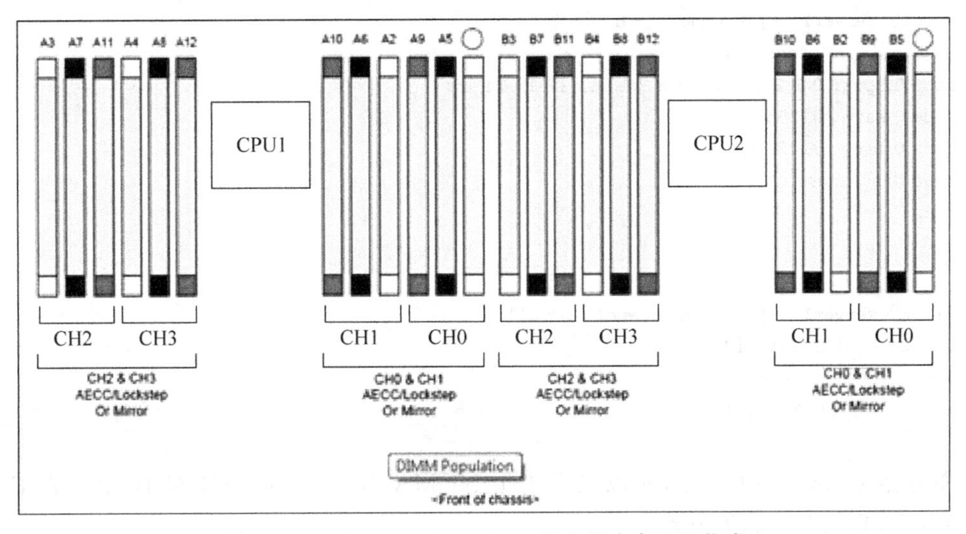


图 4-10 Dell PowerEdge R720 服务器内存配置指南

查看服务器 1 的 CPU 配置，示例命令如下：

```
[root@server1 ~]# lscpu
Architecture:          x86_64
CPU op-mode(s):        32-bit, 64-bit
Byte Order:            Little Endian
CPU(s):                16
On-line CPU(s) list:   0-15
Thread(s) per core:    2
Core(s) per socket:    8
Socket(s):             1
NUMA node(s):          1
Vendor ID:             GenuineIntel
CPU family:            6
Model:                 45
Model name:            Intel(R) Xeon(R) CPU E5-2670 0 @ 2.60GHz
Stepping:              7
CPU MHz:               3211.787
CPU max MHz:           3300.0000
CPU min MHz:           1200.0000
BogoMIPS:              5187.47
Virtualization:        VT-x
L1d cache:             32K
L1i cache:             32K
L2 cache:              256K
L3 cache:              20480K
NUMA node0 CPU(s):     0-15
```

```
[root@server1 ~]# numactl -H
available: 1 nodes (0)
node 0 cpus: 0 1 2 3 4 5 6 7 8 9 10 11 12 13 14 15
node 0 size: 31255 MB
node 0 free: 31061 MB
node distances:
node   0
  0:  10
```

```
[root@server1 ~]# virsh freecell --all
    0:   31805356 KiB
--------------------
Total:   31805356 KiB
```

服务器 1 只有 1 颗 CPU,所以只有 1 个 NUMA 节点 0。服务器有 16 个逻辑 CPU、32GB 内存,它们全部属于节点 0。

在服务器 2 上执行相同的命令以便查看 CPU 配置:

```
[root@server2 ~]# lscpu
Architecture:          x86_64
CPU op-mode(s):        32-bit, 64-bit
Byte Order:            Little Endian
CPU(s):                24
On-line CPU(s) list:   0-23
Thread(s) per core:    2
Core(s) per socket:    6
Socket(s):             2
NUMA node(s):          2
Vendor ID:             GenuineIntel
CPU family:            6
Model:                 62
Model name:            Intel(R) Xeon(R) CPU E5-2630 v2 @ 2.60GHz
Stepping:              4
CPU MHz:               1200.000
BogoMIPS:              5199.25
Virtualization:        VT-x
L1d cache:             32K
L1i cache:             32K
L2 cache:              256K
L3 cache:              15360K
NUMA node0 CPU(s):     0-5,12-17
NUMA node1 CPU(s):     6-11,18-23
```

```
[root@server2 ~]# numactl -H
available: 2 nodes (0-1)
```

```
node 0 cpus: 0 1 2 3 4 5 12 13 14 15 16 17
node 0 size: 16331 MB
node 0 free: 9065 MB
node 1 cpus: 6 7 8 9 10 11 18 19 20 21 22 23
node 1 size: 16384 MB
node 1 free: 9174 MB
node distances:
node   0   1
  0:  10  20
  1:  20  10
```

```
[root@server2 ~]# virsh freecell --all
    0:     9283100 KiB
    1:     9393116 KiB
---------------------
Total:   18676216 KiB
```

服务器2有两颗CPU，所以就有两个NUMA节点。编号0～5、12～17共12逻辑CPU属于节点0，编号6～11、18～23共12个逻辑CPU属于节点1。由于宿主机上4条8GB内存条安装位置正确，所以每个节点有16GB内存。

服务器1只有1个NUMA节点，所有的32GB内存、16个逻辑CPU都属于这个节点。这台服务器不存在由于NUMA跨节点访问而影响性能的问题。

服务器2有两个NUMA节点，每个节点有12个逻辑CPU、16GB。虚拟机的vCPU是线程，有可能会被分配给不同的NUMA节点，如果访问的内存数据在远程节点，就会产生跨节点访问，这对于运行内存密集的应用的虚拟机，对性能的影响会很大。下面我们查看一下默认NUMA策略。

## 4.5.5 查看默认的NUMA策略

在很多Linux发行版本中，由numad守护程序来协助构建最佳的NUMA平衡策略，所以通常不需要手工进行调整。

查看numad软件的信息，示例命令如下：

```
[root@server2 ~]# rpm -qi numad
Name        : numad                          Relocations: (not relocatable)
...
Summary     : NUMA user daemon
Description :
Numad, a daemon for NUMA (Non-Uniform Memory Architecture) systems, that monitors NUMA
characteristics and manages placement of processes and memory to minimize memory latency and
thus provide optimum performance.
```

查看默认的NUMA策略，示例命令如下：

```
[root@server2 ~]# numactl -H
available: 2 nodes (0-1)
node 0 cpus: 0 1 2 3 4 5 12 13 14 15 16 17
node 0 size: 16331 MB
node 0 free: 9072 MB
node 1 cpus: 6 7 8 9 10 11 18 19 20 21 22 23
node 1 size: 16384 MB
node 1 free: 9207 MB
node distances:
node   0   1
  0:  10  20
  1:  20  10
```

有一台正在运行的虚拟机 tomvm1,查看其配置文件中的 vCPU 信息,示例命令如下:

```
[root@server2 ~]# virsh dumpxml tomvm1 | grep vcpu
  <vcpu placement='static'>16</vcpu>
```

它拥有 16 个 vCPU。vCPU 还有 3 个可选的属性 placement、cpuset 和 current,用于配置 vCPU,例如:<vcpu placement='static' cpuset="1-4,^3,6" current="1">2</vcpu>。

查询虚拟机的进程 ID,示例命令如下:

```
[root@server2 ~]# ps aux | grep tomvm1
USER       PID %CPU %MEM    VSZ   RSS TTY      STAT START   TIME COMMAND
chentao  14062 16.6  0.0 3565392 31076 ?       Sl   17:10   0:17 /usr/libexec/qemu-kvm -name tomvm1 -S -M rhel6.6.0 -enable-kvm -m 2048 -realtime mlock=off -smp 16,sockets=16,cores=1,threads=1 ...
```

进程 ID 是 14062,通过它查询线程,示例命令如下:

```
[root@server2 ~]# ps -T -p 14062
  PID  SPID TTY          TIME CMD
14062 14062 ?        00:00:00 qemu-kvm
14062 14067 ?        00:00:16 qemu-kvm
14062 14068 ?        00:00:00 qemu-kvm
14062 14069 ?        00:00:00 qemu-kvm
14062 14070 ?        00:00:00 qemu-kvm
14062 14071 ?        00:00:00 qemu-kvm
14062 14072 ?        00:00:00 qemu-kvm
14062 14073 ?        00:00:00 qemu-kvm
14062 14074 ?        00:00:00 qemu-kvm
14062 14075 ?        00:00:00 qemu-kvm
```

```
14062 14076 ?          00:00:00 qemu-kvm
14062 14077 ?          00:00:00 qemu-kvm
14062 14078 ?          00:00:00 qemu-kvm
14062 14079 ?          00:00:00 qemu-kvm
14062 14080 ?          00:00:00 qemu-kvm
14062 14081 ?          00:00:00 qemu-kvm
14062 14082 ?          00:00:00 qemu-kvm
```

16 个 vCPU 就有 16 个线程，查询它们的放置情况，示例命令如下：

```
[root@server2 ~]# virsh vcpuinfo tomvm1
VCPU:           0
CPU:            0
State:          running
CPU time:       17.1s
CPU Affinity:   yyyyyyyyyyyyyyyyyyyyyyyy

VCPU:           1
CPU:            0
State:          running
CPU time:       0.1s
CPU Affinity:   yyyyyyyyyyyyyyyyyyyyyyyy

VCPU:           2
CPU:            2
State:          running
CPU time:       0.1s
CPU Affinity:   yyyyyyyyyyyyyyyyyyyyyyyy
...

VCPU:           15
CPU:            0
State:          running
CPU time:       0.1s
CPU Affinity:   yyyyyyyyyyyyyyyyyyyyyyyy
```

通过 virsh vcpuinfo 命令可以了解这 16 个 vCPU 被自动放置到哪个逻辑 CPU 上，例如：vCPU 0 放置在 CPU 0 上，vCPU 1 也放置在 CPU 0 上，vCPU 2 放置在 CPU 2 上……另外，CPU Affinity 后面的字符串是 24 个字母 y，表示这个 vCPU 进程可以在 24 个逻辑 CPU 中的任何一个上面运行。

通过 grep 和 sort 命令处理一下输出结果，会更清晰地显示 vCPU 与逻辑 CPU 的对应关系，示例命令如下：

```
[root@server2 ~]# virsh vcpuinfo tomvm1 | grep ^CPU: | sort
CPU:            1
CPU:            1
CPU:            10
CPU:            10
CPU:            10
CPU:            10
CPU:            10
CPU:            10
CPU:            10
CPU:            10
CPU:            14
CPU:            2
CPU:            2
CPU:            22
CPU:            6
CPU:            7
```

我们会发现：这 16 个 vCPU 被放置到两个 NUMA 节点上，但不是很平均。

查询 NUMA 的状态，示例命令如下：

```
[root@server2 ~]# numastat -c tomvm1
Per-node process memory usage (in MBs) for PID 14062 (qemu-kvm)
           Node 0   Node 1   Total
           ------   ------   -----
Huge       0        0        0
Heap       10       1        11
Stack      0        0        0
Private    10       9        19
-------    ------   ------   -----
Total      20       10       30
```

由于这是一个实验环境，虚拟机 tomvm1 使用的内存并不多。通过 numastat 命令可以看出，它使用 NUMA 节点 0 的内存是 20MB，使用 NUMA 节点 1 的内存是 10MB。

查询每个 NUMA 节点可用的内存数量，示例命令如下：

```
[root@server2 ~]# virsh freecell --all
    0:     9264036 KiB
    1:     9410436 KiB
--------------------
Total:     18674472 KiB
```

从输出可以看出：整体上还是比较平均的。

通过查看，我们知道默认的策略是在任何可用的逻辑 CPU 上运行虚拟机，所以有可能会出现跨 NUMA 节点进行访问的现象。如果虚拟机运行的是内存敏感型应用(需要大量的内存),则显式地设置 vCPU 的放置会更有优势。如果将虚拟机固定在指定的逻辑 CPU 上,则可以让其要访问的内存位于相同的 NUMA 节点上,从而避免跨节点进行访问。另外,也可以增加缓存的命中率。下面将讲解 vCPU 的固定。

### 4.5.6　vCPU 的固定

使用 Linux 中的 taskset 命令可以查询或设置进程(线程)并绑定 CPU(亲和性)。虚拟机是进程、vCPU 是线程,所以我们可以通过 taskset 命令将虚拟机 vCPU 固定到某个逻辑 CPU 上,示例命令如下:

```
[root@server2 ~]# whatis taskset
taskset                   (1)   - retrieve or set a process's CPU affinity
```

对于 KVM 虚拟化来讲,推荐使用自己的方法进行 vCPU 固定,因为这可以进行更细粒度的设置,而且操作更方便。

#### 1. 确认 CPU 和 NUMA 拓扑

第一步是先确认宿主机的内存和 CPU 拓扑,示例命令如下:

```
[root@server2 ~]# virsh nodeinfo
CPU model:           x86_64
CPU(s):              24
CPU frequency:       1200 MHz
CPU socket(s):       1
Core(s) per socket:  6
Thread(s) per core:  2
NUMA cell(s):        2
Memory size:         32827592 KiB
```

宿主机具有 24 个逻辑 CPU,位于两个插槽中,每个处理器具有 6 个核,每个核有两个线程。宿主机有两个 NUMA 节点。

使用 virsh capabilities 命令来获得与 CPU 配置相关的信息,示例命令如下:

```
[root@server2 ~]# virsh capabilities
<capabilities>
...
    <topology>
      <cells num='2'>
        <cell id='0'>
          <cpus num='12'>
            <cpu id='0' socket_id='0' core_id='0' siblings='0,12'/>
            <cpu id='1' socket_id='0' core_id='1' siblings='1,13'/>
```

```
          <cpu id='2' socket_id='0' core_id='2' siblings='2,14'/>
          <cpu id='3' socket_id='0' core_id='3' siblings='3,15'/>
          <cpu id='4' socket_id='0' core_id='4' siblings='4,16'/>
          <cpu id='5' socket_id='0' core_id='5' siblings='5,17'/>
          <cpu id='12' socket_id='0' core_id='0' siblings='0,12'/>
          <cpu id='13' socket_id='0' core_id='1' siblings='1,13'/>
          <cpu id='14' socket_id='0' core_id='2' siblings='2,14'/>
          <cpu id='15' socket_id='0' core_id='3' siblings='3,15'/>
          <cpu id='16' socket_id='0' core_id='4' siblings='4,16'/>
          <cpu id='17' socket_id='0' core_id='5' siblings='5,17'/>
        </cpus>
      </cell>
      <cell id='1'>
        <cpus num='12'>
          <cpu id='6' socket_id='1' core_id='0' siblings='6,18'/>
          <cpu id='7' socket_id='1' core_id='1' siblings='7,19'/>
          <cpu id='8' socket_id='1' core_id='2' siblings='8,20'/>
          <cpu id='9' socket_id='1' core_id='3' siblings='9,21'/>
          <cpu id='10' socket_id='1' core_id='4' siblings='10,22'/>
          <cpu id='11' socket_id='1' core_id='5' siblings='11,23'/>
          <cpu id='18' socket_id='1' core_id='0' siblings='6,18'/>
          <cpu id='19' socket_id='1' core_id='1' siblings='7,19'/>
          <cpu id='20' socket_id='1' core_id='2' siblings='8,20'/>
          <cpu id='21' socket_id='1' core_id='3' siblings='9,21'/>
          <cpu id='22' socket_id='1' core_id='4' siblings='10,22'/>
          <cpu id='23' socket_id='1' core_id='5' siblings='11,23'/>
        </cpus>
      </cell>
    </cells>
  </topology>
...
```

宿主机有两个 CPU 插槽，所以有两个 NUMA 节点。每个节点包含 12 个逻辑 CPU。对于具有 4 个 vCPU 的虚拟机，最好将虚拟机固定到属于 NUMA 节点 0 的 0～5、12～17，或者属于 NUMA 节点 1 的 6～11、18～23，这样可以避免访问非本地内存。

还可以通过 numactl 命令确认逻辑 CPU 与 NUMA 节点的对照关系，示例命令如下：

```
[root@server2 ~]# numactl -H
available: 2 nodes (0-1)
node 0 cpus: 0 1 2 3 4 5 12 13 14 15 16 17
node 0 size: 16331 MB
node 0 free: 6573 MB
```

```
node 1 cpus: 6 7 8 9 10 11 18 19 20 21 22 23
node 1 size: 16384 MB
node 1 free: 9180 MB
node distances:
node   0   1
  0:  10  20
  1:  20  10
```

将一台虚拟机运行在多个 NUMA 节点上不是最佳选择。例如：如果某个业务需要 16 个 vCPU，由于每个 NUMA 节点只有 12 个逻辑 CPU，所以运行两台 8 个 vCPU 虚拟机是较好的选择，而不是使用单台 16 个 vCPU 虚拟机。

**2. 确认 NUMA 节点的可用内存**

在固定虚拟机之前，需要检查节点是否有足够的可用内存，否则会引起跨节点访问。

除了使用 numactl -H 命令之外，还可以通过 virsh freecell 命令来获得可用内存，示例命令如下：

```
[root@server2 ~]# virsh freecell --all
    0:      6731208 KiB
    1:      9400128 KiB
--------------------
Total:      16131336 KiB
```

如果虚拟机需要分配 8GB 的内存，就应该在 NUMA 节点 1 运行。节点 0 仅具有 6.4GB 的可用空间，对于某些虚拟机来讲可能是不足的。

**3. 将虚拟机固定到 NUMA 节点或逻辑 CPU 集**

一台有 4 个 vCPU 的虚拟机 tomvm2，查看其 vCPU 拓扑，示例命令如下：

```
[root@server2 ~]# virsh vcpuinfo tomvm2
VCPU:           0
CPU:            0
State:          running
CPU time:       21.6s
CPU Affinity:   yyyyyyyyyyyyyyyyyyyyyyyy

VCPU:           1
CPU:            12
State:          running
CPU time:       0.9s
CPU Affinity:   yyyyyyyyyyyyyyyyyyyyyyyy

VCPU:           2
CPU:            0
```

```
State:           running
CPU time:        0.9s
CPU Affinity:    yyyyyyyyyyyyyyyyyyyyyyyy

VCPU:            3
CPU:             0
State:           running
CPU time:        0.7s
CPU Affinity:    yyyyyyyyyyyyyyyyyyyyyyyy
```

从输出信息可以看出：vCPU 被默认自动分配策略分配到的 ID 是 0、1、2、3 的逻辑 CPU。根据 NUMA 拓扑，它们都属于 NUMA 节点 0。

假设这不是最佳配置，我们需要将其固定到 NUMA 节点 1，即放置到 ID 是 6～11、18～23 CPU 集中。这需要编辑虚拟机的配置文件，修改 vCPU 的 cpuset 属性，将其锁定到一组 CPU，例如参数 6-9，示例命令如下：

```
[root@server2 ~]# virsh shutdown tomvm2

[root@server2 ~]# virsh edit tomvm2
```

将

```
<vcpu placement = 'static'> 4 </vcpu>
```

修改为

```
<vcpu cpuset = '6 - 9'> 4 </vcpu>
```

保存配置文件并重新启动虚拟机，通过 virsh vcpupin 命令查看 vCPU 的亲和性(Affinity)，示例命令如下：

```
[root@server2 ~]# virsh start tomvm2

[root@server2 ~]# virsh vcpupin tomvm2
VCPU: CPU Affinity
----------------------------------
   0: 6 - 9
   1: 6 - 9
   2: 6 - 9
   3: 6 - 9
```

再通过 virsh vcpuinfo 命令查看实际的分配，示例命令如下：

```
[root@server2 ~]# virsh vcpuinfo tomvm2
VCPU:          0
CPU:           9
State:         running
CPU time:      16.5s
CPU Affinity:  ------yyyy--------------

VCPU:          1
CPU:           9
State:         running
CPU Affinity:  ------yyyy--------------

VCPU:          2
CPU:           6
State:         running
CPU time:      0.0s
CPU Affinity:  ------yyyy--------------

VCPU:          3
CPU:           9
State:         running
CPU Affinity:  ------yyyy--------------
```

从输出中可以看到：CPU Affinity 后面的字符串，仅第 6～9 位的字符是字母 y。有 3 个 vCPU 运行在 ID 是 9 的逻辑 CPU 上，有 1 个 vCPU 运行在 ID 是 6 的逻辑 CPU 上。

能否再进行更细粒度的调整？可以，这需要通过 cputune 属性实现，示例命令如下：

```
[root@server2 ~]# virsh shutdown tomvm2

[root@server2 ~]# virsh edit tomvm2
  #将 vCPU 的部署修改为原状，并且添加<cputune>属性
  <vcpu placement = 'static'> 4 </vcpu>
  <cputune>
    <vcpupin vcpu = '0' cpuset = '6'/>
    <vcpupin vcpu = '1' cpuset = '7'/>
    <vcpupin vcpu = '2' cpuset = '8'/>
    <vcpupin vcpu = '3' cpuset = '9'/>
  </cputune>
```

将 vCPU 的 0～3 和逻辑 CPU 的 6～9 进行一一对应。重新启动虚拟机，检查调整是否成功，示例代码如下：

```
[root@server2 ~]# virsh start tomvm2

[root@server2 ~]# virsh vcpupin tomvm2
VCPU: CPU Affinity
----------------------------------
```

```
   0: 6
   1: 7
   2: 8
   3: 9

[root@server2 ~]# virsh vcpuinfo tomvm2
VCPU:          0
CPU:           6
State:         running
CPU time:      11.3s
CPU Affinity:  ------y-----------------

VCPU:          1
CPU:           7
State:         running
CPU Affinity:  -------y----------------

VCPU:          2
CPU:           8
State:         running
CPU Affinity:  --------y---------------

VCPU:          3
CPU:           9
State:         running
CPU Affinity:  ---------y--------------
```

从输出的结果可以看出调整成功了。

还可以使用 virsh vcpupin 命令调整正在运行的虚拟机上的 vCPU 的关联性，这比修改配置文件再重新引导虚拟机更简单，示例命令如下：

```
[root@server2 ~]# virsh vcpupin --help
  NAME
    vcpupin - control or query domain vcpu affinity

  SYNOPSIS
    vcpupin <domain> [--vcpu <number>] [<cpulist>] [--config] [--live] [--current]

  DESCRIPTION
    Pin domain VCPUs to host physical CPUs.

  OPTIONS
    [--domain] <string>  domain name, id or uuid
    --vcpu <number>  vcpu number
    [--cpulist] <string>  host cpu number(s) to set, or omit option to query
    --config         affect next boot
    --live           affect running domain
    --current        affect current domain
```

将 ID 号是 3 的 vCPU 运行在 ID 是 10 的逻辑 CPU 上,示例命令如下:

```
[root@server2 ~]# virsh vcpupin tomvm2
VCPU: CPU Affinity
----------------------------------
   0: 6
   1: 7
   2: 8
   3: 9

[root@server2 ~]# virsh vcpupin tomvm2 3 10

[root@server2 ~]# virsh vcpupin tomvm2
VCPU: CPU Affinity
----------------------------------
   0: 6
   1: 7
   2: 8
   3: 10
```

使用 virsh vcpuinfo 命令确认亲缘关系的更改,示例命令如下:

```
[root@server2 ~]# virsh vcpuinfo tomvm2
VCPU:           0
CPU:            6
State:          running
CPU time:       18.4s
CPU Affinity:   ------y-----------------

VCPU:           1
CPU:            7
State:          running
CPU time:       0.5s
CPU Affinity:   -------y----------------

VCPU:           2
CPU:            8
State:          running
CPU time:       0.6s
CPU Affinity:   --------y---------------

VCPU:           3
CPU:            10
State:          running
CPU time:       0.6s
CPU Affinity:   ----------y-------------
```

当使用 virt-install 创建虚拟机时,也可以使用--cpuset 选项配置 NUMA 策略。示例命令如下:

```
[root@server2 ~]# man virt - install
...
-- cpuset = CPUSET
Set which physical cpus the guest can use. "CPUSET" is a comma separated list of numbers, which
can also be specified in ranges or cpus to exclude.
Example:

0,2,3,5          : Use processors 0,2,3 and 5
1 - 5,^3,8       : Use processors 1,2,4,5 and 8

If the value 'auto' is passed, virt - install attempts to automatically determine an optimal cpu
pinning using NUMA data, if available.
...
```

CPUSET 既可以是逗号分隔的数字列表,也可以使用范围或排除表示法。

(1) 0,2,3,5:使用处理器 0、2、3 和 5。

(2) 1-5,^3,8:使用处理器 1、2、4、5 和 8

如果传递了值 auto,则 virt-install 会尝试使用 NUMA 数据(如果有)自动确定最佳的 CPU 固定。

**注意**:错误的 vCPU 固定会严重影响宿主机和虚拟机的性能。

## 4.6 内存优化技术

监视内存使用情况和诊断的工具主要包括:

(1) top 命令。

(2) vmstat 命令。

(3) numastat 命令

(4) /proc/目录下的文件。

在对 KVM 虚拟化环境的内存进行优化时,要考虑以下 3 方面:

(1) 内存分配。

(2) 内存调整。

(3) 内存支持。

### 4.6.1 内存分配

宿主机的内存分为两部分,一部分内存用于自身的应用程序,另外一部分提供给虚拟机使用。分配给虚拟机的内存会有多少呢?可以通过两个简单经验公式进行计算:

(1) 物理内存≤64GB：内存数量－2GB。

(2) 物理内存>64GB：内存数量－(2GB+5×(内存数量/64))。

假设宿主机有32GB内存，则一共可以配置32－2＝30GB的内存给虚拟机。

假设宿主机有256GB内存，则一共可以配置256－(2+0.5×(256/64))＝252GB的内存给虚拟机。

优化KVM内存性能的第一条准则是不要向虚拟机分配过多内存。

在Cockpit和virt-manager中可以设置虚拟机内存的两个参数，如图4-11所示。

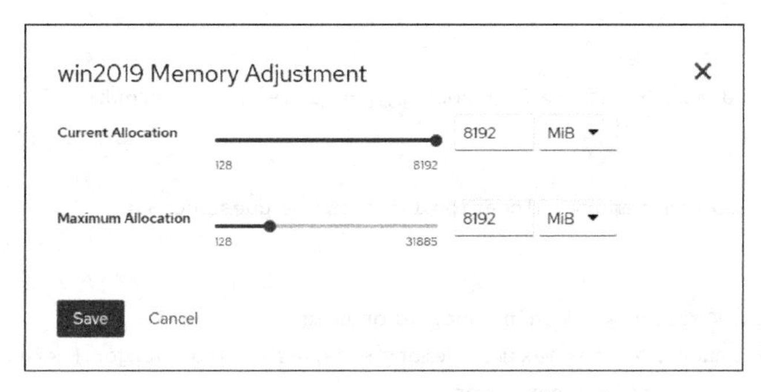


图4-11　虚拟机内存设置

(1) 当前分配：虚拟机的实际内存分配。该值可以小于最大分配，以允许迅速增加虚拟机的内存。

(2) 最大分配：虚拟机的运行时最大内存分配。这是虚拟机运行时可以分配给虚拟机的最大内存。

对应的XML文件的配置如下：

```
<memory unit='KiB'>8388608</memory>
<currentMemory unit='KiB'>8388608</currentMemory>
```

除了修改配置文件，还可以使用virsh的setmem和setmaxmem子命令调整这些参数，示例命令如下：

```
# virsh setmem --help
  NAME
    setmem - change memory allocation

  SYNOPSIS
    setmem <domain> <size> [--config] [--live] [--current]

  DESCRIPTION
    Change the current memory allocation in the guest domain.

  OPTIONS
```

```
    [--domain] <string>   domain name, id or uuid
    [--size] <number>   new memory size, as scaled integer (default KiB)
    --config          affect next boot
    --live            affect running domain
    --current         affect current domain

# virsh setmaxmem --help
  NAME
    setmaxmem - change maximum memory limit

  SYNOPSIS
    setmaxmem <domain> <size> [--config] [--live] [--current]

  DESCRIPTION
    Change the maximum memory allocation limit in the guest domain.

  OPTIONS
    [--domain] <string>   domain name, id or uuid
    [--size] <number>   new maximum memory size, as scaled integer (default KiB)
    --config          affect next boot
    --live            affect running domain
    --current         affect current domain
```

### 4.6.2 内存调整

既可以通过命令行也可以通过配置文件对虚拟机的内存进行调整。下面是命令行的帮助：

```
# virsh memtune --help
  NAME
    memtune - Get or set memory parameters

  SYNOPSIS
    memtune <domain> [--hard-limit <number>] [--soft-limit <number>] [--swap-hard-
limit <number>] [--min-guarantee <number>] [--config] [--live] [--current]

  DESCRIPTION
    Get or set the current memory parameters for a guest domain.
    To get the memory parameters use following command:

    virsh # memtune <domain>

  OPTIONS
    [--domain] <string>   domain name, id or uuid
    --hard-limit <number>   Max memory, as scaled integer (default KiB)
```

```
--soft-limit <number>   Memory during contention, as scaled integer (default KiB)
--swap-hard-limit <number>   Max memory plus swap, as scaled integer (default KiB)
--min-guarantee <number>   Min guaranteed memory, as scaled integer (default KiB)
--config        affect next boot
--live          affect running domain
--current       affect current domain
```

内存调整有以下 4 个选项。

(1) hard-limit：虚拟机可以使用的最大内存数。

(2) soft-limit：在内存争用期间要强制执行的内存限制。

(3) swap-hard-limit：虚拟机可以使用的最大内存加上交换空间。该值必须大于提供的硬限制值。

(4) min-guarantee：虚拟机的最小保证内存。

如果不提供选项，则显示当前的值，示例命令如下：

```
# virsh memtune win2019
hard_limit      : unlimited
soft_limit      : unlimited
swap_hard_limit : unlimited
```

从输出中可以看出：虚拟机默认为没有限制。

XML 配置文件中的内存调整选项在<memtune>下，代码如下：

```
...
  <memtune>
    <hard_limit unit='G'>1</hard_limit>
    <soft_limit unit='M'>128</soft_limit>
    <swap_hard_limit unit='G'>2</swap_hard_limit>
    <min_guarantee unit='Bytes'>67108864</min_guarantee>
</memtune>
...
```

如果存在 NUMA，则内存调整时需要注意每个节点可用内存的数量。

### 4.6.3 内存气球技术

内存气球(Ballooning)是一种内存回收技术，可以在虚拟机运行时动态地调整它所占用的宿主机的内存资源，而不需要关闭虚拟机。

KVM、VMware、Hyper-V 和 Xen 都有内存气球技术，其工作原理基本相似。KVM 是通过给虚拟机配置 virtio_balloon 设备实现的。默认情况下，通过 virt-install、Cockpit 和 virt-manager 创建的虚拟机都有 virtio_balloon 设备，代码如下：

```
...
<devices>
...
    <memballoon model='virtio'>
      <address type='pci' domain='0x0000' bus='0x05' slot='0x00' function='0x0'/>
    </memballoon>
...
<devices>
```

如果虚拟机采用的是 RHEL/CentOS 6 以后的发行版本，则会自动安装 virtio_balloon 设备的驱动程序，示例命令如下：

```
[root@guest ~]# cat /etc/redhat-release
CentOS Linux release 8.3.2011

[root@guest ~]# lsmod | grep virtio
virtio_balloon        20480  0
virtio_console        36864  1
virtio_blk            20480  3
virtio_net            53248  0
net_failover          24576  1 virtio_net
```

如果虚拟机采用的是 Windows 操作系统，则需要手工安装 virtio_balloon 设备的驱动程序。安装完成后，在设备管理器中会看到 VirtIO Balloon Driver，如图 4-12 所示。

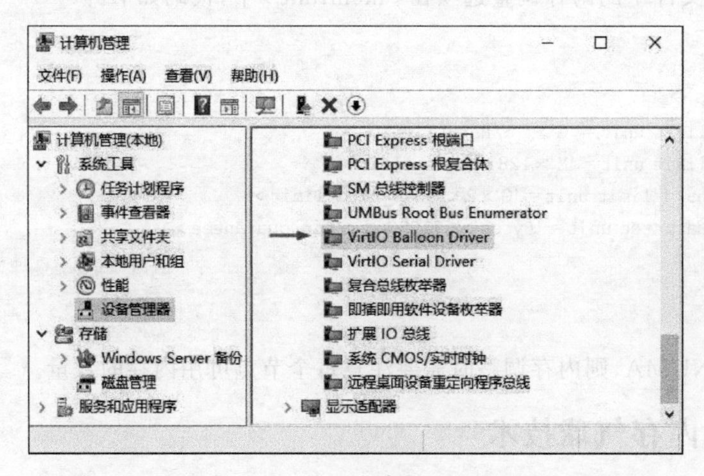


图 4-12　Windows Server 2019 中的 VirtIO Balloon Driver

内存气球技术使用气球很形象地描述了其工作原理，Hypervisor 可以对其进行充气和放气操作，如图 4-13 所示。

气球中的内存只能供宿主机使用，而不能被虚拟机访问或使用。当宿主机内存紧张时，

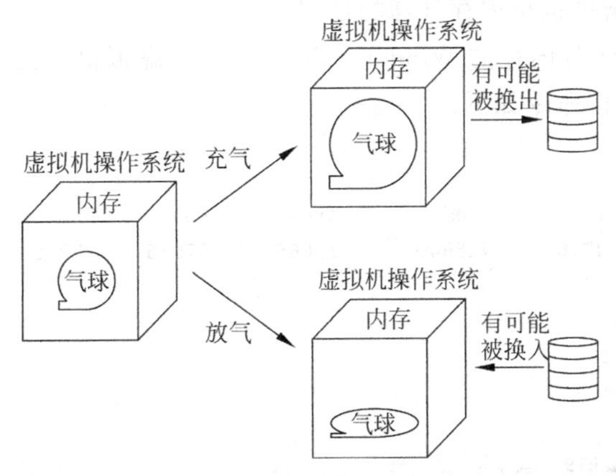


图 4-13 内存气球技术原理

可以通过充气操作增加气球设备所占用的虚拟机内存,从而将虚拟机的空闲内存回收,宿主机可以将回收到的内存再分配给其他虚拟机。具体的充气过程如下：

(1) Hypervisor 给虚拟机操作系统发送请求,让其归还一定数量的内存给宿主机。

(2) virtio_balloon 驱动接收请求后,会使内存气球膨胀。如果此时虚拟机剩余内存小于请求的数量,virtio_balloon 驱动有可能会将一部分使用中的内存换出(swap)到虚拟机的交换分区(文件),从而尽可能地满足请求,但是并不一定能完全满足。

(3) 虚拟机操作系统将气球中的内存归还给宿主机,宿主机可以将从气球中得来的内存分配到任何需要的地方。

当从气球中获得的内存不再被使用时,Hypervisor 可以通过放气操作将其返还给虚拟机。具体的放气过程如下：

(1) Hypervisor 给虚拟机的 virtio_balloon 驱动发送请求。

(2) virtio_balloon 驱动接收到请求后,会将气球中的内存释放出来,重新让虚拟机使用更多的内存。

内存气球技术在节约内存和灵活分配内存方面有明显的优势,但是也有如下一些缺点：

(1) 如果有大量内存从虚拟机系统中被回收,则内存气球技术可能会降低虚拟机操作系统运行的性能。一方面,内存的减少可能会将虚拟机中作为磁盘缓存的内存被放到气球中,从而使虚拟机中的磁盘 I/O 访问增加。另一方面,如果处理机制不够好,则可能让虚拟机中正在运行的进程由于内存不足而执行失败。

(2) 内存的动态增加或减少,可能会造成内存过度碎片化,从而降低内存使用时的性能。另外,内存的变化会影响虚拟机内核对内存使用的优化,例如：内核起初根据当前状态对内存的分配采取了某种策略,而突然由于内存气球的效果让可用内存减少了很多,这时原有的内存策略有可能就不是最佳策略了。

(3) 需要在虚拟机操作系统中安装、加载 virtio_balloon 驱动,Windows 操作系统需要手工安装。

下面通过实验理解虚拟机内存气球的使用。

启动一台配置为8GB内存的Windows Server 2019虚拟机，对比宿主机在启动前后剩余内存的变化情况，示例命令如下：

```
# free
              total       used        free      shared  buff/cache   available
Mem:          16184716    745840      13665724  57536   1773152      15062184
Swap:         4141052     0           4141052

# virsh start win2k19
Domain win2k19 started

# virsh dominfo win2k19
...
Max memory:      8388608 KiB
Used memory:      8388608 KiB
...

# free
              total       used        free      shared  buff/cache   available
Mem:          16184716    9301808     5104260   57588   1778648      6505740
Swap:         4141052     0           4141052

# echo "(13665724 - 5104260)/1024" | bc
8360
```

这台虚拟机大约消耗了宿主机8360KB物理内存。

通过virsh中的qemu-monitor-command子命令来查看内存气球的信息，示例命令如下：

```
# virsh qemu - monitor - command win2k19  -- hmp  -- cmd info balloon
balloon: actual = 8192
```

选项--hmp表示使用了human monitor protocol命令，--cmd命令指定为info，查看当前实际可以使用内存为8192，即目前气球中的内存为0。

在虚拟机中查看可用内存，当前为6.8GB，如图4-14所示。

开始进行内存充气操作，即使用balloon命令将虚拟机可以使用的内存设置为4096MB，示例命令如下：

```
# virsh qemu - monitor - command win2k19  -- hmp  -- cmd balloon 4096
```

内存充气需要一段时间，可以通过多次执行查看命令来观察可用内存的变化，示例命令如下：

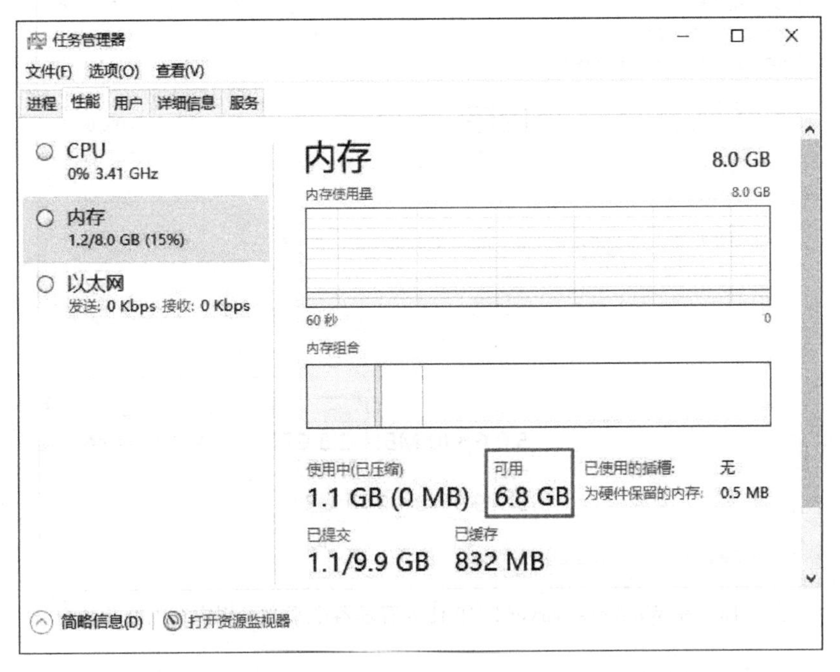


图 4-14　Windows Server 2019 任务管理器中的性能视图

```
#virsh qemu-monitor-command win2k19 --hmp --cmd info balloon
balloon: actual=7422

#virsh qemu-monitor-command win2k19 --hmp --cmd info balloon
balloon: actual=6410
...
#virsh qemu-monitor-command win2k19 --hmp --cmd info balloon
balloon: actual=4096
```

在虚拟机中查看现在的可用内存，会看到它是逐渐减小的，最终达到 2.8GB，如图 4-15 所示。

在宿主机上查看可用内存，示例命令如下：

```
#free
            total       used       free       shared   buff/cache   available
Mem:        16184716    5788748    8614676    57588    1781292      10018724
Swap:       4141052     0          4141052

#echo "(8614676-5104260)/1024" | bc
3428
```

当前可用内存增加了 3428MB。

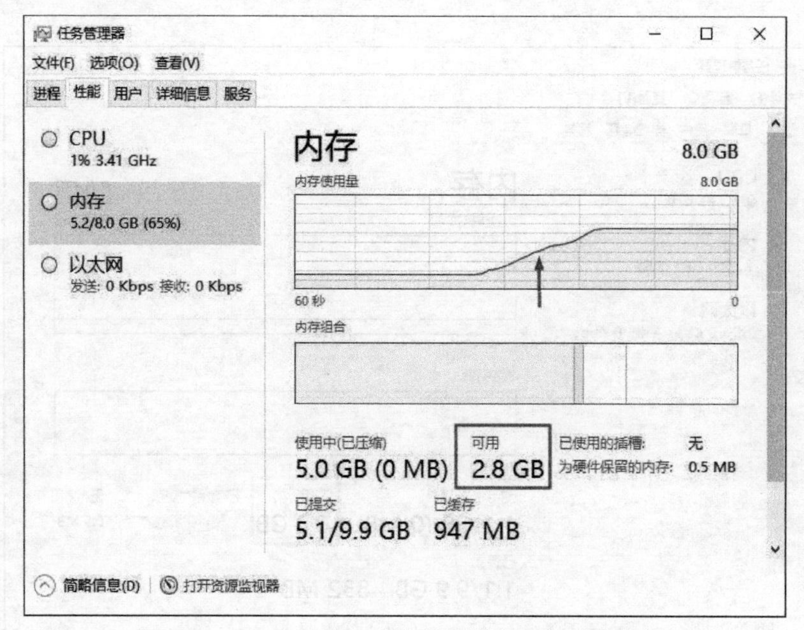


图 4-15 在 Windows Server 2019 任务管理器中观察可用内存的变化情况

恢复原有的值，即进行放气操作，示例命令如下：

```
# virsh qemu - monitor - command win2k19 -- hmp -- cmd balloon 4096
```

最后虚拟机的可用内存恢复为 6.8GB，如图 4-16 所示。

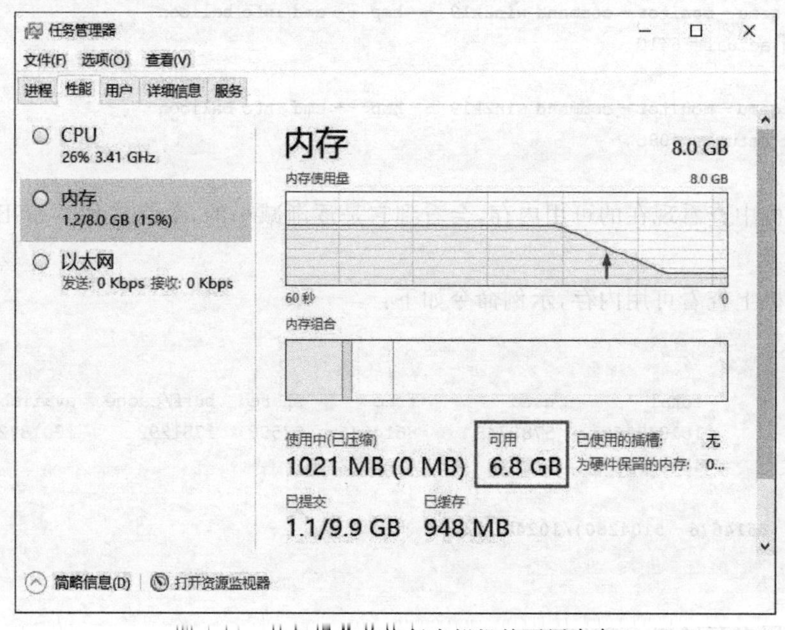


图 4-16 放气操作将恢复虚拟机的可用内存

### 4.6.4 内存虚拟化与大页的原理

为了正确理解内存管理及优化技术，我们还需要再掌握一些内存管理的概念。

KVM 内存虚拟化的目标是保障内存的合理分配、管理、隔离及高效可靠地使用，这需要将虚拟机中应用程序使用的内存与宿主机上的物理内存相关联。

早期的虚拟化技术需要维护一个三层关系的映射，如图 4-17 所示。这种 Guest Virtual Address(GVA)→Guest Physical Address(GPA) →Host Physical Address(HPA)效率很差。后来，通过软件实现了从 GVA 到 HPA 的直接映射，也就是减少了 GPA 这个中间层，这是通过为每个虚拟机维护一份影子页表(Shadow Page Table)实现的。由于是通过软件实现的，所以效率不是很高，同时影子页表也会有额外的内存消耗。

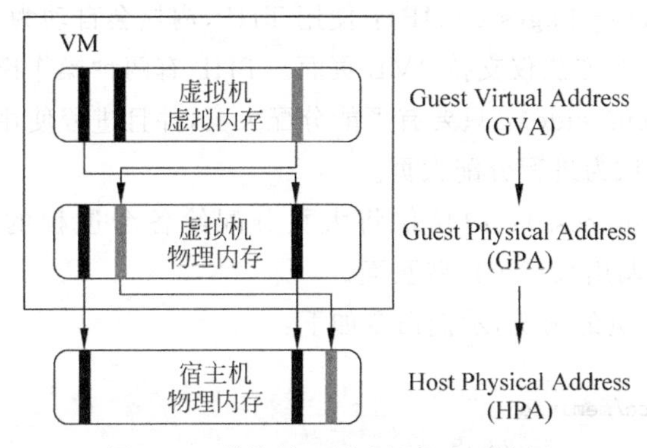


图 4-17 传统的三层虚拟机内存访问

在硬件辅助虚拟化中，Intel 使用扩展页表(Extend Page Table，EPT)技术，而 AMD 使用的是嵌入页表(Nested Page Table，NPT)技术。它们都作为 CPU 内存管理单元(Memory Management Unit，MMU)的扩展，通过硬件实现 GVA、GPA 到 HPA 的转换。从而降低了内存转换复杂度并且提升了转换效率。

另外，CPU 使用转换后备缓冲器(Translation Lookaside Buffer，TLB)缓存了 GVA 到 HPA 的映射，也提高了效率。在地址转换时，CPU 首先在 TLB 缓存中查找对应的 HPA，如果找到就返回结果；如果未找到，则在页表进行查找并将映射关系保存在 TLB 缓存中。

宿主机上通常有多台虚拟机，为了提高效率，Intel 使用了虚拟机标识符(Virtual-Processor Identifier，VPID)技术，在硬件上为 TLB 增加一个标志，每个 TLB 表项都与一个 VPID 关联，并且只对应一个 vCPU。当 vCPU 切换时，可根据 VPID 找到并保留已有的 TLB 表项，从而减少 TLB 刷新，提高了效率。

内存是以称为页(Page)的块进行管理的。不同的体系结构支持不同的页大小，x86 CPU 通常是 4KB，这在早期内存普遍比较小的环境中工作得很好，适用于大多数类型工作负载。

如果应用程序需要经常处理数百兆相对固定的数据集，使用 4KB 大小的内存页则有可能会有比较大的开销。使用大页(Huge Page)可以解决这个问题,它可以减少内存页数与页表项数,节省了页表所占用的CPU缓存空间,同时也可以减少内存地址转换次数、TLB失效和刷新的次数,从而提升内存使用效率与性能。

内存页的大小类似于纸质书籍的开本,32开的开本适合于文艺书籍,而编程类图书多数采用的是16开的开本。

在RHEL/CentOS 8中支持两种实现大页的方法。

(1) HugeTLB:也称为静态大页。在系统中预留HugeTLB有两种方法。

- 引导时预留:由于此时内存还未被大量使用,所以比较容易实现,而且产生碎片的可能性低。如果是NUMA计算机,则会自动地将大页在NUMA节点之间分配。
- 运行时预留:通过sysctl命令进行设置。可以为每个NUMA节点预留大页。

(2) Transparent HugePages (THP):使用THP,内核会自动为进程分配大页,因此无须手动预留静态大页。此方法仅支持2MB页面。THP有两种操作模式。

- 系统级别(System-wide):只要有可能分配大页,并且进程使用大的连续虚拟内存区域,内核就会尝试为进程分配大页。
- 每个进程(Per-process):内核仅将大页分配给各个进程的内存区域,进程使用madvise()系统调用来指定这些页面。

查看当前内存中大页的情况,示例命令如下:

```
# grep -i huge /proc/meminfo
AnonHugePages:          176128 KB
ShmemHugePages:         0 KB
HugePages_Total:        0
HugePages_Free:         0
HugePages_Rsvd:         0
HugePages_Surp:         0
Hugepagesize:           2048 KB
Hugetlb:                0 KB

# sysctl -a | grep -i huge
vm.hugetlb_shm_group = 0
vm.nr_hugepages = 0
vm.nr_hugepages_mempolicy = 0
vm.nr_overcommit_hugepages = 0

# cat /proc/sys/vm/nr_hugepages
0
```

/proc/meminfo中与大页有关属性含义如下。

(1) AnonHugePages:表示当前的THP大小。

(2) ShmemHugePages:Sh是Share的缩写,它是共享内存或tmpfs分配的大页数量。

(3) HugePages_Total：大页池的大小，使用/proc/sys/vm/nr_hugepages 进行配置。

(4) HugePages_Free：池中尚未使用的大页数。

(5) HugePages_Rsvd：Rsvd 是 reserved 的缩写，它是已做出承诺要从池中分配但尚未进行分配的大页数。保留的大页可确保应用程序在故障时能够从大页池中获得大页。

(6) HugePages_Surp：Surp 是 surplus（剩余）的缩写，它是池中超出/proc/sys/vm/nr_hugepages 所设定值的大页数量。多余的大页的最大数量由/proc/sys/vm/nr_overcommit_hugepages 控制。

(7) Hugepagesize：默认的大页的尺寸。

RHEL/CentOS 8 默认启动了对 THP 的支持，AnonHugePages 176128KB 表示当前的 THP 大小为 176128KB(172MB)。默认没有预留静态大页(HugeTLB)，所以 HugePages_Total 为 0。Hugepagesize：2048KB 表示大页的大小为 2MB。通过 sysctl 查看与大页有关的内核参数，vm. nr_hugepages=0 表示没有设置静态大页。

#### 1）配置 HugeTLB

在系统运行时，可以通过 sysctl 命令来配置 HugeTLB。

如果系统有多个 NUMA 节点，则需要先查看一下当前的拓扑情况。

下面示例中的宿主机有 4 个 NUMA 节点，我们将向节点 2 添加 20 个 2048KB 大小的大页，示例代码如下：

```
# numastat -cm | egrep 'Node|Huge'
                  Node 0 Node 1 Node 2 Node 3   Total add
AnonHugePages         0       2      0      8    10
HugePages_Total       0       0      0      0    0
HugePages_Free        0       0      0      0    0
HugePages_Surp        0       0      0      0    0
HugePages_Free        0       0      0
HugePages_Surp        0       0      0
```

从内存统计信息可以看出节点 2 没有预留的大页。

修改内核参数 nr_hugepages，设置 20 个要预留的大页，node2 是预留页的节点，示例命令如下：

```
# echo 20 > /sys/devices/system/node/node2/hugepages/hugepages-2048kB/nr_hugepages
```

验证是否成功：

```
# numastat -cm | egrep 'Node|Huge'
                  Node 0 Node 1 Node 2 Node 3   Total
AnonHugePages         0       2      0      8    10
HugePages_Total       0       0     40      0    40
HugePages_Free        0       0     40      0    40
HugePages_Surp        0       0      0      0    0
```

node2 在大页池中显示为 40MB(2048KB×20)。由于未使用,所以 HugePages_Free 也显示为 40MB。

恢复原有的值,示例命令如下:

```
# echo 0 > /sys/devices/system/node/node2/hugepages/hugepages - 2048kB/nr_hugepages
```

还可以在引导时预留 HugeTLB 页。在启动时,可以传递给内核的 HugeTLB 的参数如表 4-1 所示。

表 4-1 HugeTLB 的参数

| 参 数 | 描 述 | 默 认 值 |
|---|---|---|
| hugepages | 定义启动时内核中配置的持久性大页的数量。在 NUMA 系统中,定义了此参数的大页在节点之间平均分配 | 默认值为 0 |
| hugepagesz | sz 是 size 的意思。定义启动时在内核中配置的持久性大页的大小 | 有效值为 2MB 和 1GB,默认值为 2MB |
| default_hugepagesz | 定义启动时在内核中配置的持久性大页的默认大小 | 有效值为 2MB 和 1GB,默认值为 2MB |

HugeTLB 子系统支持的页面大小取决于体系结构。x86_64 体系结构支持 2MB 和 1GB 的大页。

下面示例将配置单 NUMA 节点的宿主机,在引导时预留 1GB 页面。

首先编辑/etc/default/grub. bak 文件,找到内核命令行选项 GRUB_CMDLINE_LINUX,添加参数 default_hugepagesz 和 hugepagesz,为 1GB 页面创建一个 HugeTLB 池,示例命令如下:

```
# cp /etc/default/grub ~/grub.bak

# vi /etc/default/grub
# 在 GRUB_CMDLINE_LINUX 行的尾部添加参数:
GRUB_CMDLINE_LINUX = "crashKernel = auto resume = /dev/mapper/cl - swap rd. lvm. lv = cl/root rd.
lvm. lv = cl/swap rhgb quiet intel_iommu = on default_hugepagesz = 1G hugepagesz = 1G"
```

使用编辑后的默认文件重新生成 grub2 配置。如果系统使用 BIOS 固件,则示例命令如下:

```
# grub2 - mkconfig - o /boot/grub2/grub.cfg
  Generating grub configuration file ...
    WARNING: lvmlockd process is not running.
    Reading without shared global lock.
    WARNING: lvmlockd process is not running.
```

```
    Reading without shared global lock.
    WARNING: lvmlockd process is not running.
    Reading without shared global lock.
    WARNING: lvmlockd process is not running.
    Reading without shared global lock.
  done
```

如果系统使用 UEFI 框架，则执行命令如下：

```
#grub2-mkconfig -o /boot/efi/EFI/redhat/grub.cfg
```

在系统中添加一个自动启动的新服务，由它来创建静态大页。方法是在/usr/lib/systemd/system/目录中创建一个名为 hugetlb-gigantic-pages. service 的新文件，并添加以下内容：

```
#vi /usr/lib/systemd/system/hugetlb-gigantic-pages.service
#添加以下内容
[Unit]
Description=HugeTLB Gigantic Pages Reservation
DefaultDependencies=no
Before=dev-hugepages.mount
ConditionPathExists=/sys/devices/system/node
ConditionKernelCommandLine=hugepagesz=1G

[Service]
Type=oneshot
RemainAfterExit=yes
ExecStart=/usr/lib/systemd/hugetlb-reserve-pages.sh

[Install]
WantedBy=sysinit.target
```

服务启动时会执行 Bash 脚本/usr/lib/systemd/hugetlb-reserve-pages. sh，其内容如下：

```
#vi /usr/lib/systemd/hugetlb-reserve-pages.sh
#添加以下内容

#!/bin/sh
nodes_path=/sys/devices/system/node/
if [ ! -d $nodes_path ]; then
    echo "ERROR: $nodes_path does not exist"
    exit 1
fi

reserve_pages()
```

```
{
    echo $1 > $nodes_path/$2/hugepages/hugepages-1048576kB/nr_hugepages
}

#reserve_pages number_of_pages node
reserve_pages 2 node0
```

需要将 number_of_pages 替换为要预留的 1GB 页面的数量，将 node 替换为节点名称。在本示例中，需要在 node0 上预留两个 1GB 页面。

修改脚本权限，增加可执行许可，示例命令如下：

```
#chmod +x /usr/lib/systemd/hugetlb-reserve-pages.sh
```

设置为自动启动，示例命令如下：

```
#systemctl status hugetlb-gigantic-pages.service
● hugetlb-gigantic-pages.service - HugeTLB Gigantic Pages Reservation
   Loaded: loaded (/usr/lib/systemd/system/hugetlb-gigantic-pages.service; disabled;
vendor preset: di>
   Active: inactive (dead)

#systemctl enable hugetlb-gigantic-pages.service
Created symlink /etc/systemd/system/sysinit.target.wants/hugetlb-gigantic-pages.service
→ /usr/lib/systemd/system/hugetlb-gigantic-pages.service.
```

重新引导，检查配置结果，示例命令如下：

```
#reboot

#grep -i huge /proc/meminfo
AnonHugePages:          190464 KB
ShmemHugePages:         0 KB
HugePages_Total:        2
HugePages_Free:         2
HugePages_Rsvd:         0
HugePages_Surp:         0
Hugepagesize:           1048576 KB
Hugetlb:                2097152 KB

#numastat -cm | egrep 'Node|Huge'
Token Node not in hash table.
                  Node 0 Total
AnonHugePages        174    174
ShmemHugePages       0      0
HugePages_Total      2048   2048
HugePages_Free       2048   2048
HugePages_Surp       0      0
```

当前 Hugepagesize 为 1048576KB(1GB)，HugePages_Total 数量为 2,则目前预留的大页为 2GB。

虽然在运行时或引导时都可以预留大页,但是推荐在引导时预留,因为引导时内存碎片少,预留 1GB 大页成功率高。

尽管预留的大页对有些应用程序是有益的,但预留的大页的池过大或未使用也不利于整体系统性能。

2）配置 THP

在 RHEL/CentOS 8 中默认启用了透明页 THP。查看相关的内核参数,示例命令如下：

```
# cat /sys/Kernel/mm/transparent_hugepage/enabled
[always] madvise never
```

方括号中是当前选中的值。这 3 个值的含义如下。

(1) always：在系统级别启用。

(2) madvise：仅适用于使用 madvise(MADV_HUGEPAGE)的区域。

(3) never：从不,等于禁用。

禁用 THP,示例命令如下：

```
# echo never > /sys/Kernel/mm/transparent_hugepage/enabled
```

启用 THP,示例命令如下：

```
# echo always > /sys/Kernel/mm/transparent_hugepage/enabled
```

为了防止应用程序分配超出必要的内存资源,可以不使用系统范围(system-wide)内的透明大页,此时可使用 madvise,这样只有当应用程序显式地请求时才启用透明大页,示例命令如下：

```
# echo madvise > /sys/Kernel/mm/transparent_hugepage/enabled
```

### 4.6.5 内存支持的子元素

XML 配置中有一个可选的元素 memoryBacking,它可能包含几个子元素,这些子元素设置了宿主机如何为虚拟机内存提供支持,示例代码如下：

```
<domain>
  …
  <memoryBacking>
    <hugepages>
      <page size="1" unit="G" nodeset="0-3,5"/>
```

```
      <page size="2" unit="M" nodeset="4"/>
    </hugepages>
    <nosharepages/>
    <locked/>
    <source type="file|anonymous|memfd"/>
    <access mode="shared|private"/>
    <allocation mode="immediate|ondemand"/>
    <discard/>
  </memoryBacking>
  ...
</domain>
```

**提示**：通过 virt-install、Cockpit、virt-manager 创建虚拟机都没有 memoryBacking 元素，所以需要手工设置这些元素。

### 1. hugepages 元素

推荐在宿主机和虚拟机启用透明大页 THP(RHEL/CentOS 8 默认已启用)。Red Hat 公司 2010 年发布的一个测试结果：在启用 EPT 的宿主机上的虚拟机中运行特定工作负荷，如果虚拟机启用了 THP，则它仅仅比裸机慢 5.67%；如果虚拟机不启用 THP，则慢 12.71%。差异明显，如图 4-18 所示。

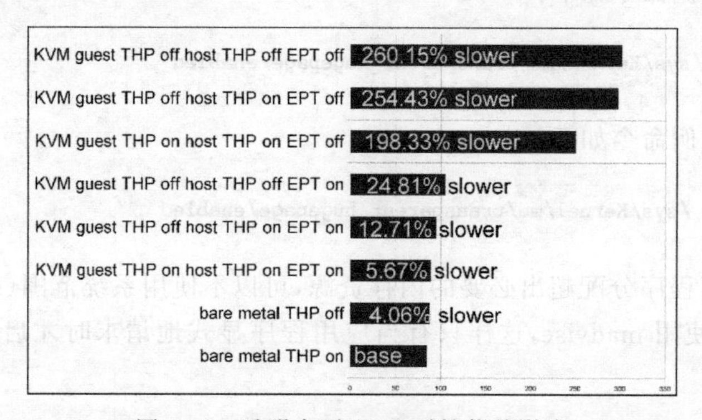


图 4-18　透明大页 THP 对性能的影响

对于运行内存要求特别高的虚拟机，如果希望进一步控制大页，就可以在虚拟机配置中通过 hugepages 元素来使用静态大页，示例命令如下：

```
# virsh edit centos8.3
...
  <memory unit='KiB'>4194304</memory>
  <currentMemory unit='KiB'>4194304</currentMemory>
  <memoryBacking>
    <hugepages/>
  </memoryBacking>
...
```

这表示宿主机将使用大页而不是使用默认页面大小为虚拟机分配内存。

启动虚拟机,会发现报错,因为此时在宿主机上还没有静态大页,示例命令如下:

```
# virsh start centos8.3
error: Failed to start domain centos8.3
error: internal error: process exited while connecting to monitor: 2021-02-07T09:28:28.
134446Z qemu-kvm: unable to map backing store for guest RAM: Cannot allocate memory
```

查看当前的大页的值,示例命令如下:

```
# grep -i huge /proc/meminfo
AnonHugePages:        184320 KB
ShmemHugePages:       0 KB
HugePages_Total:      0
HugePages_Free:       0
HugePages_Rsvd:       0
HugePages_Surp:       0
Hugepagesize:         2048 KB
Hugetlb:              0 KB
```

当前静态大页为2MB,虚拟机内存为4GB,所以至少需要2048个大页。修改内核参数以便创建3000个大页(适当多创建几个),示例命令如下:

```
# echo "4194304/2048" | bc
2048

# echo 3000 > /proc/sys/vm/nr_hugepages

# grep -i huge /proc/meminfo
AnonHugePages:        184320 KB
ShmemHugePages:       0 KB
HugePages_Total:      3000
HugePages_Free:       3000
HugePages_Rsvd:       0
HugePages_Surp:       0
Hugepagesize:         2048 KB
Hugetlb:              6144000 KB
```

启用虚拟机。虚拟机内存会消耗宿主机2048个大页,还剩余3000－2048＝952个大页,示例命令如下:

```
# virsh start centos8.3
Domain centos8.3 started

# grep -i huge /proc/meminfo
AnonHugePages:        204800 KB
```

```
ShmemHugePages:        0 KB
HugePages_Total:       3000
HugePages_Free:        952
HugePages_Rsvd:        0
HugePages_Surp:        0
Hugepagesize:          2048 KB
Hugetlb:               6144000 KB
```

要想这个配置永久生效，则需要修改/etc/sysctl.conf 文件，示例命令如下：

```
# echo 'vm.nr_hugepages = 2048' >> /etc/sysctl.conf
# sysctl -p
```

将预期的大页数设置为 2048。

### 2. nosharepages 元素

KSM(Kernel Samepage Merging，内核相同页面合并)是一种节省内存的重复数据删除功能。一台宿主机上通常会有多个虚拟机，如果虚拟机中运行相同的操作系统、应用程序，就一定会有相同的内存页。KSM 扫描这些相同的内存页，将其合并成 COW(Copy On Write，写时复制)的共享页，当某个虚拟机进程尝试更改其内容时，该进程就会获得包含更改的新副本。通过使用 KSM，可以减少宿主机物理内存的消耗，实现虚拟机的超量分配，从而达到更高的虚拟机密度，特别是桌面虚拟化应用。

KSM 并不是虚拟化专用技术，也可以为其他业务节省内存。RHEL/CentOS 8 是通过 qemu-kvm-common 软件包提供的。

查看版本信息，示例命令如下：

```
# cat /etc/redhat-release
CentOS Linux release 8.3.2011

# uname -r
4.18.0-240.1.1.el8_3.x86_64

# rpm -ql qemu-kvm-common | grep ksm
/etc/ksmtuned.conf
/etc/sysconfig/ksm
/usr/lib/systemd/system/ksm.service
/usr/lib/systemd/system/ksmtuned.service
/usr/libexec/ksmctl
/usr/sbin/ksmtuned
```

查看 RHEL/CentOS 8 对 KSM 的支持，示例命令如下：

```
# grep -i ksm /boot/config-4.18.0-240.1.1.el8_3.x86_64
CONFIG_KSM=y
```

RHEL/CentOS 8 内核默认启动对 KSM 的支持，由 ksm 和 ksmtuned 两个守护程序来完成相同内存页的合并，这个过程对于应用程序是透明的。查看守护程序的状态，示例命令如下：

```
# systemctl status ksm
● ksm.service - Kernel Samepage Merging
   Loaded: loaded (/usr/lib/systemd/system/ksm.service; enabled; vendor preset: enabled)
   Active: active (exited) since Mon 2021-02-08 11:05:38 CST; 25min ago
  Process: 1144 ExecStart=/usr/libexec/ksmctl start (code=exited, status=0/SUCCESS)
Main PID: 1144 (code=exited, status=0/SUCCESS)
    Tasks: 0 (limit: 203267)
   Memory: 0B
   CGroup: /system.slice/ksm.service

Feb 08 11:05:38 tomkvm1 systemd[1]: Starting Kernel Samepage Merging...
Feb 08 11:05:38 tomkvm1 systemd[1]: Started Kernel Samepage Merging.

# systemctl status ksmtuned
● ksmtuned.service - Kernel Samepage Merging (KSM) Tuning Daemon
   Loaded: loaded (/usr/lib/systemd/system/ksmtuned.service; enabled; vendor preset: enabled)
   Active: inactive (dead)

Feb 08 11:38:15 kvm1 systemd[1]: Starting Kernel Samepage Merging (KSM) Tuning Daemon...
Feb 08 11:38:15 kvm1 systemd[1]: Started Kernel Samepage Merging (KSM) Tuning Daemon.
Feb 08 11:48:38 kvm1 systemd[1]: Stopping Kernel Samepage Merging (KSM) Tuning Daemon...
Feb 08 11:48:38 kvm1 systemd[1]: ksmtuned.service: Succeeded.
Feb 08 11:48:38 kvm1 systemd[1]: Stopped Kernel Samepage Merging (KSM) Tuning Daemon
```

KSM 会消耗一些 CPU 资源来识别重复的页并执行合并任务。

虚拟机之间有部分共享内存，这可能不符合某些安全规定。如果通过测试发现 KSM 不能改善工作负载的性能，则可以禁用 KSM。

如果想全面禁用 KSM，则可以通过禁用 ksmtuned 和 ksm 守护程序实现，示例命令如下：

```
# systemctl stop ksm

# systemctl stop ksmtuned

# systemctl disable ksm

# systemctl disable ksmtuned
```

如果仅仅想禁用某台虚拟机的内存页共享，则可以在虚拟机配置文件中添加 nosharepages 元素来防止宿主机合并此虚拟机的内存，示例代码如下：

```
<memoryBacking>
    <nosharepages/>
</memoryBacking>
```

3. locked 元素

Linux 主机的内存管理模块有可能会将进程中内存页从物理内存换到磁盘(swap 文件或分区)中,这会影响性能。在 KVM 环境中,虚拟机内存其实位于宿主机中 qemu-kvm 进程的地址空间中。为了避免虚拟机的内存页被宿主机的内存页换出,可以使用 locked 锁定选项,示例代码如下:

```
<memoryBacking>
    <locked/>
</memoryBacking>
```

启用此选项后,宿主机系统内存中的虚拟内存页将被锁定。

## 4.7 网络优化技术

### 4.7.1 常用优化技术

不要将虚拟机的所有网络流量都放置到一条网络路径上,要通过合理隔离网络流量来避免网络拥塞。作为网络优化的第一步,建议按业务划分网络流量,将备份、迁移等流量放置到专用的网络中。

如果宿主机有两块或两块以上的网卡连接到同一个子网中,则要避免出现 ARP Flux 现象,即从多个网卡响应同一个 ARP 请求。可以通过使用 arp_filter 实现,示例命令如下:

```
#sysctl -a | grep -i all.arp
net.ipv4.conf.all.arp_accept = 0
net.ipv4.conf.all.arp_announce = 0
net.ipv4.conf.all.arp_filter = 0
net.ipv4.conf.all.arp_ignore = 0
net.ipv4.conf.all.arp_notify = 0

#echo 1 > /proc/sys/net/ipv4/conf/all/arp_filter

#sysctl -a | grep -i all.arp
net.ipv4.conf.all.arp_accept = 0
net.ipv4.conf.all.arp_announce = 0
net.ipv4.conf.all.arp_filter = 1
net.ipv4.conf.all.arp_ignore = 0
net.ipv4.conf.all.arp_notify = 0
```

可以编辑/etc/sysctl.conf文件，让重启后也生效，示例命令如下：

```
# vi /etc/sysctl.conf
# 添加
net.ipv4.conf.all.arp_filter = 1

# sysctl -p
net.ipv4.conf.all.arp_filter = 1
```

如果发现宿主机网络接口的带宽是瓶颈，则可以将多块网卡绑定为一个逻辑网卡，实现带宽扩容和负载均衡。Linux 2.4.12及以后的内核均支持绑定，这在生产环境中是一种常用的技术。

还可以调整MTU（最大传输单元）的值来减少碎片。如果改变了MTU，则路径中所有的设备都应支持新的MTU值。

### 4.7.2　VirtIO和vhost_net

VirtIO比仿真设备有更好的性能，所以在虚拟机中应尽可能使用VirtIO驱动程序。在使用VirtIO的网卡驱动程序VirtIO_net时，它在QEMU中有一个后端驱动程序。每次宿主机接收数据包的时候，数据包从Linux bridge经过TAP设备发送到用户空间的QEMU，这是一层数据的复制并且伴有从内核空间到用户空间的切换，而在用户空间QEMU中VirtIO后端驱动把数据写入虚拟机内存后还需要退到KVM中，从KVM进入虚拟机，又增加了一次模式的切换。在IO切换比较频繁的情况下，会导致模式切换次数过多从而使性能降低，如图4-19(a)所示。

为了获得更好的性能，又出现了vhost技术。它把后端驱动从QEMU中迁移到内核中，作为一个独立的内核模块，这样在数据包到来的时候，该模块直接监听TAP设备，在内核中直接把数据写入虚拟机内存中，然后通知虚拟机。这样就和QEMU解耦了，减少了模式切换次数和数据复制次数，降低延迟和CPU资源的消耗，从而提高了性能，如图4-19(b)所示。

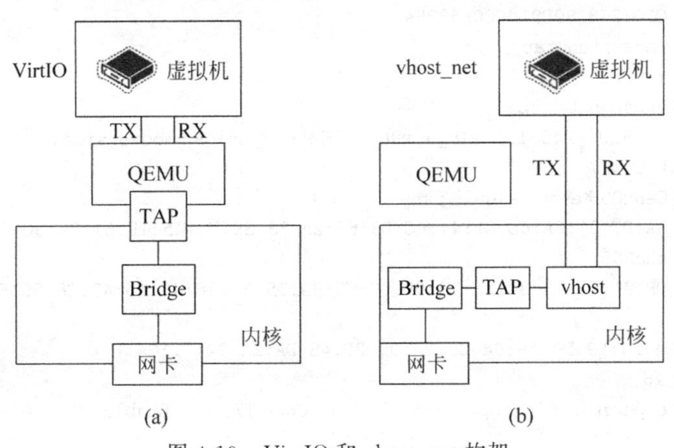


图4-19　VirtIO和vhost_net构架

从对比图 4-19 可以看出：由于前端驱动程序没有变化，所以虚拟机中不需要额外的配置，仅需要在宿主机上支持 vhost。示例命令如下：

```
# virsh list
Id   Name       State
--------------------------

# lsmod | grep vhost
无输出

# virsh start centos8
Domain centos8 started

# lsmod | grep vhost
vhost_net              28672  1
vhost                  49152  1 vhost_net
vhost_iotlb            16384  1 vhost
tap                    28672  1 vhost_net
tun                    53248  4 vhost_net
```

宿主机上的 vhost_net 是动态加载的内核模块。当有虚拟机运行时，会自动加载。查看这个内核模块，示例命令如下：

```
# modinfo vhost_net
filename:      /lib/modules/4.18.0-240.1.1.el8_3.x86_64/Kernel/drivers/vhost/vhost_net.ko.xz
alias:         devname:vhost-net
alias:         char-major-10-238
description:   Host Kernel accelerator for virtio net
author:        Michael S. Tsirkin
license:       GPL v2
version:       0.0.1
rhelversion:   8.3
srcversion:    C1D2278189D910C80F532FA
depends:       vhost,tun,tap
intree:        Y
name:          vhost_net
vermagic:      4.18.0-240.1.1.el8_3.x86_64 SMP mod_unload modversions
sig_id:        PKCS#7
signer:        CentOS Kernel signing key
sig_key:       3A:DF:91:6A:6D:C1:47:4C:D3:65:2A:38:39:E3:45:1F:61:99:8C:DB
sig_hashalgo:  sha256
signature:     DF:91:C5:13:87:13:73:9D:02:C5:9D:55:57:38:48:45:42:C8:56:F3:
...
73:6B:E8:54:24:F4:D9:D9:79:04:C8:74:03:2B:48:DA:B2:94:C1:CA:
    1A:86:1A:16
parm:          experimental_zcopytx:Enable Zero Copy TX; 1 - Enable; 0 - Disable (int)
```

vhost-net 是宿主机系统中的字符设备，示例命令如下：

```
#file /dev/vhost-net
/dev/vhost-net: character special (10/238)
```

宿主机通过-netdev tap，vhost=on 选项启动虚拟机。它将打开/dev/vhost-net 并使用多个 ioctl()调用初始化 vhost-net 实例，示例命令如下：

```
#ps aux | grep qemu-kvm | grep vhost
qemu        2641  6.2  5.0 6040896 818252 ?      Sl   15:39   0:44 /usr/libexec/qemu-kvm
-name guest=centos8,debug-threads=on
...
-netdev tap,fd=39,id=hostnet0,vhost=on,vhostfd=40
-device virtio-net-pci,netdev=hostnet0,id=net0,mac=52:54:00:cd:cd:9e, bus=pci.1,
addr=0x0
...
```

## 4.7.3　桥接零复制传输

vhost_net 模块有一个名为 experimental_zcopytx 的内核参数，默认为 0。如果将其设置为 1，则启用桥接零复制传输(Bridge Zero Copy Transmit)模式，它对大尺寸的数据包较为有效。通常在虚拟机网络和外部网络间的大数据包传输中，宿主机 CPU 的负荷可减少 15%左右，而且不会影响吞吐量。

查看内核参数 experimental_zcopytx 当前的值，示例命令如下：

```
#cat /sys/module/vhost_net/parameters/experimental_zcopytx
0
```

在/etc/modprobe.d 的目录中创建新文件 vhost-net.conf，配置 experimental_zcopytx 参数，示例命令如下：

```
#vi /etc/modprobe.d/vhost-net.conf
添加以下内容
options vhost_net   experimental_zcopytx=1
```

关闭虚拟机并移除 vhost-net 模块，重新启动虚拟机，检查新参数是否生效，示例命令如下：

```
#virsh shutdown centos8

#modprobe -r vhost-net

#virsh start centos8
```

```
#cat /sys/module/vhost_net/parameters/experimental_zcopytx
1
```

提示：对大数据包的环境，修改此参数配置可以产生明显的效果，但不会对虚拟机之间、虚拟机到宿主机或小数据包网络产生影响。

### 4.7.4 多队列 virtio-net

使用 virtio-net 的瓶颈之一是其单一的 RX 和 TX 队列，因此虚拟机无法并行发送和接收数据包，即使有多个 vCPU，网络吞吐量也受到此限制的影响。

多队列(Multi-Queue) virtio-net 提供了随着 vCPU 数量的增加而提升网络性能的方法，即允许一次通过一组以上的 virtqueue 传输数据包。多队列 virtio-net 在以下场景可以提供最佳性能：

(1) 流量相对较大。

(2) 虚拟机同时有很多处于活动状态的网络连接，包括虚拟机之间、虚拟机与宿主机或与外部系统之间。

(3) 队列数量与 vCPU 相同。

提示：多队列 virtio-net 在入站流量中运行良好，但在少数情况下可能会影响出站流量的性能，具体来讲是通过 TCP 流(stream)发送 1500 字节以下的数据包时。

可以编辑虚拟机 XML 配置文件实现多队列 virtio-net，示例命令如下：

```
#virsh edit centos8
…
  <vcpu placement = 'static'> 4 </vcpu>
…
    <interface type = 'network'>
      <mac address = '52:54:00:cd:cd:9e'/>
      <source network = 'default'/>
      <model type = 'virtio'/>
      <driver name = 'vhost' queues = '4'/>
      <address type = 'pci' domain = '0x0000' bus = '0x01' slot = '0x00' function = '0x0'/>
    </interface>
```

因为内核最多支持 256 个队列用于多队列 TAP 设备，所以 queues 的取值范围是 1～256。还需要通过 ethtool 命令启用网卡对多队列的支持，示例命令如下：

```
#ethtool -L ens32 combined 4
```

为网卡 ens32 启用 4 个队列。

提示：并非所有的网卡都支持多队列。如果不支持，则会收到提示：Channel parameters for 网卡名称：Cannot get device channel parameters：Operation not supported.

### 4.7.5 直接设备分配和 SR-IOV

其他网络调整选项还包括直接设备分配(Device Assignment)和单 Root IO 虚拟化(Single Root I/O Virtualization,SR-IOV)。

直接设备分配可以将宿主机上物理 PCI 网卡直接分配给虚拟机,从而提高性能,如图 4-20(a)所示,但是这也带来紧耦合的缺点,而且物理网卡仅能供一台虚拟机独占使用。

如果宿主机上的物理网卡支持单 Root IO 虚拟化技术,则可以将其划分为多个虚拟功能(Virtual Functions,VF)模块。每个 VF 就是一个独立的虚拟网卡,可以直接分配给虚拟机使用,如图 4-20(b)所示。单 Root IO 虚拟化技术既可以提高性能,又可以避免虚拟机对物理设备的独占。

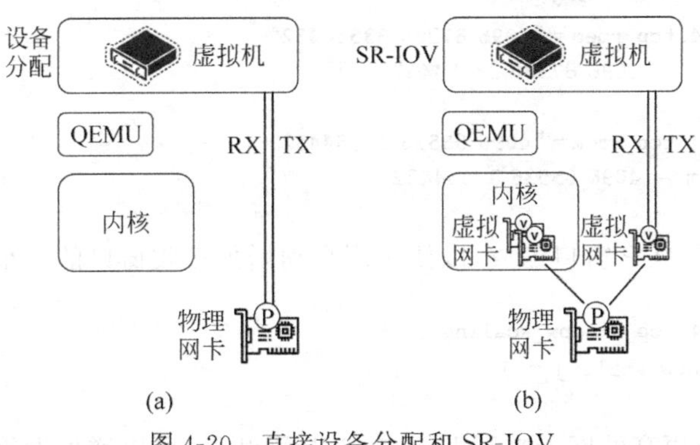


图 4-20 直接设备分配和 SR-IOV

### 4.7.6 调整内核参数以提高网络性能

目前大多数 Linux 发行版本针对各种工作负载进行了充分优化,而且 RHEL/CentOS 还可以通过 Tuned 使用一组配置文件简化操作。本节将介绍一些在宿主机上调整 Linux 内核的最佳实践,它们通常可以提高网络性能。

还是建议在更改之前先进行测量,然后进行小幅度更改,通过再次测量来评估这些更改对性能的影响。

(1) 增加最大 TCP 发送和接收套接字缓冲区的大小,为 TCP 协议栈分配更多的内存,示例命令如下:

```
# sysctl net.core.rmem_max
net.core.rmem_max = 212992

# sysctl net.core.wmem_max
net.core.wmem_max = 212992

# sysctl net.core.rmem_max = 33554432
net.core.rmem_max = 33554432
```

```
# sysctl net.core.wmem_max = 33554432
net.core.wmem_max = 33554432
```

(2) 增加 TCP 缓冲区限制，包括最小、默认和最大字节数。对于千兆网卡，将最大字节数设置为 16MB；对于万兆网卡，将最大字节数设置为 32MB 或 54MB。TCP 根据默认值动态调整发送缓冲区的大小，示例命令如下：

```
# sysctl net.ipv4.tcp_rmem
net.ipv4.tcp_rmem = 4096 87380 6291456

# sysctl net.ipv4.tcp_wmem
net.ipv4.tcp_wmem = 4096 16384 4194304

# sysctl net.ipv4.tcp_rmem = "4096 87380 33554432"
net.ipv4.tcp_rmem = 4096 87380 33554432

# sysctl net.ipv4.tcp_wmem = "4096 65536 33554432"
net.ipv4.tcp_wmem = 4096 65536 33554432
```

(3) 确保启用了 TCP 窗口缩放，以便实现自动增加接收窗口的大小，示例命令如下：

```
# sysctl net.ipv4.tcp_window_scaling
net.ipv4.tcp_window_scaling = 1
```

(4) 使用千兆或更高的网卡来帮助提高 TCP 吞吐量，可以增加网络接口的传输队列的长度，示例命令如下：

```
# ifconfig eth0 txqueuelen 5000
```

(5) 减少 tcp_fin_timeout 的值，指定在强制关闭套接字之前最终等待 FIN 数据包的时间，示例命令如下：

```
# sysctl net.ipv4.tcp_fin_timeout
net.ipv4.tcp_fin_timeout = 60

# sysctl net.ipv4.tcp_fin_timeout = 30
net.ipv4.tcp_fin_timeout = 30
```

(6) 减少 tcp_keepalive_intvl 的值，示例命令如下：

```
# sysctl net.ipv4.tcp_keepalive_intvl
net.ipv4.tcp_keepalive_intvl = 75

# sysctl net.ipv4.tcp_keepalive_intvl = 30
net.ipv4.tcp_keepalive_intvl = 30
```

(7) 启用 TIME_WAIT 套接字的快速回收，默认为禁用(值为 0)，示例命令如下：

```
# sysctl net.ipv4.tcp_tw_recycle
net.ipv4.tcp_tw_recycle = 0

# sysctl net.ipv4.tcp_tw_recycle = 1
net.ipv4.tcp_tw_recycle = 1
```

(8) 为新连接启用 TIME_WAIT 状态的套接字重用。默认为禁用(值为 0)。在具有大量虚拟机的宿主机上，这可能会对建立新连接的速度产生重大影响，示例命令如下：

```
# sysctl net.ipv4.tcp_tw_reuse
net.ipv4.tcp_tw_reuse = 0

# sysctl net.ipv4.tcp_tw_reuse = 1
net.ipv4.tcp_tw_reuse = 1
```

(9) 从内核版本 2.6.13 开始，Linux 支持可插拔的拥塞控制算法(Pluggable Congestion Control Algorithms)。执行以下命令获取内核中可用的拥塞控制算法列表：

```
# sysctl net.ipv4.tcp_available_congestion_control
net.ipv4.tcp_available_congestion_control = reno cubic
```

cubic 是许多 Linux 发行版的默认值。

```
# sysctl net.ipv4.tcp_congestion_control
net.ipv4.tcp_congestion_control = cubic
```

(10) 在有大量 SYN 请求的系统上，使用 tcp_max_syn_backlog 和 tcp_synack_retries 选项进行调整，可能会有所帮助，示例命令如下：

```
# sysctl net.ipv4.tcp_max_syn_backlog
net.ipv4.tcp_max_syn_backlog = 512

# sysctl net.ipv4.tcp_max_syn_backlog = 16384
net.ipv4.tcp_max_syn_backlog = 16384

# sysctl net.ipv4.tcp_synack_retries
net.ipv4.tcp_synack_retries = 5

# sysctl net.ipv4.tcp_synack_retries = 1
net.ipv4.tcp_synack_retries = 1
```

(11) UNIX/Linux 中一切皆文件，每个网络连接都使用文件描述符/套接字，所以具有足够数量的可用文件描述符非常重要。查看当前的最大文件描述符数量和可用文件描述符的示例命令如下：

```
# sysctl fs.file-nr
fs.file-nr = 2976    0    1609724
```

要增加最大文件描述符，示例命令如下：

```
# sysctl fs.file-max=10000000
fs.file-max = 10000000

# sysctl fs.file-nr
fs.file-nr = 3040    0    10000000
```

## 4.8 存储优化技术

虚拟机的存储可以是块设备或者映像文件。单纯从性能上来考虑，使用块设备不受宿主机文件系统的影响，所以通常比映像文件性能更好一些，但是映像文件易于管理、特性丰富。使用稀疏映像文件可实现存储的超量分配，但会降低虚拟磁盘的性能。

VirtIO 驱动程序可以提供更好的性能，因此建议使用 VirtIO 磁盘总线，而不是 IDE 总线。

常用的映像文件格式为 RAW 和 QCOW2。使用 RAW 格式会获得最佳性能，但是功能及特性不如 QCOW2 格式丰富。

当前版本的 Cockpit 不能设置磁盘性能选项，而 virt-manager 有 4 个选项，如图 4-21 所示。

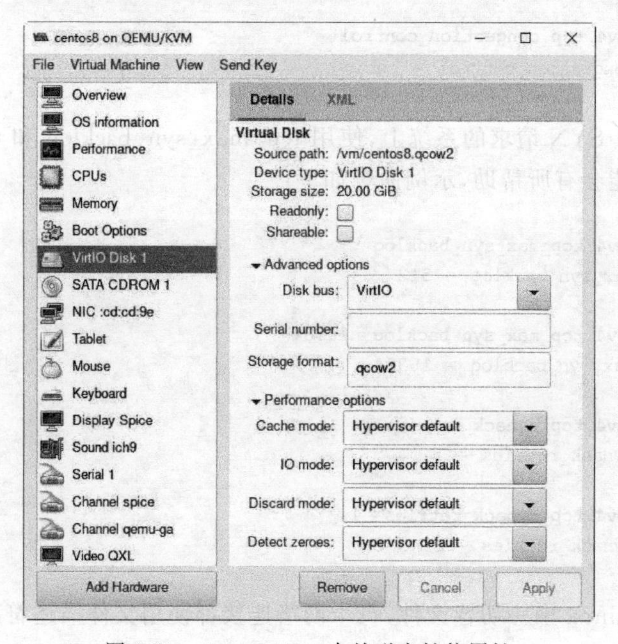


图 4-21 virt-manager 中的磁盘性能属性

### 4.8.1 缓存模式

缓存模式(Cache Mode)对性能影响很大，在 virt-manager 中可选项共有 6 种，如图 4-22 所示。

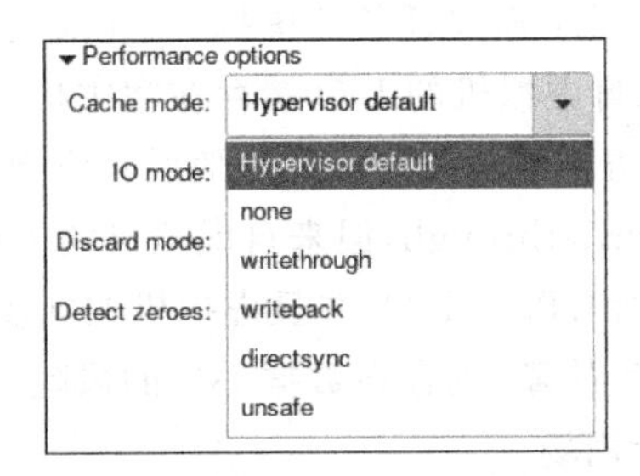


图 4-22　磁盘性能中的缓存模式

虚拟机配置文件中的缓存选项如下：

```
...
    <disk type='file' device='disk'>
      <driver name='qemu' type='qcow2' cache='writeback'/>
      <source file='/vm/vm1.qcow2'/>
      <target dev='vda' bus='virtio'/>
    </disk>
...
```

不同的缓存模式的工作原理如图 4-23 所示。

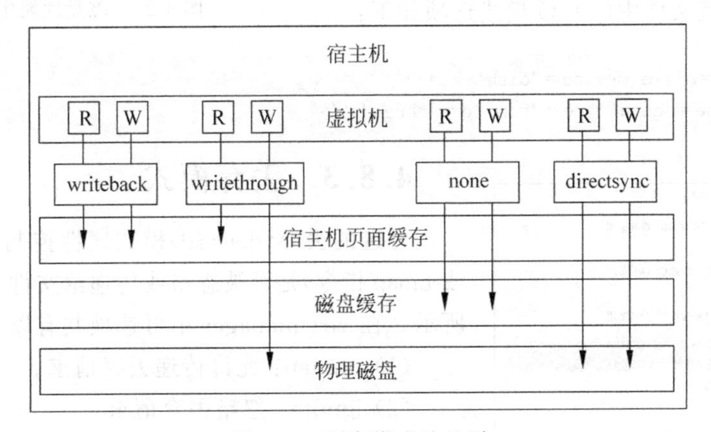


图 4-23　缓存模式的差别

(1) none：无页面缓存。此选项将跳过宿主机页面缓存，并且 I/O 直接在 qemu-kvm 用户空间缓冲区与基础存储设备之间发生。也就是说，来自虚拟机的 I/O 不会缓存在宿主机上，而是可以保留在磁盘缓存中。对于具有大量 I/O 要求的虚拟机可使用此选项。cache=none

被认为是最佳选择，这也是支持实时迁移的唯一选择。

（2）writethrough：直写。直接写入物理磁盘，仅当数据写入物理磁盘后，才会报告写入已完成。来自虚拟机的I/O会被缓存在宿主机上。虽然可以确保数据的完整性，但此模式速度较慢并且容易出现扩展问题。如果有少量具有较低I/O要求的虚拟机，则可以选择此模式。对于不支持回写缓存且不需要迁移的虚拟机，建议使用此缓存模式。

（3）writeback：回写。来自虚拟机的I/O缓存在宿主机上。写入宿主机页的缓存成功后，就会报告给虚拟机写入成功。页面缓存管理器随后再将数据写入底层储设备。

（4）directsync：类似于writethrough，但来自虚拟机的I/O绕过了宿主机页面缓存。

（5）unsafe：宿主机缓存所有磁盘I/O，并且虚拟机的同步请求将被忽略。这种模式是不安全的，因为如果宿主机发生故障，则存在数据丢失的风险，但是，这在执行虚拟机安装或类似非关键任务时可能会派上用场。

（6）Hypervisor default：未设置。

## 4.8.2 I/O模式

如图4-24所示，在virt-manager中，I/O模式（I/O Mode）的可选项共有以下3种。

（1）native：通常在使用块设备的虚拟机上表现更好。这需要将缓存模式设置为none或directsync，否则相当于threads模式。

（2）threads：通常在使用映像文件的虚拟机上表现更好。

（3）Hypervisor default：未设置。

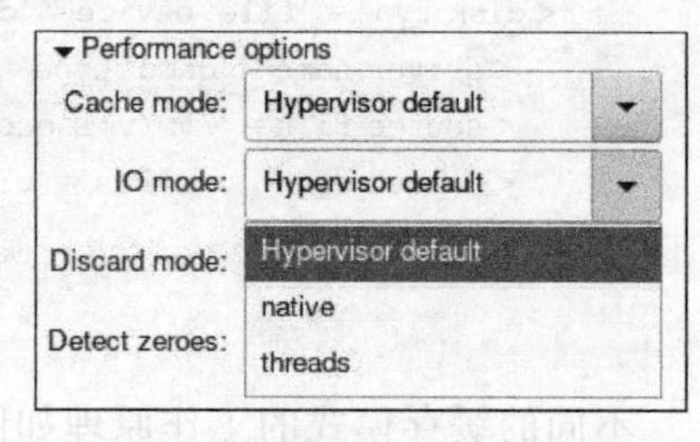


图4-24　磁盘性能中的I/O模式

虚拟机配置文件中的I/O模式选项如下：

```
<disk type='file' device='disk'>
<driver name='qemu' type='raw' io='threads'/>
```

## 4.8.3 丢弃模式

可选的丢弃（discard）模式属性控制丢弃请求（trim或umap指令）是否被忽略或传递给文件系统，如图4-25所示。在virt-manager中可选项共有以下3种。

（1）unmap：允许传递丢弃请求。

（2）ignore：忽略丢弃请求。

（3）Hypervisor default：未设置。

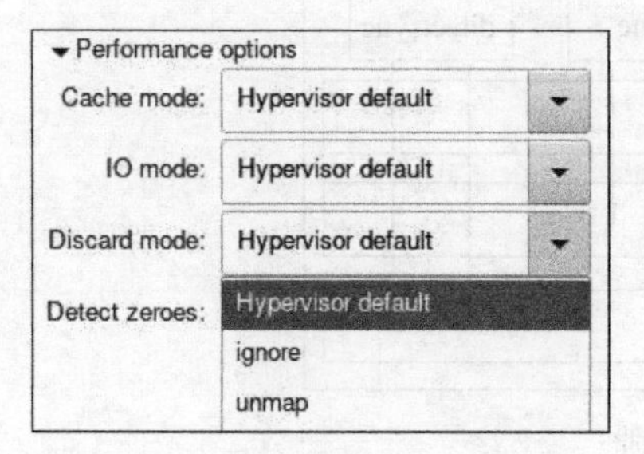


图4-25　磁盘性能中的丢弃模式

虚拟机配置文件中的丢失模式选项如下：

```
<driver name='qemu' type='qcow2' discard='unmap' />
```

### 4.8.4 检测零模式

可选的detect_zeroes属性用于控制是否检测零写入请求，它是一项计算密集型操作，但可以减少文件空间占用和慢速介质的访问时间，如图4-26所示。在virt-manager中可选项共有以下4种。

(1) off：关闭检测。

(2) on：打开检测。

(3) unmap：开启检测功能，并根据丢失模式的值尝试从映像文件中丢弃这些区域。

(4) Hypervisor default：未设置。

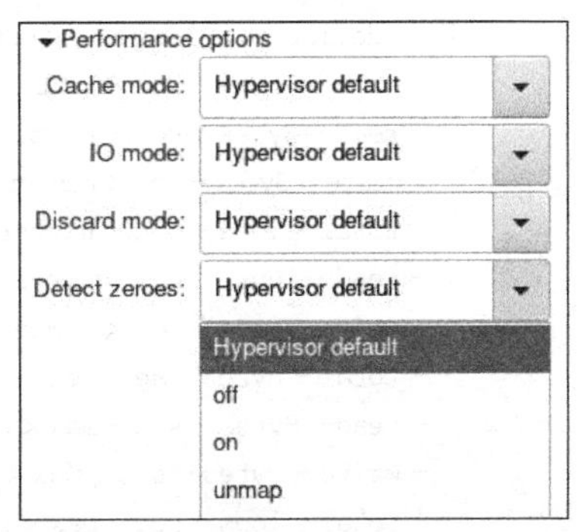


图4-26 磁盘性能中的检测零

虚拟机配置文件中的检测零选项如下：

```
<driver name = 'qemu' type = 'qcow2' detect_zeroes = 'unmap' />
```

### 4.8.5 I/O调整

当宿主机中有多个虚拟机时，为了防止某个虚拟机过度使用共享资源，需要限制每个虚拟机的磁盘I/O以确保所有虚拟机都获得足够的资源进行工作，换句话说，就是实现服务质量(QoS)。

KVM可以对每个虚拟机上的每个块设备进行I/O节流，并支持吞吐量和I/O操作的限制。这可以通过virsh的blkdeviotune子命令实现，示例命令如下：

```
# virsh blkdeviotune --help
  NAME
    blkdeviotune - Set or query a block device I/O tuning parameters.

  SYNOPSIS
    blkdeviotune <domain> <device> [--total-Bytes-sec <number>] [--read-Bytes-sec <number>] [--write-Bytes-sec <number>] [--total-iops-sec <number>] [--read-iops-sec <number>] [--write-iops-sec <number>] [--total-Bytes-sec-max <number>] [--read-Bytes-sec-max <number>] [--write-Bytes-sec-max <number>] [--total-iops-sec-max <number>] [--read-iops-sec-max <number>] [--write-iops-sec-max <number>] [--size-iops-sec <number>] [--group-name <string>] [--total-Bytes-sec-max-length <number>] [--read-Bytes-sec-max-length <number>] [--write-Bytes-sec-max-length <number>] [--total-iops-sec-max-length <number>] [--read-iops-sec-max-length <number>] [--write-iops-sec-max-length <number>] [--config] [--live] [--current]

  DESCRIPTION
```

```
    Set or query disk I/O parameters such as block throttling.

  OPTIONS
    [--domain] <string>  domain name, id or uuid
    [--device] <string>  block device
    --total-Bytes-sec <number>  total throughput limit, as scaled integer (default Bytes)
    --read-Bytes-sec <number>  read throughput limit, as scaled integer (default Bytes)
    --write-Bytes-sec <number>  write throughput limit, as scaled integer (default Bytes)
    --total-iops-sec <number>  total I/O operations limit per second
    --read-iops-sec <number>  read I/O operations limit per second
    --write-iops-sec <number>  write I/O operations limit per second
    --total-Bytes-sec-max <number>  total max, as scaled integer (default Bytes)
    --read-Bytes-sec-max <number>  read max, as scaled integer (default Bytes)
    --write-Bytes-sec-max <number>  write max, as scaled integer (default Bytes)
    --total-iops-sec-max <number>  total I/O operations max
    --read-iops-sec-max <number>  read I/O operations max
    --write-iops-sec-max <number>  write I/O operations max
    --size-iops-sec <number>  I/O size in Bytes
    --group-name <string>  group name to share I/O quota between multiple drives
    --total-Bytes-sec-max-length <number>  duration in seconds to allow total max Bytes
    --read-Bytes-sec-max-length <number>  duration in seconds to allow read max Bytes
    --write-Bytes-sec-max-length <number>  duration in seconds to allow write max Bytes
    --total-iops-sec-max-length <number> duration in seconds to allow total I/O
operations max
    --read-iops-sec-max-length <number> duration in seconds to allow read I/O operations max
    --write-iops-sec-max-length <number> duration in seconds to allow write I/O
operations max
    --config          affect next boot
    --live            affect running domain
    --current         affect current domain
```

例如，要将虚拟机 test1 的磁盘 vdb 限制为每秒 2000 I/O 操作和每秒 50MB 的吞吐量，可以执行的命令如下：

```
# virsh blkdeviotune test1 vdb --total-iops-sec 2000 \
    --total-Bytes-sec 52428800
```

## 4.9 本章小结

本章讲解了性能监视与优化的思路和工具，然后介绍了如何使用 Tuned 进行优化配置，复习了 VirtIO 驱动程序，最后又介绍针对 CPU、内存、网络和存储的优化技术。

第 5 章将讲解 P2V 和 V2V 迁移。

第5章

# P2V 和 V2V 迁移

很多企业在部署新的虚拟化平台之前，总会有一些现存的应用系统。在新环境重新部署应用系统并导入原有数据是一个推荐的选项，因为通过重建可以对原有业务系统进行梳理、整合，并且解决一些历史遗留问题，但这要求技术人员对原有应用系统的配置比较熟悉。如果达不到此要求，就需要采用迁移的方案。

针对现存应用系统的运行平台，迁移方案分为物理到虚拟(Physical to Virtual，P2V)和虚拟到虚拟(Virtual to Virtual，V2V)两种。

P2V 将物理计算机的所有数据迁移到虚拟化平台中，从而获得虚拟化所带来的灵活性和降低成本的好处。V2V 则是将在一种虚拟化平台上的虚拟机迁移到另外一种平台中。本章将介绍两个用于迁移的开源工具软件 virt-p2v 和 virt-v2v。

**本章要点**

- ❑ 使用 virt-v2v。
- ❑ 使用 virt-p2v。
- ❑ 使用磁盘映像工具 libguestfs。

## 5.1 V2V 迁移工具 virt-v2v

virt-p2v 依赖于 virt-v2v，所以先介绍 virt-v2v。

### 5.1.1 virt-v2v 实用程序简介

在专业迁移工具出现之前，很多工程师会使用 dd 加 nc 命令，或者使用 Ghost、再生龙之类的硬盘克隆、映像制作工具，它们也可以实现进行 P2V、V2V 转移，但是这都涉及大量的手动工作，因此失败的可能性很高。

为了使迁移过程自动化，需要使用专门的工具。因为这种转换不仅是以位的形式将数据从一个磁盘(卷、映像文件)复制到另一个映像文件，而且还涉及添加、移动和修改数据，例如注入半虚拟化驱动程序、修改虚拟机操作系统配置等。

libguestfs 是一个用于访问和修改虚拟机磁盘映像的工具集。使用它可以进行 P2V 和

V2V 迁移、查看和编辑虚拟机中的文件、获得磁盘状态、创建虚拟机、执行备份、克隆虚拟机、格式化磁盘、调整磁盘大小、对虚拟机进行脚本更改等。这个工具集的官方网站是 www. libguestfs. org。

virt-v2v 是 libguestfs 工具集中的一个命令行实用程序。使用它可以将外部 Hypervisor 中的虚拟机转换为由 libvirt、oVirt、Red Hat Enterprise Virtualization（RHEV）和 OpenStack 管理的虚拟机。

virt-p2v 是 virt-v2v 的附加工具。它以 ISO 或 CD 映像的形式提供，通过调用 virt-p2v 的核心功能，可以将物理机转换为虚拟机。我们可以简单地将 virt-p2v 理解为它是在 virt-v2v 基础上加了一个“壳”。

virt-v2v 可以自动完成在系统转换中涉及的所有手动工作。当前支持将在 VMware ESXi、Xen 虚拟化平台上运行的 RHEL/CentOS 的 4/5/6/7、Windows XP、Windows Vista、Windows 7、Windows Server 2003、Windows Server 2008 等转换为 KVM 的虚拟机。

virt-v2v 当前支持以下源 Hypervisor：

（1）VMware vSphere ESX / ESXi-版本 3. 5、4. 0、4. 1、5. 0、5. 1 和 5. 5。

（2）libvirt 管理的 Xen。

**提示：** 虽然 libguestfs. org 近期没有更新 virt-v2v、virt-p2v 所支持的操作系统的版本列表，但是作者的实践证明可以正常转换 RHEL/CentOS 8、Windows 7、Windows 10、Windows Server 2016 和 Windows Server 2019。

## 5. 1. 2 virt-v2v 的工作原理

安装了 virt-v2v 软件包的系统称为 virt-v2v 转换服务器，可以将其安装在虚拟机或物理机上，如图 5-1 所示。

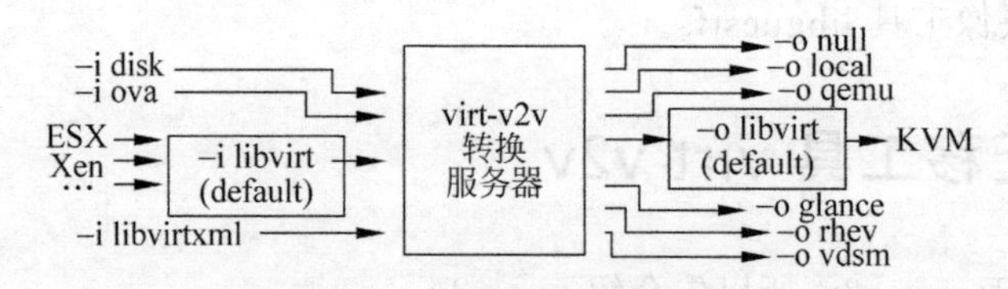


图 5-1 virt-v2v 转换服务器

在迁移时，可以将 virt-v2v 的执行过程简化理解成 4 个步骤：

（1）virt-v2v 使用 libvirt 连接到源 Hypervisor，从中检索虚拟机配置，生成新虚拟机的 XML 配置文件。

（2）读取源虚拟机磁盘中的数据，导出到新的磁盘映像。新文件可以保存在 virt-v2v 转换服务器的本地，也可以是远程的目标 Hypervisor。

（3）使用 libguestfs 工具修改磁盘映像。针对不同的操作系统进行修改会有所差别，主要包括：安装 VirtIO 驱动程序、更新/etc/fstab 和 xorg. conf 文件、重建 initrd 文件等。

（4）有了虚拟机的配置与映像文件，就可以在目标 Hypervisor 上创建新虚拟机了。

真实的过程要比这复杂很多，首先看一下示例命令：

```
#virt - v2v --help
virt - v2v: convert a guest to use KVM

virt - v2v  - ic vpx://vcenter.example.com/Datacenter/esxi  - os imported esx_guest

virt - v2v  - ic vpx://vcenter.example.com/Datacenter/esxi esx_guest  - o rhv  - os rhv.nfs:/
export_domain  -- network ovirtmgmt

virt - v2v  - i libvirtxml guest - domain.xml  - o local  - os /var/tmp

virt - v2v  - i disk disk.img  - o local  - os /var/tmp

virt - v2v  - i disk disk.img  - o glance
...
```

virt-v2v 是一个命令行工具，最重要的选项是输入(-i 或-ic)和输出(-o)。输入是源 Hypervisor、磁盘映像，而输出则是目标 Hypervisor、磁盘映像。例如：如果要从 VMware ESXi 转换虚拟机，则 VMware ESXi 或 VMware vCenter 是输入，转换后要在其中运行虚拟机的平台就是目标，它可以是独立的 KVM 宿主机或 oVirt Glance(oVirt 的存储)。

根据源 Hypervisor、磁盘映像的不同，在转换期间会自动执行以下操作：

(1) 调用 virt-inspector 命令来获得源操作系统的信息。

(2) 如果是 Xen 虚拟机，则删除内核软件包，以及删除 Xen 驱动程序，例如 xen_net、xen_blk 等。

(3) 如果虚拟机安装了多个内核，则它将找出包含 virtio 驱动程序的最新内核。

(4) 如果新内核支持 VirtIO，则将网络驱动程序替换为 virtio_net，将块驱动程序替换为 virtio_blk，否则使用非半虚拟驱动程序。

(5) 用 cirrus 替换显卡驱动程序。

(6) 更新/etc/fstab 配置。

(7) 确保 initrd 可以引导真正的根设备。

(8) 执行通过选项设置的操作。例如：-of 所指定的新的磁盘格式。

**注意**：目前尚不支持实时的 v2v，所以需要在迁移之前将虚拟机关闭。

### 5.1.3　virt-v2v 的安装

由于 RHEL/CentOS 8 的软件仓库提供了 virt-v2v 软件包，所以可以很方便地使用 dnf 或 yum 软件包管理器进行安装，示例命令如下：

```
#cat /etc/redhat - release
CentOS Linux release 8.3.2011

#dnf  - y install virt - v2v virtio - win libguestfs - winsupport
```

virt-v2v 是将虚拟机从非 KVM 的 Hypervisor 迁移到 KVM 的核心程序。virtio-win 包括了适用于 Windows 操作系统的 VirtIO 半虚拟化驱动程序。libguestfs-winsupport 在 virt-v2v 和 virt-p2v 程序中增加了对 NTFS 文件系统的支持。后两个软件包对迁移运行 Windows 操作系统的虚拟机和物理机特别重要。

### 5.1.4 V2V 的准备工作

使用 virt-v2v 实用程序进行 V2V 迁移，需要满足一些先决条件。

**1. 源虚拟机操作系统的先决条件**

(1) Linux：确保运行的 Linux 操作系统支持 virt-io 驱动程序。

(2) Windows：临时禁用防病毒等安全软件，因为它们有可能会使 virt-v2v 安装的新驱动程序无法正常工作。检查 Windows 组策略，特别是软件安装、驱动程序数据签名类的策略，防止引起 virt-v2v 安装的新驱动程序无法正常工作。

**2. 源 Hypervisor 的先决条件**

(1) KVM 虚拟化：确保 root 用户可以通过 SSH 登录，例如：在/etc/ssh/sshd_config 中将 PermitRootLogin 指令设置为 yes。

(2) VMware vSphere/ESXi 虚拟化：

- 删除虚拟机中安装的 VMware-Tools 软件。
- 如果需要与 VMware vCenter 对话以执行转换，则需要管理员用户凭据具有访问数据存储区和虚拟机配置的最小权限集。
- 目前不支持快照虚拟机。如果有快照，则可按 P2V 进行迁移。

(3) XEN 虚拟化：确保 root 用户可以通过 SSH 登录。因为需要下载新内核和驱动程序，转换虚拟机必须有权访问存储仓库。

### 5.1.5 示例：迁移 VMware 虚拟机

下面的示例是将虚拟机从 VMware vSphere 平台迁移至 KVM 平台。

VMware vCenter Server 的 IP 地址是 192.168.1.20，有一个名为 ZZ 的数据中心，虚拟机所在的 ESXi 服务器的 IP 地址是 192.168.1.21，要迁移的虚拟机是一台名为 TEMP 的 Windows 2003 虚拟机。迁移的目标是 v2v 服务器本地的/var/tmp/目录。

首先通过 virsh 命令验证源 Hypervisor 是否可以正常访问，示例命令如下：

```
# virsh -c 'vpx://192.168.1.20/ZZ/192.168.1.21?no_verify=1'\
    list --all
Enter username for 192.168.1.20 [administrator]: tomtrain%5cchentao
Enter tomtrain%5cchentao's password for 192.168.1.20: 输入密码
Id    Name                   State
----------------------------------------
-     CT_Test_CentOS6.81     shut off
-     TEMP                   shut off
-     TESTWIN7               shut off
```

其中 vpx 的语法格式为 vpx://vCenter 的名称或 IP/数据中心名称/主机名或 IP。

在此实验环境中，VMware vCenter Server 使用的是自签名证书，所以需要在上述命令的 URI 的最后添加?no_verify=1，这可以显式地禁用证书检查。

由于实验中使用的是域账户 tomtrain\chentao，所以需要将域名与账户名之间的正斜线(\)进行 URL 转码(结果为%5c)，输入的用户名为 tomtrain%5cchentao。

测试通过后，就可以进行迁移操作，示例命令如下：

```
# virt - v2v \
    - ic vpx://tomtrain % 5cchentao@192.168.1.20/Datacenter/esxi "TEMP" \
    - o local - os /var/tmp - oa sparse

[    0.0] Opening the source - i libvirt - ic vpx://tomtrainchentao@192.168.1.20/Datacenter/
esxi TEMP
virt - v2v: error: because of libvirt bug
https://bugzilla.redhat.com/show_bug.cgi?id = 1134592 you must set this
environment variable:

export LIBGUESTFS_BACKEND = direct and then rerun the virt - v2v command.

If reporting bugs, run virt - v2v with debugging enabled and include the
complete output:

  virt - v2v - v - x [...]
```

命令的选项参数的含义如下。

(1) -ic uri：libvirt 的 URI。

(2) -o <glance|json|libvirt|local|null|openstack|qemu|rhv|rhv-upload|vdsm>：设置输出模式。默认为 libvirt，本实验是本地目录(local)。

(3) -os <storage>：设置输出存储位置。

(4) -oa <sparse|preallocated>：设置输出磁盘映像的分配模式，本实验是稀疏模式(sparse)。

这次执行会出错。根据信息提示，需要先设置一个环境变量，然后次执行 virt-v2v 命令，示例命令如下：

```
# export LIBGUESTFS_BACKEND = direct

# virt - v2v \
    - ic vpx://tomtrain % 5cchentao@192.168.1.20/Datacenter/esxi "TEMP" \
    - o local - os /var/tmp - oa sparse

[    0.0] Opening the source - i libvirt - ic vpx://tomtrain % 5cchentao@192.168.1.20/ZZ/192.
168.1.21?no_verify = 1 TEMP
Enter tomtrain % 5cchentao's password for 192.168.1.20:
```

```
Enter host password for user 'tomtrain%5cchentao': 输入密码
cURL -q --insecure --user '<hidden>' --head --silent --URL 'https://192.168.1.20/
folder/TEMP/TEMP-000001-flat.vmdk?dcPath=ZZ&dsName=local21-2'
HTTP/1.1 404 Not Found
Date: Thu, 8 Dec 2016 05:20:42 GMT
Set-Cookie: vmware_soap_session="52c57ad2-0133-5004-b5b0-f7debad220bd"; Path=/;
HttpOnly; Secure;
Connection: close
Content-Type: text; charset=plain
Content-Length: 0

virt-v2v: error: vcenter: URL not found:
https://192.168.1.20/folder/TEMP/TEMP-000001-flat.vmdk?dcPath=ZZ&dsName=local21-2

The '--dcpath' parameter may be useful.  See the explanation in the
virt-v2v(1) man page OPTIONS section.

If reporting bugs, run virt-v2v with debugging enabled and include the
complete output:

  virt-v2v -v -x [...]
```

virt-v2v 命令不支持带快照的虚拟机，所以会出现上述错误。在 VMware vSphere 中删除这个虚拟机的快照，然后再次执行 virt-v2v 命令，示例命令如下：

```
#virt-v2v \
    -ic vpx://tomtrain%5cchentao@192.168.1.20/Datacenter/esxi "TEMP" \
    -o local -os /var/tmp -oa sparse

[   0.0] Opening the source -i libvirt -ic vpx://tomtrain%5cchentao@192.168.1.20/ZZ/192.
168.1.21?no_verify=1 TEMP
Enter tomtrain%5cchentao's password for 192.168.1.20: 输入密码
Enter host password for user 'tomtrain%5cchentao':
[   8.0] Creating an overlay to protect the source from being modified
[   9.0] Opening the overlay
[  78.0] Initializing the target -o local -os /var/tmp
[  78.0] Inspecting the overlay
[ 141.0] Checking for sufficient free disk space in the guest
[ 141.0] Estimating space required on target for each disk
[ 141.0] Converting Microsoft Windows Server 2003 to run on KVM
virt-v2v: This guest has virtio drivers installed.
[ 149.0] Mapping filesystem data to avoid copying unused and blank areas
[ 150.0] Closing the overlay
[ 150.0] Checking if the guest needs BIOS or UEFI to boot
[ 150.0] Copying disk 1/1 to /var/tmp/TEMP-sda (raw)
(3.00/100%)
```

这次命令执行成功后，开始进行迁移。

在迁移时，可以在控制台上看到进度和当前正在进行的转换步骤，例如安装 virt-io 驱动程序。

迁移时间取决于虚拟机的磁盘大小和网络带宽。可以使用 sar 查看网络的使用情况，示例命令如下：

```
#sar -n DEV 1 4
...
Avg:     IFACE          rxpck/s    txpck/s   rxkB/s     txkB/s   rxcmp/s   txcmp/s   rxmcst/s
Avg:     eth0           2955.25    2553.00   44537.24   166.55   0.00      0.00      0.00
Avg:     eth1           0.50       0.00      0.03       0.00     0.00      0.00      0.00
Avg:     eth2           0.50       0.00      0.03       0.00     0.00      0.00      0.00
Avg:     lo             0.75       0.75      1.36       1.36     0.00      0.00      0.00
Avg:     virbr0-nic     0.00       0.00      0.00       0.00     0.00      0.00      0.00
Avg:     virbr0         0.00       0.00      0.00       0.00     0.00      0.00      0.00
```

迁移完成后，virt-v2v 将在/var/tmp 目录中创建一个新的 XML 文件和磁盘映像文件。

检查 XML 的内容，然后可以使用 virsh define 定义新虚拟机，从而完成本次转换。

**提示**：virt-v2v 会将一些与迁移相关的信息保存在系统日志/var/log/messages 中，例如：打开虚拟机的磁盘文件、向其中添加文件、修改注册表等操作。

### 5.1.6 导入 OVF/OVA 格式的文件

开源虚拟化格式(Open Virtualization Format，OVF)文件是一种开源的文件规范，其描述了一个安全、可扩展的便携式虚拟机打包及分发格式。它一般由多个文件组成，包括 OVF 文件、MF 文件、CERT 文件、VMDK 文件和 ISO 文件等。

开放虚拟化设备(Open Virtualization Appliance，OVA)文件与 OVF 文件的作用类似，不过它是打包后的文件(TAR 格式)，包含了 OVF 中的所有文件。与 OVF 相比，OVA 是单一文件，所以更易于管理。

virt-v2v 支持导入 OVA/OVF 格式的文件，这就给虚拟化的迁移带来了更大的灵活性，例如这个场景：

如果需要将在 VirtualBox、Oracle VM 或任何其他不受 virt-v2v 支持的虚拟化平台上的虚拟机迁移到 KVM 中，则该怎么办?

有两种解决方案：

(1) 把此虚拟机当作物理系统，然后使用 virt-p2v 进行迁移。

(2) 先导出开放式虚拟化格式的文件(OVF 或 OVA)，然后将其复制到 virt-v2v 转换服务器，使用-i ova 选项进行导入。示例命令如下：

```
#export pool=vmdata

#virt-v2v -i ova -os $pool /tmp/erptest.ova
```

上述命令将读取保存在 OVA 文件中的清单，并在本地创建虚拟机。生成的磁盘映像存储在名为 vmdata 的存储池中。

### 5.1.7 转换 OVF 格式的文件

对于有些 OVF 格式的文件，甚至可以不用 virt-v2v 的导入功能，就可以直接在 KVM 平台下使用。下面就以某厂商发布的 OVF 格式的文件演示一下，如图 5-2 所示。

| 名称 | 修改日期 | 类型 | 大小 |
|---|---|---|---|
| FS6U5-NSSVA-800-MK.mf | 2015/5/2 0:14 | MF 文件 | 1 KB |
| FS6U5-NSSVA-800-MK.ovf | 2015/5/2 0:14 | 开放虚拟化格式程序包 | 33 KB |
| FS6U5-NSSVA-800-MK-disk1.vmdk | 2015/5/2 0:09 | VMware 虚拟磁盘文件 | 1,339,020... |
| FS6U5-NSSVA-800-MK-disk2.vmdk | 2015/5/2 0:10 | VMware 虚拟磁盘文件 | 680 KB |

图 5-2 OVF 格式文件

首先，通过阅读 OVF 文件(XML 格式)可以知道虚拟机有两个 VMDK 文件，然后将它们上传到 KVM 服务器上，示例命令如下：

```
# ls -l
total 1339700
-rw-r--r--. 1 root root 1371156480 May  2  2015 FS6U5-NSSVA-800-MK-disk1.vmdk
-rw-r--r--. 1 root root     696320 May  2  2015 FS6U5-NSSVA-800-MK-disk2.vmdk
```

查看这两个映像文件的信息，示例命令如下：

```
# qemu-img info FS6U5-NSSVA-800-MK-disk1.vmdk
image: FS6U5-NSSVA-800-MK-disk1.vmdk
file format: vmdk
virtual size: 70G (75161927680 Bytes)
disk size: 1.3G
cluster_size: 65536
Format specific information:
    cid: 1031567622
    parent cid: 4294967295
    create type: streamOptimized
    extents:
        [0]:
            compressed: true
            virtual size: 75161927680
            filename: FS6U5-NSSVA-800-MK-disk1.vmdk
            cluster size: 65536
            format:
```

```
# qemu-img info FS6U5-NSSVA-800-MK-disk2.vmdk
image: FS6U5-NSSVA-800-MK-disk2.vmdk
file format: vmdk
virtual size: 1.0G (1073741824 Bytes)
disk size: 680K
cluster_size: 65536
Format specific information:
   cid: 2716224742
   parent cid: 4294967295
   create type: streamOptimized
   extents:
      [0]:
         compressed: true
         virtual size: 1073741824
         filename: FS6U5-NSSVA-800-MK-disk2.vmdk
         cluster size: 65536
         format:
```

出于性能的考虑，可以把它们转换为 qcow2 格式，示例命令如下：

```
# qemu-img convert -O qcow2 FS6U5-NSSVA-800-MK-disk1.vmdk \
  FS6U5-NSSVA-800-MK-disk1.qcow2

# qemu-img convert -O qcow2 FS6U5-NSSVA-800-MK-disk2.vmdk \
  FS6U5-NSSVA-800-MK-disk2.qcow2
```

**提示**：转换磁盘映像格式并不是必需的。

根据 OVF 文件中的配置信息，确定新虚拟机的主要参数如下：

(1) 两个 vCPU。

(2) 4GB 内存。

(3) 4 个 virtio 网卡。

(4) 磁盘接口为 SCSI 接口。

(5) 显示协议是 VNC。

(6) OS 类型为 RHEL6。

使用 virt-install 命令利用现在的映像文件创建新的虚拟机，示例命令如下：

```
# virt-install \
  --import \
  --name=demo8 \
  --vcpus=2 --ram=4096 \
  --disk bus=scsi,path=/vm/falc/FS6U5-NSSVA-800-MK-disk1.qcow2 \
  --disk bus=scsi,path=/vm/falc/FS6U5-NSSVA-800-MK-disk2.qcow2 \
  --network type=bridge,source=br0 \
  --network type=bridge,source=br0 \
```

```
--network type=bridge,source=br0 \
--network type=bridge,source=br0 \
--graphics vnc,listen=0.0.0.0 \
--os-type=Linux \
--os-variant=rhel6   \
--noautoconsole
```

### 5.1.8 与 virt-v2v 相关的故障排除

在对 virt-v2v 进行故障排除时,应注意以下几点:

(1) 确保在 virt-v2v 转换服务器上安装了必需的 V2V 软件包,例如 libguestfs-winsupport 和 virtio-win。

(2) 确保在源主机上启用了 SSH。

(3) 确保目标宿主机具有足够的空间来容纳新的虚拟机。

(4) 确认所使用的 virt-v2v 命令语法正确。

(5) 使用最新版本的 virt-v2v。

**提示**:virt-v2v 手册页中有每个参数的详细说明和丰富示例。

另外,还可以通过设置环境变量来启用 virt-v2v 调试日志,示例命令如下:

```
LIBGUESTFS_TRACE=1
LIBGUESTFS_DEBug=1
```

如下所示,这行命令会在屏幕上输出详细的执行过程:

```
# LIBGUESTFS_TRACE=1 LIBGUESTFS_DEBug=1 virt-v2v -ic \
  vpx://yunhedata%5cchentao@192.168.1.20/ZZ/192.168.1.21?no_verify=1 \
  "TEMP"  -o local  -os /var/tmp  -oa sparse
```

## 5.2 P2V 迁移工具 virt-p2v

virt-v2v 可以与源 Hypervisor 进行对话,以获取虚拟机的硬件信息和元数据,但是,当源是物理系统时,virt-v2v 则无法收集有关硬件的信息。为了解决此问题,libguestfs 项目提供了一个 Live CD/USB,它是一个小的可引导的 Linux 操作系统,其中包括 virt-p2v 的工具。用它启动物理主机,virt-p2v 会通过 SSH 将物理系统的数据发送到 virt-v2v 主机,然后将其转换为目标 Hypervisor 上的虚拟机。

virt-p2v 可以将大多数物理计算机转换为虚拟机,但也对物理机有一些要求:

(1) 至少具有 512MB 的 RAM。

(2) 不支持大于 2TB 的卷。

(3) 仅支持对基于 x86 或 x86_64 体系的计算机进行 P2V 转换。

(4) 不支持软件 RAID 的根文件系统。

(5) 必须具有网络连接。

### 5.2.1　创建或下载 virt-p2v 可启动映像

获得 virt-p2v 可启动映像有两种方法：自己制作和下载 libguestfs 之前推荐的非官方的 ISO 二进制版本。

自己制作需要使用 virt-p2v-make-disk 或 virt-p2vmake-kickstart 的实用程序，它们都是 virt-p2v-maker 软件包的一部分，示例命令如下：

```
# dnf -y install virt-p2v-maker

# rpm -qi virt-p2v-maker
Name         : virt-p2v-maker
Epoch        : 1
Version      : 1.42.0
Release      : 5.el8
...
Summary      : Convert a physical machine to run on KVM
Description :
Virt-p2v converts (virtualizes) physical machines so they can be run
as virtual machines under KVM.

This package contains the tools needed to make a virt-p2v boot CD or
USB key which is booted on the physical machine to perform the
conversion.  You also need virt-v2v installed somewhere else to
complete the conversion.

To convert virtual machines from other hypervisors, see virt-v2v.
```

制作用于启动的 U 盘，示例命令如下：

```
# virt-p2v-make-disk -o /dev/sdX fedora-33
```

**注意**：此操作会删除 USB 驱动器/dev/sdX 中所有的数据。

制作启动 ISO 并进行刻录，示例命令如下：

```
# virt-p2v-make-kickstart fedora

# Live CD-creator p2v.ks
```

这将为 Fedora 在当前目录中创建一个 p2v.ks 的 kickstart 文件。有了 kickstart 文件后，就可以使用 Live CD-creator 制作 Live CD 了。

**提示**：virt-p2v-make-disk 和 Live CD-creator 需要连接到互联网上的在线软件仓库，有时可能会失败。

libguestfs 之前推荐过一个非官方的二进制版本，包含 ISO 文件和 PXE 引导映像，目前最新的版本是通过 RHEL-7.3 构建的。其下载网址为 http://oirase.annexia.org/virt-p2v/。

将下载的 ISO 文件(virt-p2v-1.32.7-2.el7.iso)刻录成 CD,或使用 Rufus(https://rufus.ie/)类的工具制作一个启动 U 盘。

下面通过 virt-p2v-1.32.7-2.el7.iso 进行 P2V 的转换。

## 5.2.2 示例:迁移 Windows 2008 R2 服务器

使用 virt-p2v 可引导的 CD 或 U 盘引导物理系统,就会出现启动菜单,如图 5-3 所示。

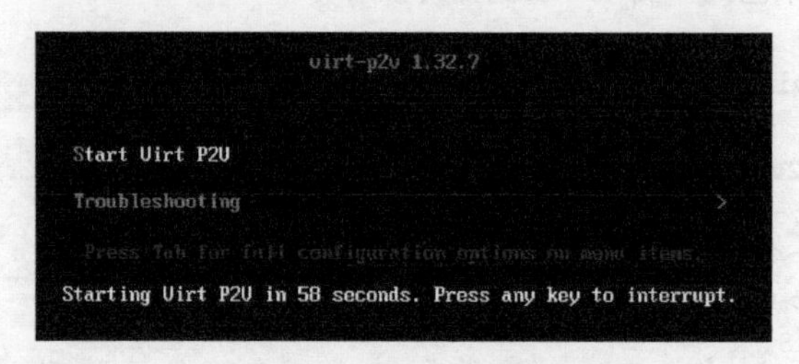


图 5-3 virt-p2v 启动菜单

通过键盘的上下键选中 Start Virt P2V,然后按 Enter 键。virt-p2v 提供了一个 GUI 界面,输入 virt-v2v 转换服务器的 IP 地址、root 账号及密码,如图 5-4 所示。

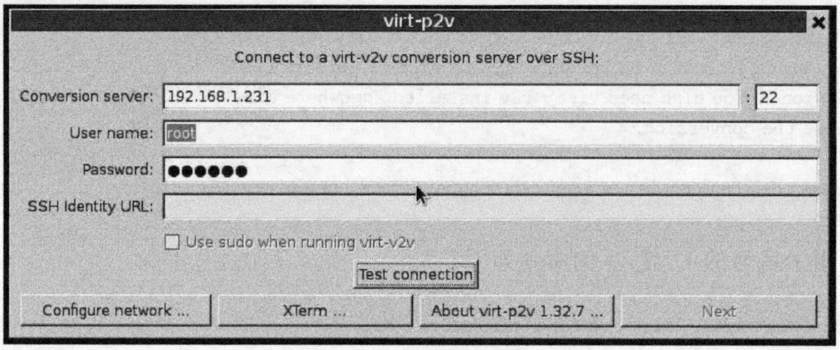

图 5-4 连接到 virt-v2v 转换服务器

如果网络中没有 DHCP 服务器,则需要单击 Configure network 按钮,然后进行网络配置,设置静态 IP 地址。

单击 Test connection 按钮,测试与转换服务器的 SSH 连接。如果成功,则可单击 Next 按钮继续。

这时会出现 virt-p2v 的主配置界面,如图 5-5 所示。界面分为 4 部分:

(1) 在 Target Properties 中,可以设置新虚拟机的名称、vCPUs 的数量和内存大小。

(2) 右侧有 3 个窗格,用于控制将在新虚拟机中创建哪些硬盘、可移动媒体设备和网络接口。通常保持默认设置即可。

(3) 在左中部的 Virt-v2v output options 中,可以指定输出目标、输出连接、输出存储位置、输出格式、存储分配方式。默认情况下,存储分配格式为稀疏格式。

(4) 左下角区域会显示 virt-p2v 和 virt-v2v 的角色及版本信息。

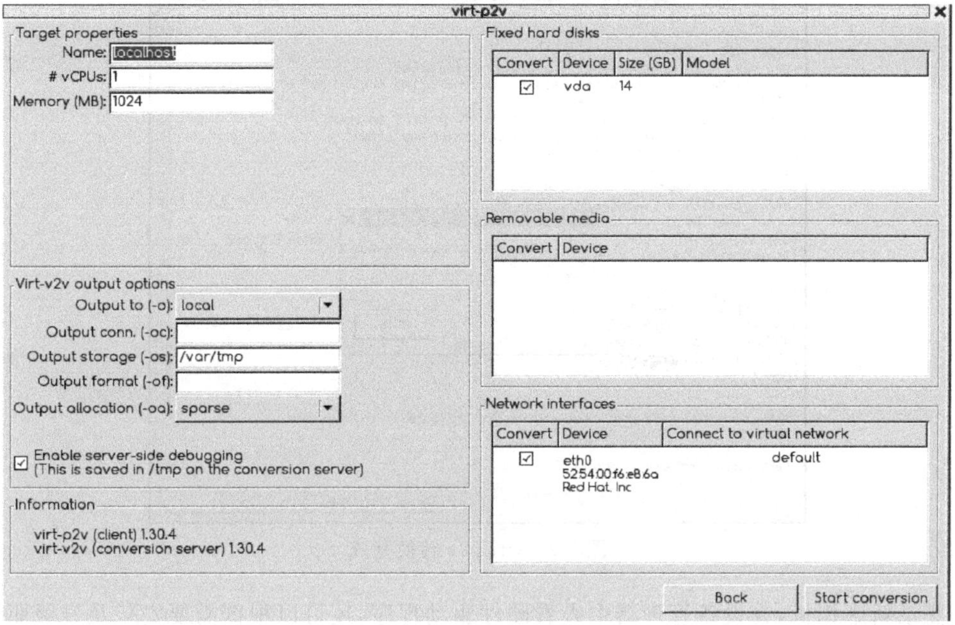

图 5-5　virt-p2v 主要配置界面

检查确认无误后，单击 Start conversion 按钮启动转换。在转换过程中，会显示动作及进度，如图 5-6 所示。

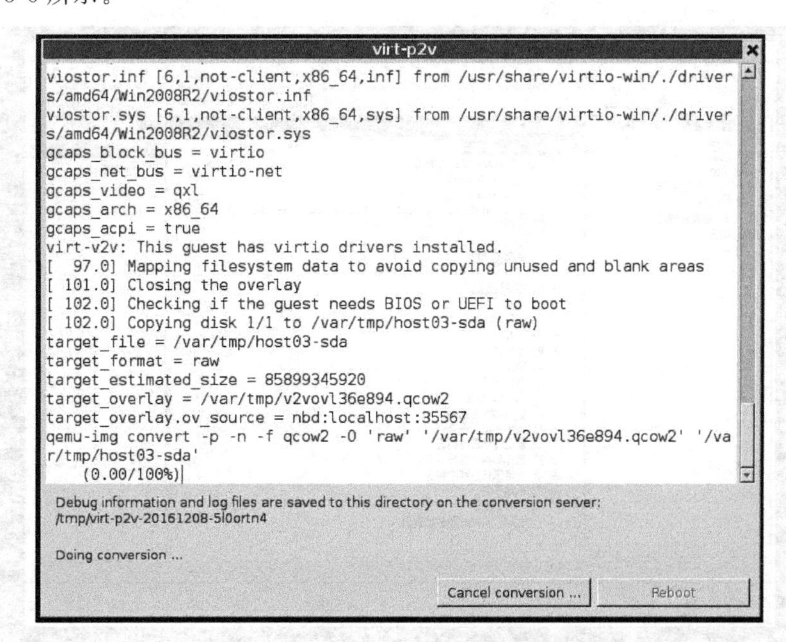


图 5-6　virt-p2v 转换进度窗口

转换完成后会弹出转换成功窗口，单击 OK 按钮，如图 5-7 所示。

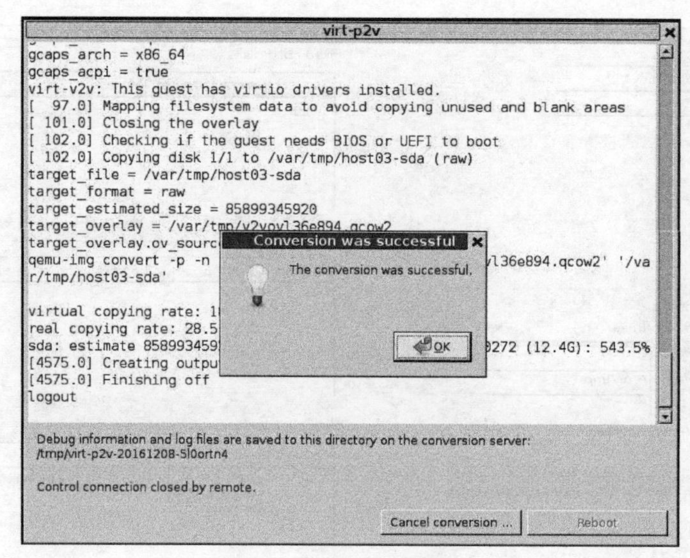


图 5-7 virt-p2v 转换完成

登录新虚拟机，在设备管理器中查看硬件驱动程序，还可以根据需要安装及升级驱动程序，如图 5-8 所示。

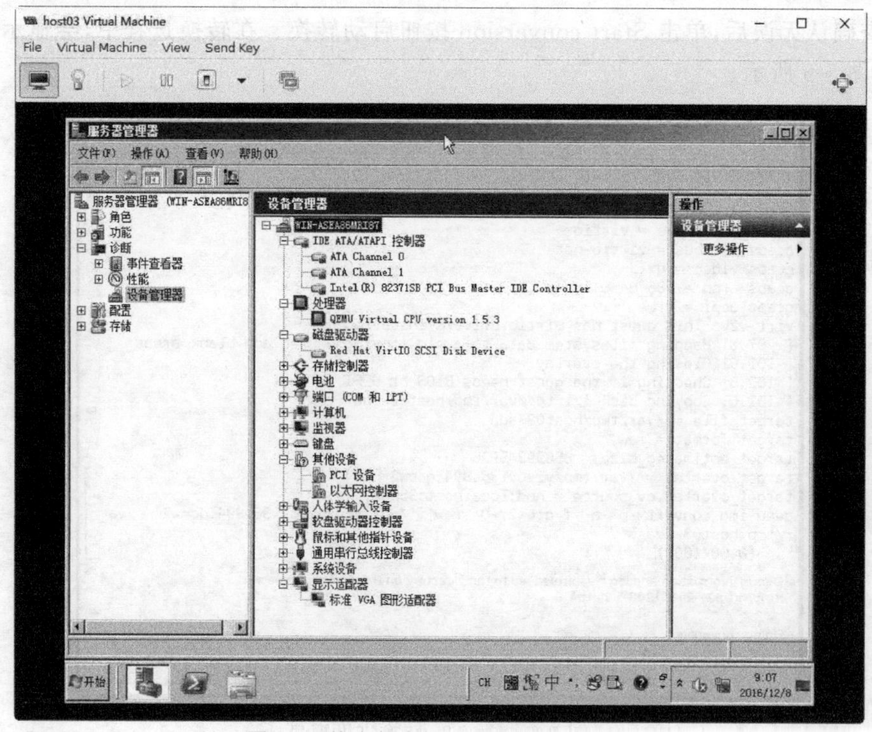


图 5-8 检查新虚拟机的硬件驱动程序

## 5.2.3　故障排错及杂项

如果运行 virt-p2v 时出现故障，则需要根据屏幕提示和日志的信息进行线索查找，例如蓝屏错误。

在有些环境下，转换 Windows Sever 2003 操作系统成功后，新虚拟机在启动时会出现蓝屏错误，如图 5-9 所示。

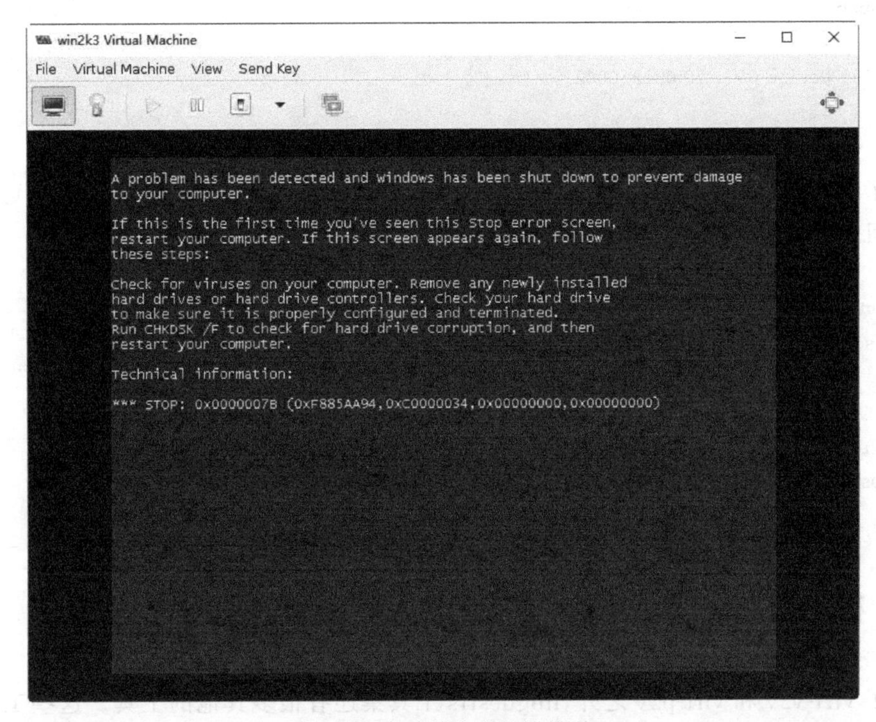


图 5-9　启动时出现蓝屏错误

查看屏幕上显示的错误号码及摘要信息，可以知道这是一个 Windows Server 2003 经常出现的 IDE 接口 0x0000007B 错误。微软的一篇知识库文件对此有详细的说明：

http://support. microsoft. com/kb/314082/en-us
Booting a virtual clone (IDE) of a physical Windows system partition may fail with a BSOD referring to the problem
STOP: 0x0000007B (0xF741B84C,0xC0000034,0x00000000,0x00000000)
INACCESSIBLE_BOOT_DEVICE
this means that the source physical Windows machine had no support for IDE controller, or at least the one virtually replaced by kvm (see Microsoft KB article article for details): as Microsoft suggests, create a mergeide. reg file (File: Mergeide. zip) file on the physical machine and merge that in the registry, 'before the P2V migration. Btw, it may not be necessary but should be no harm, anyway, and save you lots of time and headaches.

知道了原因，就可以很容易地找到解决方案了。在本示例中，需要在源主机上修改注册表来启用标准 IDE 驱动程序。可参考 https://pve.proxmox.com/wiki/File：Mergeide.zip。

在进行 P2V 转换的时候，还有可能会遇到 USB 接口的软件加密狗。

解决方案是先将 USB 的加密狗插入宿主机的 USB 接口，然后获得 USB 设备的 ID，示例命令如下：

```
# lsusb
…
Bus 005 Device 007: ID 08e2:0002
…
```

有了设备 ID，就可以使用 USB"透传"功能，将 USB 加密狗分配给虚拟机。这可以通过在虚拟机的配置文件中添加以下内容实现：

```
<hostdev mode='subsystem' type='usb' managed='no'>
  <source>
    <vendor id='0x08e2'/>
    <product id='0x0002'/>
  </source>
</hostdev>
```

## 5.3 磁盘映像工具 libguestfs

除了 virt-v2v 和 virt-p2v 之外，libguestfs 工具集还有很多其他的工具。这些工具的简单说明如下。

- guestfs：主要的 API 文档。
- guestfish：交互式 Shell 程序。
- guestmount：在宿主机上挂载虚拟机文件系统。
- guestunmount：卸载虚拟机文件系统。
- virt-alignment-scan：检查虚拟机分区的对齐方式。
- virt-builder：快速生成映像文件。
- irt-builder-repository：创建 virt-builder 库。
- virt-cat：显示映像中的文件内容。
- virt-copy-in：将宿主机的文件和目录复制到虚拟机。
- virt-copy-out：从虚拟机中将文件和目录复制到宿主机。
- virt-customize：自定义虚拟机。

- virt-df：显示可用空间。
- virt-dib：安全的 diskimage-builder。
- virt-diff：比较映像文件的差异。
- virt-edit：编辑映像文件。
- virt-filesystems：显示有关文件系统、设备、LVM 信息。
- virt-format：擦除并制作空白磁盘。
- virt-get-Kernel：从磁盘获取内核。
- virt-inspector：探查映像文件。
- virt-list-filesystems：列出文件系统。
- virt-list-partitions：列出分区。
- virt-log：显示日志文件。
- virt-ls：列出文件。
- virt-make-fs：创建文件系统。
- virt-p2v：将物理机转换到 KVM 平台上。
- virt-p2v-make-disk：制作用于 P2V 的 ISO 文件或 U 盘。
- virt-p2v-make-kickstart：制作 kickstart 文件。
- virt-rescue：救援模式 Shell。
- virt-resize：调整映像文件的大小。
- virt-sparsify：将映像文件设置为稀疏格式(精简配置)。
- virt-sysprep：克隆前删除虚拟机的个性化配置。
- virt-tail：查看日志文件尾部。
- virt-tar：归档文件。
- virt-tar-in：归档和上传文件。
- virt-tar-out：归档和下载文件
- virt-v2v：将其他平台的虚拟机转换到 KVM 平台上。
- virt-win-reg：导出和合并 Windows 注册表项。
- libguestfs-test-tool：测试 libguestfs。
- libguestfs-make-fixed-appliance：使 libguestfs 固定设备。
- hivex：提取 Windows 注册表配置单元。
- hivexregedit：从 regedit 格式的文件合并和导出注册表更改。
- hivexsh：Windows 注册表配置单元 Shell。
- hivexml：将 Windows 注册表配置单元转换为 XML。
- hivexget：从 Windows 注册表配置单元中提取数据。
- supermin：用于构建 supermin 设备的工具。

- guestfsd：guestfs 守护程序。

这些命令基本上要求虚拟机处于关闭状态。下面介绍几个常用的命令。

(1) virt-df 类似于 df 命令,示例命令如下：

```
# virt - df /vm/win2019.qcow2
Filesystem                  1K - blocks   Used       Available   Use %
win2019.qcow2:/dev/sda1     562172        31408      530764      6 %
win2019.qcow2:/dev/sda2     104293372     11664548   92628824    12 %
```

(2) virt-cat 类似于 cat 命令,示例命令如下：

```
# virt - cat /vm/centos8.3.qcows  /etc/hosts
127.0.0.1    localhost localhost.localdomain localhost4 localhost4.localdomain4
::1          localhost localhost.localdomain localhost6 localhost6.localdomain6
```

(3) virt-edit 类似于 vi 命令,示例命令如下：

```
# virt - edit /vm/centos8.3.qcows  /etc/hosts
```

(4) virt-ls 类似于 ls 或 dir 命令,示例命令如下：

```
# virt - ls  /vm/win2019.qcow2 '/Windows/System32/drivers/etc/'
hosts
lmhosts.sam
networks
protocol
services
```

(5) virt- inspector 用于探查映像文件中的操作系统信息,示例命令如下：

```
# virt - inspector  /vm/win2019.qcow2 | head
<?xml version = "1.0"?>
< operatingsystems >
  < operatingsystem >
    < root >/dev/sda2 </root >
    < name > Windows </name >
    < arch > x86_64 </arch >
    < distro > Windows </distro >
    < product_name > Windows Server 2019 Standard </product_name >
    < product_variant > Server </product_variant >
    < major_version > 10 </major_version >
```

## 5.4 本章小结

本章讲解了 libguestfs 工具集中两个用于转换的工具：virt-v2v、virt-p2v，它们可以实现虚拟机到虚拟机、物理机到虚拟机的转换，支持的操作系统有 Windows 和 Linux 操作系统，还讲解了如何通过启用调试日志进行排错。最后又简单了解了 libguestfs 工具集中的其他命令。

第 6 章将讲解备份与恢复。

# 第 6 章

# 备份与恢复

备份对于任何业务系统都是必需的。对 KVM 虚拟化系统的备份和备份物理服务器没有太大区别，备份虚拟机甚至比备份物理服务器更方便。

本章先介绍备份与恢复的基本概念，然后使用 rsync 之类的工具备份 KVM 虚拟机，采用不同的方法来创建和管理快照。

**本章要点**

- 数据损坏风险及备份策略。
- 虚拟机冷备。
- 虚拟机快照的基本原理。
- 虚拟机快照的管理。
- 虚拟机备份脚本示例。

## 6.1 数据损坏风险及备份策略

数据损坏、丢失对于各种规模的企业、组织和个人都是一个严重的问题，这将意味着需要花费时间和金钱进行恢复。一些丢失的数据是可以恢复的，但通常需要专业人员的帮助。在有些情况下，即使是专业人员也无法恢复丢失的文件和信息。导致数据损坏丢失的原因有很多，著名数据恢复公司 Kroll Ontrack 曾经做过一个统计，排名前 3 的原因如下：

(1) 67%是由硬盘驱动器崩溃或系统故障引起的。

(2) 14%是由人为错误引起的。

(3) 10%是由软件故障造成的。

防止数据损坏、丢失的最有效的方法是备份。备份数据意味着至少拥有所有数据的另一份副本。需要考虑具体的业务、数据重要性及成本等多种因素才能设计出适合的备份策略。

备份数据的最主要目标就是恢复数据，所谓“养兵千日，用兵一时”，我们需要从恢复的角度来设计备份方案。在恢复过程中，有 2 个最关键的衡量指标。

(1) RTO：Recovery Time Objective，指在故障发生之后，业务停止工作的最高可承受时间。例如：在 15：00 系统宕机导致业务停顿，然后进行恢复操作，在 22：00 恢复运营，将从停顿到恢复运营的时间间隔称为 RTO(7h)，如图 6-1 所示。RTO 是反映系统业务恢复及时性的指标，表示业务从中断到恢复正常所需的时间，RTO 数值越小，代表数据恢复能力越强。RTO＝0 就意味着在任何情况下都不允许业务系统有任何停顿。

(2) RPO：Recovery Point Objective，指一个过去的时间点。当故障发生时，数据可以恢复到的时间点，是业务系统所能容忍的数据丢失量。例如：每天 0：00 进行数据备份，如果今天发生了故障，数据则可以恢复到的时间点(RPO)就是今天的 0：00，如果 15：00 发生故障，损失的数据就是此 15h 间产生的数据，如果 23：59 发生故障，则损失的数据就是此约 24h 间产生的数据，所以该用户的 RPO 就是 24h，即用户最大的数据损失量是此 24h 间产生的数据。

RTO 和 RPO 指标并不是孤立的，而是从不同角度来反映数据中心的容灾能力。

RPO 指的是用户允许损失的最大数据量。这主要与数据备份的频率有关，为了改进 RPO，必然需要增加数据备份的频率，例如每 12h 备份 1 次，那么 RPO＝12h。RPO 主要反映了业务连续性管理体系下备用数据的有效性，即 RPO 值越小，表示系统对数据完整性的保证能力越强。

另外还要考虑备份介质的可靠性，如果最近一次备份介质不可用，就需要使用上一次的备份介质，图 6-1 中损失数据就是 24＋15＝39h 的数据，即 RPO＝39h。

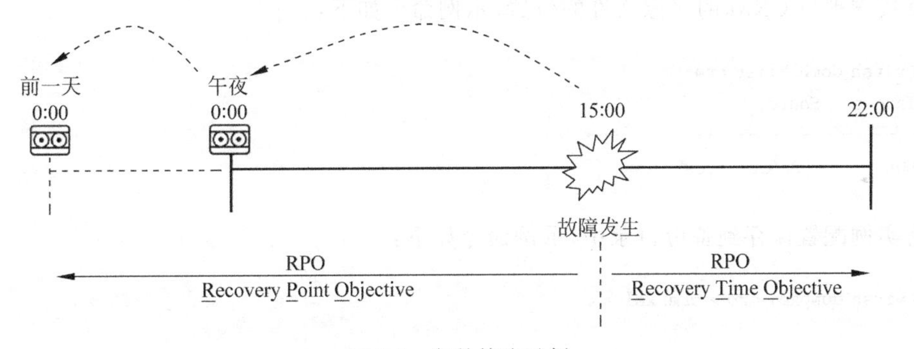


图 6-1 备份策略示例

传统的数据备份方案采用周期性备份，所以一直有备份窗口伴随着。如果特别重要的数据对 RPO 要求特别高，就无法满足需求了。持续数据保护(Continuous Data Protection，CDP)技术是对传统数据备份术的一个重大突破，它会不断监测关键数据的变化，从而不断地自动实现数据的保护，即 RPO≈0。

全球网络存储工业协会(Storage Networking Industry Association，SNIA)对 CDP 的定义为“持续数据保护是一套方法，它可以捕获或跟踪数据的变化，并将其在生产数据之外独立存放，以确保数据可以恢复到过去的任意时间点。持续数据保护系统可以基于块、文件或应用实现，可以为恢复对象提供足够细的恢复粒度，实现几乎无限多的恢复时间点。”

有一个形象的比喻：传统备份就像是照相机，只在按快门的时候产生照片；CDP 则是摄像机，打开就不停地工作，任何时间的图像都不会错过。

本章仅实现 KVM 虚拟化环境的周期备份。在进行备份时，根据虚拟机的状态分为以下 2 种备份方法。

(1) 脱机(冷)备份：在虚拟机处于关闭状态下的备份。

这种备份方法简单、维护量小、可靠性高，但在备份时业务不可用。需要备份虚拟机的配置文件、所有虚拟机的映像文件或 LVM 卷。

(2) 联机(热)备份：在虚拟机处于运行状态下的备份。

在备份时业务不中断。先创建虚拟机(磁盘文件)的外部快照文件，然后备份快照文件。对于数据库等应用还需要额外考虑数据的一致性。

## 6.2 虚拟机冷备

备份 KVM 虚拟机的最简单方法就是冷备。下面我们将使用 tar 和 rsync 创建 KVM 虚拟机实例的备份，并将其存储在远程服务器上。

创建备份目录并切换到此目录，示例命令如下：

```
# mkdir /bak_crm && cd /bak_crm
```

查找虚拟机 CRM 的映像文件的位置，示例命令如下：

```
# virsh domblklist crm
 Target   Source
---------------------------
 hda      /vm/crm.qcow2
```

将实例配置保存到备份目录中，示例命令如下：

```
# virsh dumpxml crm > crm.xml
```

停止虚拟机，然后将映像文件复制到备份目录，示例命令如下：

```
# virsh shutdown crm
Domain crm is being shutdown

# cp /vm/crm.qcow2 .

# ls -lha
total 2.3G
drwxr-xr-x.   2 root root    38 Feb 24 10:30 .
dr-xr-xr-x.  21 root root   276 Feb 24 10:24 ..
-rw-r--r--.   1 root root   2.3G Feb 24 10:30 crm.qcow2
-rw-r--r--.   1 root root   5.9K Feb 24 10:27 crm.xml
```

将配置文件和映像文件打包成一个归档文件，示例命令如下：

```
# tar zcvf crm_bakup.tar.gz .
./
./crm.xml
./crm.qcow2
tar: .: file changed as we read it

# ls -l
total 3427400
-rw-r--r--. 1 root root 1146549251 Feb 24 10:34 crm_bakup.tar.gz
-rw-r--r--. 1 root root 2363097088 Feb 24 10:30 crm.qcow2
-rw-r--r--. 1 root root       6000 Feb 24 10:27 crm.xml

# rm crm.qcow2 crm.xml
```

同步备份文件到远程服务器，示例命令如下：

```
# rsync -vaz crm_bakup.tar.gz tomkvm2:/tmp
root@192.168.1.232's password:
sending incremental file list
crm_bakup.tar.gz

sent 1,139,275,076 Bytes  received 35 Bytes  23,984,739.18 B/sec
total size is 1,146,549,251  speedup is 1.01
```

在远程服务器上模拟恢复过程。找到备份文件，提取归档中的文件，示例命令如下：

```
[root@tomkvm2 ~]# cd /tmp

[root@tomkvm2 tmp]# tar zxvf crm_bakup.tar.gz
./
./crm.xml
./crm.qcow2
```

将映像文件复制到配置的位置并定义实例，示例命令如下：

```
[root@tomkvm2 tmp]# cp crm.qcow2 /vm/

[root@tomkvm2 tmp]# virsh define crm.xml
Domain crm defined from crm.xml

[root@tomkvm2 tmp]# virsh list --all | grep crm
 -      crm                 shut off

[root@tomkvm2 tmp]# virsh start crm
Domain crm started
```

**注意**：为了保证数据的一致性和完整性，使用 tar 等工具对磁盘映像文件进行操作时，必须首先停止虚拟机。

## 6.3 快照的基本原理

与物理计算机相比，虚拟化提供了无可比拟的敏捷性优势，例如：更快的部署、快照等简单易行的备份和恢复。

KVM 虚拟机的快照用来保存虚拟机在某个时间点的内存、磁盘或者设备状态。快照有许多应用场景，例如：在对虚拟机进行重大操作之前创建快照，如果操作导致了故障，就可以通过还原快照轻松地恢复到先前的状态。

从不同的视角进行分类，KVM 快照有如下不同的分类方法：

(1) 根据对象不同，快照可以分为磁盘快照和内存快照。两者加起来构成了一个系统还原点，记录虚拟机在某个时间点的全部状态。

(2) 根据创建快照时虚拟机是否在运行，快照又可以分为实时快照和离线快照。

(3) 根据快照存储方式的不同，又分为内部快照和外部快照。

**提示**：除了 KVM 快照之外，还可以通过卷管理(如 LVM)或文件系统(如 ZFS)的快照实现对虚拟机的保护。

内部快照和外部快照都有其自身的优点和局限性。

**1) 内部快照(Internal Snapshot)**

创建快照前后的数据都存储在单个映像文件中，所以管理工作量小、灵活性高，但是仅支持 QCOW2 格式的映像文件，不支持 RAW 格式。在对运行的虚拟机创建内部快照时，虚拟机会暂停、业务会中断，快照创建完毕后会恢复运行状态。

**2) 外部快照(External Snapshot)**

外部快照使用 COW(Copy On Write)机制。创建快照时，原始映像文件将变为只读状态，我们将其称为基础(Base)映像文件，同时创建一个新的覆盖(Overlay)映像文件，随后的新增、更新、删除(标记)等写入操作会保存在其中，如图 6-2 所示。

覆盖映像文件最初创建时的长度为 0 字节。随着后续的写入操作，此文件会增长直到原始磁盘的大小。覆盖映像文件必须是 QCOW2 格式，但基础映像可以是 RAW、QCOW2 或 libvirt 支持的任何其他磁盘映像格式。

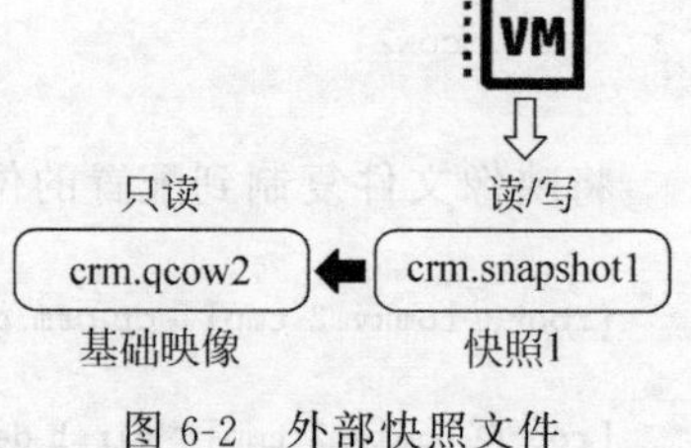

图 6-2 外部快照文件

**提示**：有些技术文档中会使用 Delta File(差异文件)，它与 Overlay File(覆盖)含义相同。

libvirt 实现的内部快照与 qemu-img 的 snapshot 子命令管理的快照所使用的是相同的机制，外部快照与 qemu-img 的派生映像的原理也相同。它们主要区别是 qemu-img 只能对未运行虚拟机中的磁盘映像进行操作。

目前版本的 Cockpit 不提供快照功能，virt-manager 仅能管理内部快照，而 virsh 快照管理功能最为丰富。

## 6.4　内存快照

virsh 的 save 子命令会将正在运行的虚拟机的内存数据（而不是磁盘状态）保存到状态文件。保存后，会将虚拟机关闭，效果大致等效于计算机的休眠。

virsh 的 restore 子命令可以从此状态文件恢复，恢复原来运行的状态，效果大致等效于从休眠中被唤醒。

内存快照的应用场景包括以下几种。

(1) 测试环境：需要不断地将虚拟机恢复为原有状态，然后重新开始测试。采用传统的删除再重建虚拟机的方式比较慢，而使用内存快照是一个比较好的选择。

(2) 备份操作的辅助手段：使用 save 子命令将运行虚拟机安全地关闭，映像文件不再有 IO 操作，复制它们以完成备份操作，完成后需重新启动虚拟机。

为正在运行的虚拟机 CRM 创建内存快照，示例命令如下：

```
# virsh list --all | grep crm
 1    crm          running

# virsh save crm state1 --verbose
Save: [100 %]
Domain crm saved to state1
```

使用--verbose 选项会显示保存进度。如果指定了--bypass-cache 选项，则将不使用文件系统缓存，从而会减慢操作速度。

save 子命令所需要的时间取决于虚拟机内存中数据的多少。可以使用 virsh 的 domjobinfo 子命令监视进度，也可以使用 domjobabort 子命令取消任务。

查看虚拟机及状态文件的信息，示例命令如下：

```
# virsh list --all | grep crm
 -    crm          shut off

# ls -lh state1
-rw-------. 1 root root 236M Mar  2 15:13 state1

# file state1
state1: Libvirt QEMU Suspend Image, version 2, XML length 5554, running
```

save 操作结束后，虚拟机会自动关闭。生成的状态文件的大小取决于内存中数据的多少。状态文件是一个 XML 格式的文件，文件的头部包含与虚拟机及状态有关的配置信息，后面是内存数据。查看生成的状态文件，示例命令如下：

```
#tail -n 1 state1 | wc
      0   35176   192507

#head state1
LibvirtQemudSave?<domain type='kvm'>
  <name>crm</name>
  <uuid>8f1149d2-9dc8-4e9b-b120-df28a0bb704a</uuid>
  <metadata>
    <libosinfo:libosinfo xmlns:libosinfo="http://libosinfo.org/xmlns/libvirt/domain/1.0">
      <libosinfo:os id="http://centos.org/centos/6.10"/>
    </libosinfo:libosinfo>
  </metadata>
  <memory unit='KiB'>1048576</memory>
  <currentMemory unit='KiB'>1048576</currentMemory>
```

启动虚拟机，示例命令如下：

```
#virsh start crm
Domain crm started
```

为了验证内存快照功能，可以登录虚拟机进行一些操作。

在回滚时，需要先将虚拟机关闭，否则会出错。示例命令如下：

```
#virsh restore state1
error: Failed to restore domain from state1
error: Requested operation is not valid: domain 'crm' is already active

#virsh shutdown crm
Domain crm is being shutdown
```

由于state文件中有虚拟机的名称等标识信息，所以virsh的restore子命令并不要求提供虚拟机的信息，示例命令如下：

```
#virsh restore state1
Domain restored from state1
```

可以使用--paused选项将恢复后的虚拟机置于暂停状态，而不是默认的运行状态。

查看虚拟机的运行状态，示例命令如下：

```
#virsh list --all | grep crm
2     crm              running
```

登录虚拟机，会发现原有的进程还在正常地运行。

## 6.5 内部快照

内部快照仅适用于 QCOW2 格式的映像文件，它是磁盘快照和虚拟机内存状态的组合，所以也被称为检查点（Checkpoint）。在需要时，可以轻松地还原指定的检查点。

**提示**：可以使用 qemu-img 命令将其他格式的映像文件转换为 QCOW2 格式。

使用 virt-manager 和 virsh 可以管理内部快照，其中 virsh 管理快照的子命令更加灵活高效，常用子命令如下。

（1）snapshot-create：使用 XML 文件创建快照。

（2）snapshot-create-as：使用参数创建快照。

（3）snapshot-current：获取或设置当前快照。

（4）snapshot-delete：删除虚拟机快照。

（5）snapshot-dumpxml：转储 XML 格式的快照配置。

（6）snapshot-edit：编辑快照的 XML 文件。

（7）snapshot-info：获取快照信息。

（8）snapshot-list：列出虚拟机快照。

（9）snapshot-parent：获取快照父级名称。

（10）snapshot-revert：将虚拟机还原到指定的快照。

### 6.5.1 创建内部快照

实验用虚拟机 CRM 的映像文件是 QCOW2 格式。查看其当前的快照列表，示例命令如下：

```
#virsh snapshot - list crm
Name    Creation Time    State
----------------------------------
```

snapshot-list 子命令会输出给定虚拟机的快照列表。默认信息包括快照名称、创建时间和状态，还可以通过其他选项列出更多的信息。

为正在运行的虚拟机 CRM 创建一个内部快照，示例命令如下：

```
#virsh list -- all | grep crm
1     crm            running

#virsh snapshot - create crm
Domain snapshot 1614752012 created
```

如果给 snapshot-create 子命令只提供虚拟机名称，则 virsh 会自动生成一个数字作为快照名称,所以推荐提供更容易理解的自定义名称和描述。

除了 snapshot-create 子命令之外,还可以使用 snapshot-create-as 子命令创建快照。它们的区别是: snapshot-create-as 需要通过参数传递设置值,而 snapshot-create 仅接受 XML 文件提供的设置值。

为快照设置名称和描述,使管理工作更方便,示例命令如下:

```
# virsh snapshot-create-as crm --name "snapshot2" \
    --description "Test snapshot2" --atomic
Domain snapshot snapshot2 created
```

通过指定--atomic 选项,可确保快照操作要么成功,要么失败且不会发生任何更改。建议始终使用--atomic 选项以避免创建快照时造成任何损坏。

查看虚拟机的快照列表,示例命令如下:

```
# virsh snapshot-list crm
Name          Creation Time                State
------------------------------------------------------------
1614752012    2021-03-03 14:13:32 +0800    running
snapshot2     2021-03-03 14:30:41 +0800    running
```

由于这两个快照是在虚拟机运行时创建的。完成快照创建的时间取决于虚拟机的内存大小及当前修改内存数据的活跃程度。

在创建快照的过程中,虚拟机会进入暂停模式,创建结束后会自动恢复运行,所以创建内部快照会导致业务系统中断一段时间,这可以通过 ping 虚拟机来验证。

(1) 如果创建快照耗时短,则暂停时间短,不太容易被察觉,但是 ping 命令会显示响应延迟,示例命令如下:

```
# ping crm
PING crm (192.168.122.230) 56(84) Bytes of data.
64 Bytes from 192.168.122.230: icmp_seq=1 ttl=64 time=0.473 ms
64 Bytes from 192.168.122.230: icmp_seq=2 ttl=64 time=1.95 ms
64 Bytes from 192.168.122.230: icmp_seq=5 ttl=64 time=1.24 ms
64 Bytes from 192.168.122.230: icmp_seq=6 ttl=64 time=1.34 ms
64 Bytes from 192.168.122.230: icmp_seq=103 ttl=64 time=1533 ms ← 返回响应慢
64 Bytes from 192.168.122.230: icmp_seq=104 ttl=64 time=510 ms ← 返回响应慢
64 Bytes from 192.168.122.230: icmp_seq=105 ttl=64 time=2.71 ms
64 Bytes from 192.168.122.230: icmp_seq=106 ttl=64 time=1.67 ms
64 Bytes from 192.168.122.230: icmp_seq=107 ttl=64 time=0.739 ms
64 Bytes from 192.168.122.230: icmp_seq=108 ttl=64 time=0.777 ms
^C
```

(2) 如果创建快照耗时长,则可导致暂停时间长,ping 命令会显示响应超时,示例命令如下：

```
# ping crm
PING crm (192.168.122.230) 56(84) Bytes of data.
64 Bytes from 192.168.122.230: icmp_seq = 168 ttl = 64 time = 0.958 ms
64 Bytes from 192.168.122.230: icmp_seq = 169 ttl = 64 time = 0.684 ms
From 192.168.122.1 icmp_seq = 175 Destination Host Unreachable ← 返回超时
From 192.168.122.1 icmp_seq = 176 Destination Host Unreachable
From 192.168.122.1 icmp_seq = 177 Destination Host Unreachable
...
From 192.168.122.1 icmp_seq = 188 Destination Host Unreachable
64 Bytes from 192.168.122.230: icmp_seq = 171 ttl = 64 time = 18918 ms
64 Bytes from 192.168.122.230: icmp_seq = 172 ttl = 64 time = 17917 ms
64 Bytes from 192.168.122.230: icmp_seq = 189 ttl = 64 time = 552 ms
64 Bytes from 192.168.122.230: icmp_seq = 173 ttl = 64 time = 16874 ms
64 Bytes from 192.168.122.230: icmp_seq = 174 ttl = 64 time = 15850 ms
64 Bytes from 192.168.122.230: icmp_seq = 190 ttl = 64 time = 0.622 ms
64 Bytes from 192.168.122.230: icmp_seq = 191 ttl = 64 time = 0.622 ms
^C
```

因此，在对正在运行的虚拟机做内部快照之前，要评估对业务的影响程度。

在生产中，可以根据需要创建多个快照。下面再为虚拟机 CRM 创建 1 个关机时的快照，这样总共拥有 3 个快照。示例命令如下：

```
# virsh shutdown crm
Domain crm is being shutdown

# virsh snapshot - create - as crm snapshot3
Domain snapshot snapshot3 created

# virsh snapshot - list crm -- parent
Name              Creation Time                State      Parent
--------------------------------------------------------------------------
 1614752012    2021 - 03 - 03 14:13:32  + 0800    running
 snapshot2     2021 - 03 - 03 14:30:41  + 0800    running   1614752012
 snapshot3     2021 - 03 - 03 22:04:30  + 0800    shutoff   snapshot2
```

使用--parent 选项可显示快照的父子关系，这有助于了解快照的顺序：第 1 个快照 1614752012 的父级为空，这意味着它是直接在磁盘映像上创建的，并且 1614752012 是 Snapshot2 的父级，而 Snapshot2 是 Snapshot3 的父级。

State 列显示了特定快照的创建时机：第 1 个、第 2 个快照是在虚拟机运行时创建的，而第 3 个快照是在虚拟机关闭时创建的。

可以使用--tree 选项获得快照的树状视图，示例命令如下：

```
# virsh snapshot - list crm -- tree
1614752012
  |
  +- snapshot2
      |
      +- snapshot3
```

可以使用 qemu-img 命令获取有关内部快照的更多信息,例如快照大小、标记、创建时间,示例命令如下:

```
# qemu - img info /vm/crm.qcow2
image: /vm/crm.qcow2
file format: qcow2
virtual size: 80 GiB (85899345920 Bytes)
disk size: 4.67 GiB
cluster_size: 65536
Snapshot list:
ID   TAG              VM SIZE                DATE       VM CLOCK
1    1614752012       193 MiB  2021 - 03 - 03  14:13:32   00:03:23.231
2    snapshot2        194 MiB  2021 - 03 - 03  14:30:41   00:20:29.200
3    snapshot3        0 B      2021 - 03 - 03  22:04:30   00:00:00.000
Format specific information:
    compat: 1.1
    lazy refcounts: true
    refcount bits: 16
    corrupt: false
```

使用 check 子命令检查 qcow2 映像文件的完整性,示例命令如下:

```
# qemu - img check /vm/crm.qcow2
No errors were found on the image.
1310720/1310720 = 100.00 % allocated, 0.02 % fragmented, 0.00 % compressed clusters
Image end offset: 86538387456
```

如果映像中发生任何损坏,则会输出错误信息。

## 6.5.2 恢复内部快照

创建快照的主要目的是在需要时恢复到指定的状态。例如:在创建快照 Snapshot3 后,错误地安装了一个应用程序并修改了系统的多个配置,在这种情况下,通过恢复快照 Snapshot3 就可以轻松恢复正常。示例命令如下:

```
# virsh list -- all | grep crm
 4    crm            running
```

```
# virsh snapshot - revert crm snapshot3

# virsh list  -- all | grep crm
 -     crm             shut off

# virsh start crm
```

如果恢复的是关闭状态的快照,则默认情况下会导致虚拟机关闭,所以还需要手工启动虚拟机。如果在恢复时使用--running 选项,则可自动启动它,示例命令如下:

```
# virsh snapshot - revert crm snapshot3  -- running

# virsh list  -- all | grep crm
 6     crm             running
```

### 6.5.3 删除内部快照

快照会影响性能,所以一旦确定不再需要,就应该及时删除。

可以使用 snapshot-delete 子命令删除虚拟机的快照。例如删除虚拟机 CRM 的第 1 个快照,示例命令如下:

```
# virsh snapshot - list crm
 Name                 Creation Time                 State
---------------------------------------------------------
 1614752012      2021 - 03 - 03 14:13:32  + 0800     running
 snapshot2       2021 - 03 - 03 14:30:41  + 0800     running
 snapshot3       2021 - 03 - 03 22:04:30  + 0800     shutoff

# virsh snapshot - delete crm 1614752012
Domain snapshot 1614752012 deleted

# virsh snapshot - list crm
 Name                 Creation Time                 State
---------------------------------------------------------
 snapshot2       2021 - 03 - 03 14:30:41  + 0800     running
 snapshot3       2021 - 03 - 03 22:04:30  + 0800     shutoff
```

### 6.5.4 使用 virt-manager 管理快照

virt-manager 提供了一个管理虚拟机内部快照的图形界面。单击工具栏上的按钮便可切换到快照视图,如图 6-3 所示。

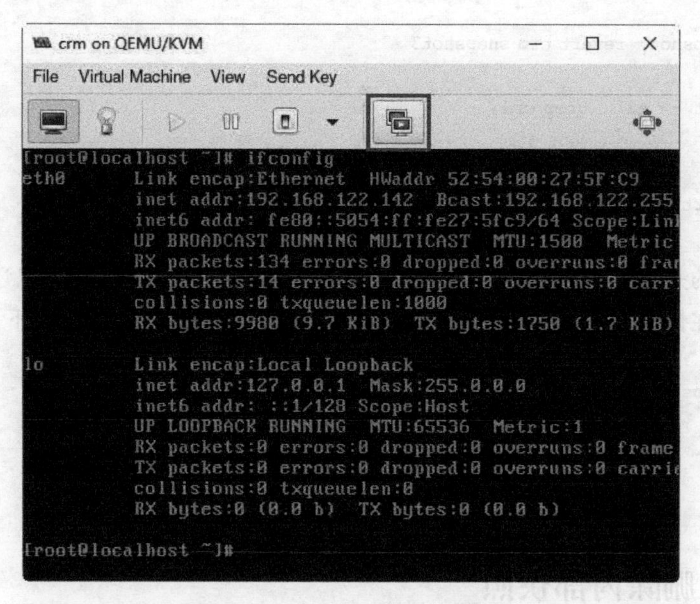


图 6-3 virt-manager 的主界面

通过图形界面可以很方便地查看、添加和删除快照，如图 6-4 所示。

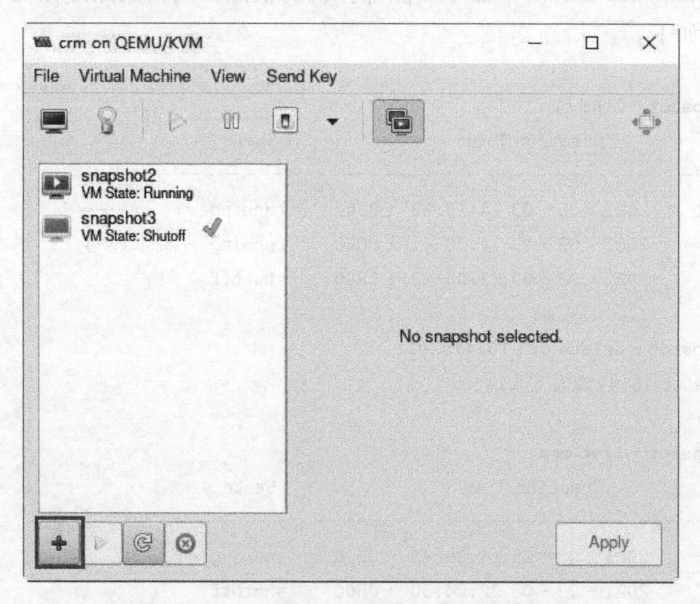


图 6-4 virt-manager 对虚拟机快照进行管理

如果要创建快照，则可以单击按钮，在新窗口中输入快照名和描述，如图 6-5 所示。单击按钮，恢复到指定的快照。单击按钮，删除指定的快照。

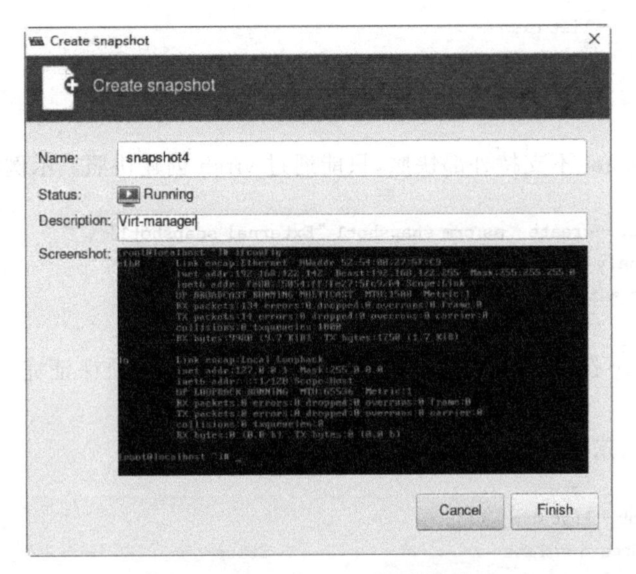


图 6-5 在 virt-manager 中创建快照

## 6.6 外部快照

外部快照的“外部”是指在外部保存快照资源。使用外部快照，需要了解以下两个术语。

(1) backing_file：后备文件，是指虚拟机的原始磁盘映像(只读)。

(2) overlay_image：覆盖映像，是指快照映像文件(可写)。

创建外部快照后，backing_file 变为只读状态，数据将写入 overlay_image 中。如果出现问题，则可以简单地丢弃 overlay_image 映像，从而恢复到原有状态。

对于外部快照来讲，backing_file 映像可以有多种映像格式，包括 RAW、QCOW、QCOW2、VMDX 等。

### 6.6.1 创建外部快照

可以在虚拟机运行或关闭时创建外部快照。

查看虚拟机的运行状态，示例命令如下：

```
# virsh list --all | grep crm
 1     crm           running

# virsh domblklist crm --details
 Type    Device    Target    Source
------------------------------------------
 file    disk      vda       /vm/crm.qcow2
 file    cdrom     hda       -
```

```
# virsh snapshot - list crm
 Name   Creation  Time       State
---------------------------------
```

目前 virt-manager 不支持外部快照,只能通过 virsh 创建快照。示例命令如下:

```
# virsh snapshot - create - as crm snapshot1 "External snapshot1" \
    -- disk - only -- atomic
Domain snapshot snapshot1 created
```

使用--disk-only 选项将创建磁盘快照。使用--atomic 选项保证完整性并避免可能的损坏。

查看快照信息,示例命令如下:

```
# virsh snapshot - list crm
 Name         Creation Time                        State
------------------------------------------------------------
 snapshot1    2021 - 03 - 04 15:36:31  + 0800     disk - snapshot

# virsh snapshot - info crm snapshot1
Name:            snapshot1
Domain:          crm
Current:         yes
State:           disk - snapshot
Location:        external
Parent:          -
Children:        0
Descendants:     0
Metadata:        yes
```

默认情况下快照文件与原始映像文件在同一个目录下。它们只是磁盘状态的快照,不包括内存中的数据。在虚拟机运行时创建外部快照,虚拟机可正常运行,业务没有中断。

查看虚拟机的所有块设备信息,示例命令如下:

```
# virsh domblklist crm
 Target       Source
------------------------------
 vda          /vm/crm.snapshot1
 hda          -
```

需要注意,在创建快照后,虚拟机磁盘的源发生了更改,变成了当前的快照。

查看一下虚拟机的 XML 配置,示例命令如下:

```
#virsh dumpxml crm
...
    <disk type='file' device='disk'>
      <driver name='qemu' type='qcow2'/>
      <source file='/vm/crm.snapshot1' index='3'/>
      <backingStore type='file' index='2'>
        <format type='qcow2'/>
        <source file='/vm/crm.qcow2'/>
        <backingStore/>
      </backingStore>
      <target dev='vda' bus='virtio'/>
      <alias name='virtio-disk0'/>
      <address type='pci' domain='0x0000' bus='0x00' slot='0x07' function='0x0'/>
    </disk>
...
```

查看虚拟机映像文件当前的状态，示例命令如下：

```
#ls /vm/crm.* -lh
-rw-r--r--. 1 qemu qemu  81G Mar  4 15:36 /vm/crm.qcow2
-rw-------. 1 qemu qemu 576K Mar  4 15:41 /vm/crm.snapshot1

#qemu-img info /vm/crm.snapshot1
qemu-img: Could not open '/vm/crm.snapshot1': Failed to get shared "write" lock
Is another process using the image [/vm/crm.snapshot1]?

#virsh shutdown crm

#qemu-img info /vm/crm.snapshot1
image: /vm/crm.snapshot1
file format: qcow2
virtual size: 80 GiB (85899345920 Bytes)
disk size: 4.81 MiB
cluster_size: 65536
backing file: /vm/crm.qcow2
backing file format: qcow2
Format specific information:
    compat: 1.1
    lazy refcounts: false
    refcount bits: 16
    corrupt: false
```

快照文件 crm.snapshot1 初始很小，随着文件增加、删除和更改，此文件会慢慢变大。

快照 crm.snapshot1 的后备文件（Backing File）指定了原始的映像文件/vm/crm.qcow2，它是只读文件，对 crm.snapshot1 所做的任何更改都不会反映在文件 crm.

qcow2 中。

在虚拟机处于关闭状态时再创建 1 个快照，示例命令如下：

```
#virsh snapshot-create-as crm snapshot2 "External snapshot2" \
    --disk-only --atomic
Domain snapshot snapshot2 created

#virsh domblklist crm
 Target     Source
--------------------------------
 vda        /vm/crm.snapshot2
 hda        -

#virsh dumpxml crm
...
    <disk type='file' device='disk'>
      <driver name='qemu' type='qcow2'/>
      <source file='/vm/crm.snapshot2'/>
      <backingStore type='file'>
        <format type='qcow2'/>
        <source file='/vm/crm.snapshot1'/>
        <backingStore type='file'>
          <format type='qcow2'/>
          <source file='/vm/crm.qcow2'/>
        </backingStore>
      </backingStore>
      <target dev='vda' bus='virtio'/>
      <address type='pci' domain='0x0000' bus='0x00' slot='0x07' function='0x0'/>
    </disk>
...
```

虚拟机的当前磁盘指向了快照 2。查看快照链的信息，示例命令如下：

```
#virsh snapshot-list crm --parent
 Name        Creation Time               State          Parent
--------------------------------------------------------------------------
 snapshot1   2021-03-04 15:36:31 +0800   disk-snapshot
 snapshot2   2021-03-04 15:53:16 +0800   shutoff        snapshot1

#virsh snapshot-list crm --tree
snapshot1
  |
  +- snapshot2
```

```
#qemu-img info /vm/crm.snapshot2 --backing-chain
image: /vm/crm.snapshot2
file format: qcow2
virtual size: 80 GiB (85899345920 Bytes)
disk size: 196 KiB
cluster_size: 65536
backing file: /vm/crm.snapshot1
backing file format: qcow2
Format specific information:
    compat: 1.1
    lazy refcounts: false
    refcount bits: 16
    corrupt: false

image: /vm/crm.snapshot1
file format: qcow2
virtual size: 80 GiB (85899345920 Bytes)
disk size: 3.88 MiB
cluster_size: 65536
backing file: /vm/crm.qcow2
backing file format: qcow2
Format specific information:
    compat: 1.1
    lazy refcounts: false
    refcount bits: 16
    corrupt: false

image: /vm/crm.qcow2
file format: qcow2
virtual size: 80 GiB (85899345920 Bytes)
disk size: 4.26 GiB
cluster_size: 65536
Format specific information:
    compat: 1.1
    lazy refcounts: true
    refcount bits: 16
    corrupt: false
```

### 6.6.2 静默选项

启动虚拟机CRM，再创建一个快照，示例命令如下：

```
#virsh start crm

#virsh snapshot-create-as crm snapshot3 "External snapshot3" \
    --disk-only --atomic --quiesce
Domain snapshot snapshot3 created
```

使用--quiesce 选项，libvirt 会通知虚拟机的 qemu-ga 执行静默操作。静默是一种文件系统冻结/释放(fsfreeze / fsthaw)机制，可以将缓存中的数据“刷入”磁盘，从而将虚拟机文件系统置于一致状态，而且可以避免映像文件的意外损坏。

创建快照时，建议尽可能使用此选项以确保安全。

如果虚拟机中 qemu-ga 没有正确地运行，则会出现错误，示例命令如下：

```
# virsh snapshot - create - as crm snapshot3 "External snapshot3" \
    -- disk - only -- atomic -- quiesce
error: Guest agent is not responding: QEMU guest agent is not connected
```

## 6.6.3 快照链

如果虚拟机有多个快照，则它们之间会相互依赖，从而形成快照链。与内部快照不同，每个外部快照都是独立的文件，所以在进行文件管理时要特别小心。

列出虚拟机的所有快照，示例命令如下：

```
# virsh snapshot - list crm
 Name           Creation Time                  State
------------------------------------------------------------
 snapshot1     2021 - 03 - 04 15:36:31  + 0800     disk - snapshot
 snapshot2     2021 - 03 - 04 15:53:16  + 0800     shutoff
 snapshot3     2021 - 03 - 04 20:54:50  + 0800     disk - snapshot
```

获得虚拟机的当前活动(读/写)磁盘/快照的信息，示例命令如下：

```
# virsh domblklist crm
 Target     Source
-------------------------------
 vda        /vm/crm.snapshot3
 hda
```

使用 qemu-img 的--backing-chain 选项枚举当前活动(读/写)快照的后备文件链，示例命令如下：

```
# qemu - img info -- backing - chain /vm/crm.snapshot3
image: /vm/crm.snapshot3
file format: qcow2
virtual size: 80 GiB (85899345920 Bytes)
disk size: 3.94 MiB
cluster_size: 65536
backing file: /vm/crm.snapshot2
backing file format: qcow2
```

```
Format specific information:
    compat: 1.1
    lazy refcounts: false
    refcount bits: 16
    corrupt: false

image: /vm/crm.snapshot2
file format: qcow2
virtual size: 80 GiB (85899345920 Bytes)
disk size: 5.69 MiB
cluster_size: 65536
backing file: /vm/crm.snapshot1
backing file format: qcow2
Format specific information:
    compat: 1.1
    lazy refcounts: false
    refcount bits: 16
    corrupt: false

image: /vm/crm.snapshot1
file format: qcow2
virtual size: 80 GiB (85899345920 Bytes)
disk size: 3.88 MiB
cluster_size: 65536
backing file: /vm/crm.qcow2
backing file format: qcow2
Format specific information:
    compat: 1.1
    lazy refcounts: false
    refcount bits: 16
    corrupt: false

image: /vm/crm.qcow2
file format: qcow2
virtual size: 80 GiB (85899345920 Bytes)
disk size: 4.26 GiB
cluster_size: 65536
Format specific information:
    compat: 1.1
    lazy refcounts: true
    refcount bits: 16
    corrupt: false
```

```
# qemu-img info --backing-chain /vm/crm.snapshot3 | grep back
backing file: /vm/crm.snapshot2
backing file format: qcow2
backing file: /vm/crm.snapshot1
backing file format: qcow2
backing file: /vm/crm.qcow2
backing file format: qcow2
```

根据所获得的信息，可以画出一个快照链，如图 6-6 所示。

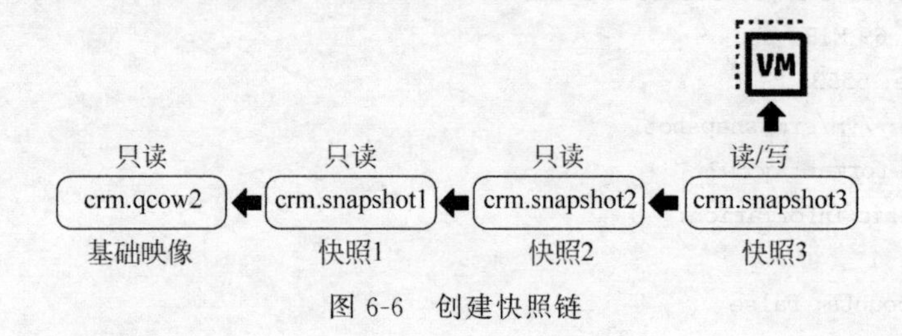


图 6-6 创建快照链

快照 3 是当前的活动快照，虚拟机可以对其进行读写。当虚拟机尝试读取虚拟磁盘数据时，必须按此次序进行读取：快照 3 将快照 2 作为其后备文件；快照 2 将快照 1 作为其后备文件；快照 1 将基础映像作为其后备文件，所以快照链如果比较长，则会严重影响 IO 性能。

libvirt 通过配置文件来存储内部快照和外部快照的信息，这些信息被称为快照的元数据(metadata)，示例命令如下：

```
# tree /var/lib/libvirt/qemu/snapshot/
/var/lib/libvirt/qemu/snapshot/
└── crm
    ├── snapshot1.xml
    ├── snapshot2.xml
    └── snapshot3.xml

1 directory, 3 files

# head /var/lib/libvirt/qemu/snapshot/crm/snapshot3.xml
<!--
WARNING: THIS IS AN AUTO-GENERATED FILE. CHANGES TO IT ARE LIKELY TO BE
OVERWRITTEN AND LOST. Changes to this xml configuration should be made using:
  virsh snapshot-edit
or other application using the libvirt API.
-->

<domainsnapshot>
  <name>snapshot3</name>
  <description>External snapshot3</description>
```

### 6.6.4　恢复外部快照

与内部快照不同，当前 RHEL/CentOS 8.3 版本还不支持直接恢复外部快照。

实验环境中发行版本及虚拟化组件的信息如下：

```
#cat /etc/redhat-release
CentOS Linux release 8.3.201

#libvirtd --version
libvirtd (libvirt) 6.0.0

#rpm -qi qemu-kvm | grep -i ^version
Version     : 4.2.0
```

虚拟机 CRM 有 3 个外部快照，尝试使用 revert 子命令进行恢复，示例命令如下：

```
#virsh snapshot-list crm
 Name          Creation Time                    State
---------------------------------------------------------------
 snapshot1     2021-03-04 15:36:31 +0800      disk-snapshot
 snapshot2     2021-03-04 15:53:16 +0800      shutoff
 snapshot3     2021-03-04 20:54:50 +0800      disk-snapshot

#virsh snapshot-info crm snapshot2
Name:           snapshot2
Domain:         crm
Current:        no
State:          shutoff
Location:       external
Parent:         snapshot1
Children:       1
Descendants:    1
Metadata:       yes

#virsh snapshot-revert crm snapshot3
error: unsupported configuration: revert to external snapshot not supported yet

#virsh snapshot-revert crm snapshot2
error: unsupported configuration: revert to external snapshot not supported yet

#virsh snapshot-revert crm snapshot1
error: unsupported configuration: revert to external snapshot not supported yet
```

执行命令后会出现错误提示。要想恢复外部快照，目前只能通过修改虚拟机的 XML 文件进行手工还原。

假设需要还原到外部快照 snapshot2，首先需要关闭虚拟机，然后编辑 XML 文件中的磁盘配置，使其使用快照的映像文件。具体操作如下：

（1）找到快照 snapshot2 对应的映像文件。最好的方法是从快照的 XML 文件获得信息，示例命令如下：

```
#virsh snapshot-dumpxml crm snapshot2 | grep "source file"
      <source file='/vm/crm.snapshot2'/>
        <source file='/vm/crm.snapshot1'/>
          <source file='/vm/crm.qcow2'/>
```

（2）/vm/crm.snapshot2 是与快照 snapshot2 关联的文件，查看此快照的信息，示例命令如下：

```
#qemu-img info /vm/crm.snapshot2
image: /vm/crm.snapshot2
file format: qcow2
virtual size: 80 GiB (85899345920 Bytes)
disk size: 5.69 MiB
cluster_size: 65536
backing file: /vm/crm.snapshot1
backing file format: qcow2
Format specific information:
   compat: 1.1
   lazy refcounts: false
   refcount bits: 16
   corrupt: false
```

（3）验证快照文件的完整性，示例命令如下：

```
#qemu-img check /vm/crm.snapshot2
No errors were found on the image.
74/1310720 = 0.01% allocated, 44.59% fragmented, 0.00% compressed clusters
Image end offset: 6029312
```

如果检测到错误，则可以使用-r leaks 或-r allt 参数进行修复。

（4）关闭虚拟机 CRM，修改其 XML 文件，从虚拟机中删除当前连接的磁盘，然后添加 /vm/crm.snapshot2 文件，示例命令如下：

```
#virsh list --all | grep crm
 -      crm            shut off

#virsh domblklist crm
 Target       Source
--------------------------------
 vda          /vm/crm.snapshot3
 hda          -
```

除了使用 virsh edit 命令修改 XML 文件外,还可以使用 virt-xml 命令修改。

首先移除当前的磁盘 vda,示例命令如下:

```
# virt-xml crm --remove-device --disk target=vda
Domain 'crm' defined successfully.
WARNING  XML did not change after domain define. You may have changed a value that libvirt is
setting by default.

# virsh domblklist crm
 Target       Source
-------------------
 hda          -
```

然后添加快照 snapshot2 作为虚拟机的磁盘,示例命令如下:

```
# virt-xml crm --add-device --disk /vm/crm.snapshot2,format=qcow2,bus=virtio
Domain 'crm' defined successfully.
WARNING  XML did not change after domain define. You may have changed a value that libvirt is
setting by default.

# virsh domblklist crm
 Target       Source
--------------------------------
 hda          -
 vda          /vm/crm.snapshot2
```

(5) 检查虚拟机 XML 配置文件中的磁盘配置,示例命令如下:

```
# virsh dumpxml crm
...
    <disk type='file' device='disk'>
      <driver name='qemu' type='qcow2'/>
      <source file='/vm/crm.snapshot2' index='1'/>
      <backingStore type='file' index='3'>
        <format type='qcow2'/>
        <source file='/vm/crm.snapshot1'/>
        <backingStore type='file' index='4'>
          <format type='qcow2'/>
          <source file='/vm/crm.qcow2'/>
          <backingStore/>
        </backingStore>
      </backingStore>
...
```

(6) 启动虚拟机,这将恢复到快照 snapshot2 的状态,示例命令如下:

```
# virsh start crm
Domain crm started
```

### 6.6.5 合并、删除外部快照

当快照不用的时候,应及时将其删除。

与内部快照不同,目前删除外部快照还有些棘手:不能直接删除外部快照,只能手工进行删除,而且在手工删除之前,还需要先将快照进行合并。

有两种快照合并机制,分别如下。

(1) blockcommit:向前合并,合并 overlay 至 backing file。overlay 通常要小于其 backing file,所以合并的速度比较快。

(2) blockpull:向后合并,将 backing file 数据合并至 overlay。

下面通过实验来查看这两种合并机制。

**1. 使用 blockcommit 合并外部快照**

在实验环境中有一个虚拟机 CRM,它具有一个名为 crm.qcow2 的基础映像(原始映像),该映像包含 4 个外部快照。/vm/crm.snapshot4 是实时快照而且是活动快照,其余快照处于只读模式。

查看虚拟机及快照的状态,示例命令如下:

```
# virsh list --all | grep crm
 2     crm            running

# virsh domblklist crm
 Target       Source
------------------------------
 hda          -
 vda          /vm/crm.snapshot4

# virsh snapshot-list crm --parent
 Name         Creation Time                    State          Parent
------------------------------------------------------------------------
 snapshot1    2021-03-04 15:36:31 +0800    disk-snapshot
 snapshot2    2021-03-04 15:53:16 +0800    shutoff        snapshot1
 snapshot3    2021-03-04 20:54:50 +0800    disk-snapshot  snapshot2
 snapshot4    2021-03-05 07:03:58 +0800    disk-snapshot  snapshot3

# virsh snapshot-list crm --tree
snapshot1
  |
  +- snapshot2
      |
      +- snapshot3
          |
          +- snapshot4
```

```
# ls - lh /vm/crm. *
-rw-r--r--. 1 qemu qemu  81G Mar  4 15:36 /vm/crm.qcow2
-rw-------. 1 qemu qemu 3.9M Mar  4 15:42 /vm/crm.snapshot1
-rw-r--r--. 1 qemu qemu  15M Mar  5 06:59 /vm/crm.snapshot2
-rw-------. 1 qemu qemu 3.9M Mar  4 21:00 /vm/crm.snapshot3
-rw-------. 1 qemu qemu 194K Mar  5 07:03 /vm/crm.snapshot4
```

现在准备收缩整个快照链，即删除与此虚拟机关联的所有快照，示例命令如下：

```
# virsh blockcommit crm vda --pivot --active --verbose
Block commit: [100 %]
Successfully pivoted
```

选项的含义如下。

(1) active：触发对顶层文件(Top File)的两阶段活动提交。

(2) wait：等待作业完成(与 active 一起使用，等待作业同步)。

(3) pivot：等同于--active --wait。

(4) verbose：与--wait 一起使用，显示进度。

合并完成之后，检查虚拟机当前的状态及活动块设备，示例命令如下：

```
# virsh list --all | grep crm
 2     crm              running

# virsh domblklist crm
 Target      Source
--------------------------
 hda         -
 vda         /vm/crm.qcow2

# virsh dumpxml crm
...
    <disk type='file' device='disk'>
      <driver name='qemu' type='qcow2'/>
      <source file='/vm/crm.qcow2' index='5'/>
      <backingStore/>
      <target dev='vda' bus='virtio'/>
      <alias name='virtio-disk0'/>
      <address type='pci' domain='0x0000' bus='0x00' slot='0x07' function='0x0'/>
    </disk>
...
```

**注意**：当前的活动块设备是基础映像 crm.qcow2，并且所有写入都已切换到该映像，这意味着已成功地将快照映像合并到基础映像中。

查看虚拟机的快照，示例命令如下：

```
# virsh snapshot - list crm
 Name               Creation Time                    State
------------------------------------------------------------
 snapshot1     2021-03-04 15:36:31 +0800     disk-snapshot
 snapshot2     2021-03-04 15:53:16 +0800     shutoff
 snapshot3     2021-03-04 20:54:50 +0800     disk-snapshot
 snapshot4     2021-03-05 07:03:58 +0800     disk-snapshot

# ls -l /var/lib/libvirt/qemu/snapshot/crm/
total 44
-rw-------. 1 root root 10867 Mar  4 15:36 snapshot1.xml
-rw-------. 1 root root  5407 Mar  4 15:53 snapshot2.xml
-rw-------. 1 root root 11485 Mar  4 20:54 snapshot3.xml
-rw-------. 1 root root 10901 Mar  5 07:03 snapshot4.xml

# ls -lh  /vm/crm.*
-rw-r--r--. 1 qemu qemu  81G Mar  5 07:18 /vm/crm.qcow2
-rw-------. 1 qemu qemu 3.9M Mar  4 15:42 /vm/crm.snapshot1
-rw-r--r--. 1 qemu qemu  15M Mar  5 06:59 /vm/crm.snapshot2
-rw-------. 1 qemu qemu 3.9M Mar  4 21:00 /vm/crm.snapshot3
-rw-------. 1 root root 194K Mar  5 07:03 /vm/crm.snapshot4
```

此时会发现系统中还残留着快照的配置信息和文件，这需要手工删除。首先删除这些快照配置信息（快照的元数据），示例命令如下：

```
# virsh snapshot - delete crm snapshot1 --children --metadata
Domain snapshot snapshot1 deleted
```

命令中使用以下两个选项。

（1）children：删除快照和所有子快照，即删除 snapshot1，同时会将 snapshot2、snapshot3、snapshot4 也删除。

（2）metadata：仅删除 libvirt 元数据，而保留快照内容。

删除元数据其实是删除快照的配置文件。查看当前的快照信息，示例命令如下：

```
# virsh snapshot - list crm
 Name   Creation Time   State
------------------------------------

# ls -l /var/lib/libvirt/qemu/snapshot/crm/
total 0
```

还需要删除残留的快照文件，示例命令如下：

```
# ls -lh /vm/crm.*
-rw-r--r--. 1 qemu qemu  81G Mar  5 07:18 /vm/crm.qcow2
-rw-------. 1 qemu qemu 3.9M Mar  4 15:42 /vm/crm.snapshot1
-rw-r--r--. 1 qemu qemu  15M Mar  5 06:59 /vm/crm.snapshot2
-rw-------. 1 qemu qemu 3.9M Mar  4 21:00 /vm/crm.snapshot3
-rw-------. 1 root root 194K Mar  5 07:03 /vm/crm.snapshot4

# rm -i /vm/crm.snapshot*
rm: remove regular file '/vm/crm.snapshot1'? y
rm: remove regular file '/vm/crm.snapshot2'? y
rm: remove regular file '/vm/crm.snapshot3'? y
rm: remove regular file '/vm/crm.snapshot4'? y
```

在此示例中，使用 blockcommit 合并、删除了所有外部快照。当然也可以合并、删除某一个快照。在创建了很长的快照链之后，使用这种方法缩短链长度十分有用。

**2. 使用 blockpull 合并外部快照**

查看实验环境中虚拟机 CRM 的信息，示例命令如下：

```
# virsh list --all | grep crm
 2     crm            running

# virsh domblklist crm
 Target      Source
--------------------------------
 hda         -
 vda         /vm/crm.snapshot4
```

虚拟机处于运行状态。查看快照的信息，示例命令如下：

```
# virsh snapshot-list crm --parent
Name              Creation Time               State           Parent
--------------------------------------------------------------------------
 snapshot1    2021-03-07 12:00:51 +0800    disk-snapshot
 snapshot2    2021-03-07 12:05:11 +0800    disk-snapshot    snapshot1
 snapshot3    2021-03-07 12:06:51 +0800    disk-snapshot    snapshot2
 snapshot4    2021-03-07 12:07:30 +0800    disk-snapshot    snapshot3

# virsh snapshot-list crm --tree
snapshot1
  |
  +- snapshot2
      |
      +- snapshot3
          |
          +- snapshot4
```

```
# ls -lh /vm/crm*
-rw-r--r--. 1 qemu qemu 81G Mar  7 12:00 /vm/crm.qcow2
-rw-------. 1 qemu qemu 64M Mar  7 12:05 /vm/crm.snapshot1
-rw-------. 1 qemu qemu 78M Mar  7 12:06 /vm/crm.snapshot2
-rw-------. 1 qemu qemu 52M Mar  7 12:07 /vm/crm.snapshot3
-rw-------. 1 qemu qemu 57M Mar  7 12:14 /vm/crm.snapshot4
```

虚拟基础映像文件为 crm.qcow2，当前共有 4 个快照，活动的快照为 snapshot4。

如果想删除 snapshot1、snapshot2、snapshot3 这 3 个快照(删除之后快照的关系如图 6-7 所示)，则应当如何删除呢？

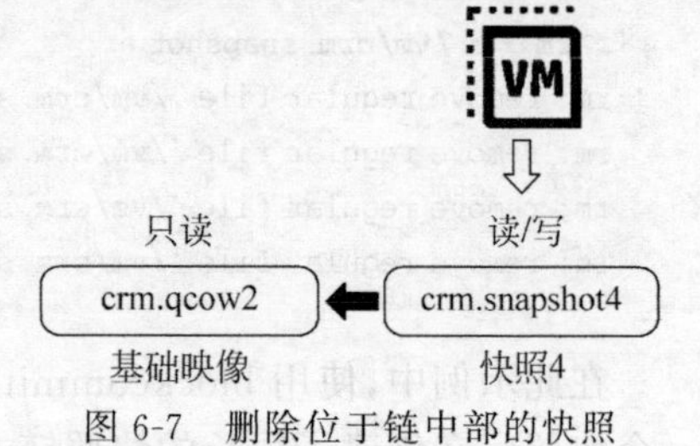


图 6-7 删除位于链中部的快照

如果使用 blockpull 子命令，则示例命令如下：

```
# virsh blockpull crm --path /vm/crm.snapshot4 --base /vm/crm.qcow2 \
    --wait --verbose
Block Pull: [100 %]
Pull complete
```

各选项含义如下。

(1) path：磁盘的绝对路径。

(2) base：如果仅对部分文件进行拉取合并，则用于指定链中 backing file 路径。

(3) wait：等待工作完成才退出。

(4) verbose：与--wait 一起显示进度。

检查相关文件的大小，示例命令如下：

```
# ls -lh /vm/crm*
-rw-r--r--. 1 qemu qemu 81G Mar  7 12:00 /vm/crm.qcow2
-rw-------. 1 qemu qemu 64M Mar  7 12:05 /vm/crm.snapshot1
-rw-------. 1 qemu qemu 78M Mar  7 12:06 /vm/crm.snapshot2
-rw-------. 1 qemu qemu 52M Mar  7 12:07 /vm/crm.snapshot3
-rw-------. 1 qemu qemu 84M Mar  7 12:18 /vm/crm.snapshot4
```

将 snapshot1、snapshot2、snapshot3 中的数据合并到 snapshot4 中，所以文件 snapshot4 变大了(不是简单的线性累加)。

查看虚拟机的磁盘信息，示例命令如下：

```
# virsh domblklist crm
 Target     Source
------------------------------
 hda        -
 vda        /vm/crm.snapshot4
```

```
# virsh dumpxml crm
...
    <disk type='file' device='disk'>
      <driver name='qemu' type='qcow2'/>
      <source file='/vm/crm.snapshot4' index='1'/>
      <backingStore type='file' index='5'>
        <format type='qcow2'/>
        <source file='/vm/crm.qcow2'/>
        <backingStore/>
      </backingStore>
...
```

当前活动快照 snapshot4 依赖于 crm.qcow2,再通过 qemu-img 命令进行验证,示例命令如下:

```
# virsh shutdown crm

# qemu-img info --backing-chain /vm/crm.snapshot4
image: /vm/crm.snapshot4
file format: qcow2
virtual size: 80 GiB (85899345920 Bytes)
disk size: 87.6 MiB
cluster_size: 65536
backing file: /vm/crm.qcow2
backing file format: qcow2
Format specific information:
   compat: 1.1
   lazy refcounts: false
   refcount bits: 16
   corrupt: false

image: /vm/crm.qcow2
file format: qcow2
virtual size: 80 GiB (85899345920 Bytes)
disk size: 4.56 GiB
cluster_size: 65536
Format specific information:
   compat: 1.1
   lazy refcounts: true
   refcount bits: 16
   corrupt: false
```

清理残留的元数据,示例命令如下:

```
# virsh snapshot-delete crm snapshot1 --metadata

# virsh snapshot-delete crm snapshot2 --metadata

# virsh snapshot-delete crm snapshot3 --metadata
```

清理残留的文件，示例命令如下：

```
# rm /vm/crm.snapshot1

# rm /vm/crm.snapshot2

# rm /vm/crm.snapshot3
```

如果希望通过 blockpull 收缩整个快照链，就不要使用 base 选项。示例命令如下：

```
# virsh blockpull crm --path /vm/crm.snapshot4  --wait --verbose
Block Pull: [100 %]
Pull complete

# virsh dumpxml crm
...
    <disk type='file' device='disk'>
      <driver name='qemu' type='qcow2'/>
      <source file='/vm/crm.snapshot4' index='1'/>
      <target dev='vda' bus='virtio'/>
      <alias name='virtio-disk0'/>
      <address type='pci' domain='0x0000' bus='0x00' slot='0x07' function='0x0'/>
    </disk>
...

# qemu-img info /vm/crm.snapshot4
image: /vm/crm.snapshot4
file format: qcow2
virtual size: 80 GiB (85899345920 Bytes)
disk size: 1.43 GiB
cluster_size: 65536
Format specific information:
    compat: 1.1
    lazy refcounts: false
    refcount bits: 16
    corrupt: false
```

现在虚拟机只使用了/vm/crm.snapshot4 这 1 个映像文件。

## 6.7 虚拟机备份脚本示例

一个完整的备份方案既要防止逻辑错误也要能够抵御物理故障，所以单纯使用快照并不是完整的备份方案，因为它仅仅防止逻辑故障。例如：如果基础映像文件发生物理损坏，则依赖它的所有快照将全部无效，而且虚拟机带着快照长时间运行，也会影响性能。

在备份虚拟机时，还要考虑业务的连续性及备份介质的存储。

下面提供了一个示例脚本，它会将虚拟机配置文件和磁盘映像文件备份到 NFS 服务器的目录。其主要思路如下：

(1) 通过 virsh 的 dumpxml 子命令备份虚拟机配置文件。

(2) 创建临时性外部快照，这样可以安全地将原有映像文件复制到备份目录。

(3) 复制时使用 qemu-img 的 convert 子命令，将备份格式设置为 QCOW2 格式以减少磁盘空间的占用。

(4) 复制完成后，将快照合并到原始映像文件。

(5) 删除临时快照的元数据和文件。

示例脚本如下：

```
#!/bin/bash
#设置参数
#备份目录
BACKUP_DIR=/nfs/backup
#时间戳,用于文件名
TIMESTAMP=`date +%Y%m%d-%H%M%S`
#要备份虚拟机及映像文件的名称
ACTIVEVM="crm"
DISK_PATH_ITEM="/vm/crm.qcow2"

#如果虚拟机没有启动,则启动它
if virsh domstate $ACTIVEVM | grep running
  then
    VMSTATE="running"
  else
    VMSTATE="shutoff"
    virsh start $ACTIVEVM
    sleep 2
  fi

#备份操作
#备份配置文件
virsh dumpxml $ACTIVEVM > $BACKUP_DIR/$ACTIVEVM-$TIMESTAMP.xml

#创建临时外部快照
```

```
virsh snapshot-create-as \
    --domain $ACTIVEVM tmp-ext-snap-$TIMESTAMP \
    --disk-only --atomic --quiesce
#增加时间间隔是为了防止写操作过于频繁
sleep 2

#将映像文件保存并收缩到备份目录
FILENAME=`basename $DISK_PATH_ITEM`
qemu-img convert -O qcow2 -c $DISK_PATH_ITEM $BACKUP_DIR/$ACTIVEVM-$FILENAME.
$TIMESTAMP.bak
sleep 2

#合并快照
DISK_PATH=`virsh domblklist $ACTIVEVM | grep -e vd -e sd | grep -e '/' | awk '{print $2}'`
TEMP_SNAPSHOT_FILE=$DISK_PATH
virsh blockcommit $ACTIVEVM $DISK_PATH --active --verbose --pivot
sleep 2

#删除临时快照的元数据
TEMP_SNAPSHOT=`virsh snapshot-list $ACTIVEVM | grep tmp-ext-snap | awk '{print $1}'`
virsh snapshot-delete $ACTIVEVM $TEMP_SNAPSHOT --metadata

#删除临时的快照文件
rm $TEMP_SNAPSHOT_FILE

#如果虚拟机的原来状态是关闭的,就关闭它
if [ $VMSTATE = "shutoff" ];
  then
    virsh shutdown $ACTIVEVM
  fi
```

**注意**：此脚本是一个简化的示例脚本，不包含错误控制机制，所以不能直接用于正式的生产环境。

检查备份效果，示例命令如下：

```
#ls -lh /nfs/backup/
total 546M
-rw-r--r--. 1 root root 6.6K Mar  7 21:21 crm-20210307-212112.xml
-rw-r--r--. 1 root root 546M Mar  7 21:22 crm-crm.qcow2.20210307-212112.bak

#head /nfs/backup/crm-20210307-212112.xml
<domain type='kvm' id='1'>
  <name>crm</name>
  <uuid>8f1149d2-9dc8-4e9b-b120-df28a0bb704a</uuid>
  <metadata>
```

```
    <libosinfo:libosinfo xmlns:libosinfo="http://libosinfo.org/xmlns/libvirt/domain/1.0">
      <libosinfo:os id="http://centos.org/centos/6.10"/>
    </libosinfo:libosinfo>
  </metadata>
  <memory unit='KiB'>1048576</memory>
  <currentMemory unit='KiB'>1048576</currentMemory>

 # qemu-img info /nfs/backup/crm-crm.qcow2.20210307-212112.bak
image: /nfs/backup/crm-crm.qcow2.20210307-212112.bak
file format: qcow2
virtual size: 80 GiB (85899345920 Bytes)
disk size: 545 MiB
cluster_size: 65536
Format specific information:
    compat: 1.1
    lazy refcounts: false
    refcount bits: 16
    corrupt: false

 # qemu-img check /nfs/backup/crm-crm.qcow2.20210307-212112.bak
No errors were found on the image.
22382/1310720 = 1.71% allocated, 92.90% fragmented, 91.97% compressed clusters
Image end offset: 571998208
```

## 6.8　本章小结

本章讲解了 RPO、RTO 等备份恢复的基本概念，然后介绍了内存快照、磁盘内部快照和磁盘外部快照的特点及应用场景，最后通过一个脚本将这些技术组合起实现了一个简单的备份功能。

# 第7章 oVirt(RHV)安装与基本管理

oVirt(www.ovirt.org)是一个开源的虚拟化管理项目。如果说KVM是汽车引擎,则oVirt就是一辆完整的汽车。管理员通过集中式的图形用户界面或REST API可以管理整个虚拟化基础架构,包括宿主机、虚拟机、网络、存储和用户。

oVirt是Red Hat Virtualization(RHV)的上游版本,这类似于Fedora和RHEL的关系。oVirt是一个社区项目,由Red Hat赞助。新功能、新特性首先会应用在oVirt中,等经过测试稳定后,这些功能会被合并到RHV中。与RHV相比,oVirt没有商业支持。

oVirt是一个比较大的项目,限于篇幅,本章仅涉及安装与基本管理。

**本章要点**

- oVirt结构。
- oVirt安装。
- 数据中心管理。
- 存储管理。
- 虚拟机管理。
- 实现迁移与高可用。
- 用户与权限管理。
- 备份与恢复。

## 7.1 oVirt结构

oVirt主要包括以下4个关键组件。

(1) oVirt管理器(Manager):提供图形用户界面和REST API来管理环境中的资源服务,它可以安装在物理机或虚拟机上。

(2) 虚拟化宿主机:既可以是RHEL/CentOS主机,也可以是专用的oVirt Node(为oVirt进行了专门裁剪和优化的RHEL/CentOS版本)。

(3) 共享存储:用于存储与虚拟机有关的数据。

(4) 数据仓库:用于存储oVirt管理器的配置信息和统计数据。

根据 oVirt 管理器部署的位置的不同，可以将 oVirt 架构分为 2 类。

**1. 自托管引擎(Self-Hosted Engine)架构**

这是推荐的架构，oVirt 管理器运行在一台虚拟机中，如图 7-1 所示。这种架构至少需要两台宿主机，应在每台宿主机上安装代理程序 VDSM(Virtual Desktop and Server Manager)和高可用服务。这种架构无须额外为管理器配置高可用。

这种架构需要有共享存储，所有宿主机都必须可以访问该存储。

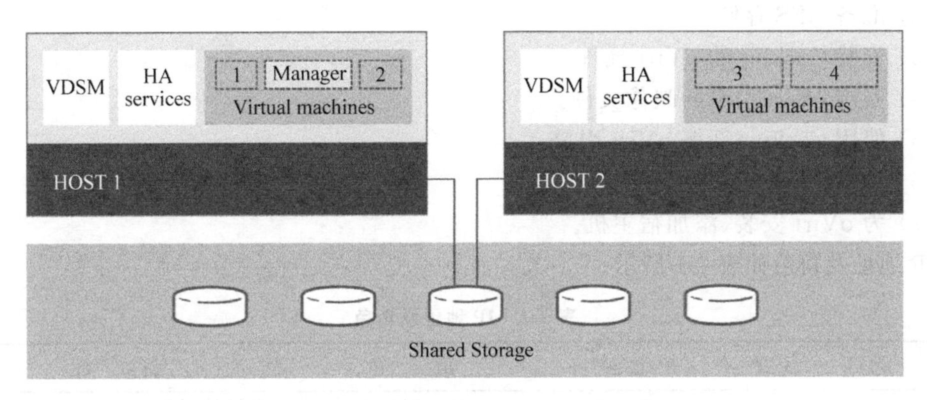


图 7-1 自托管引擎架构(来自 oVirt 网站)

**2. 独立的管理器架构**

oVirt 管理器运行在一台物理服务器之上，或者单独的虚拟化环境中所托管的虚拟机上，如图 7-2 所示。这种独立的管理器易于部署和管理，但需要额外的物理服务器，同时还需要保证其高可用。

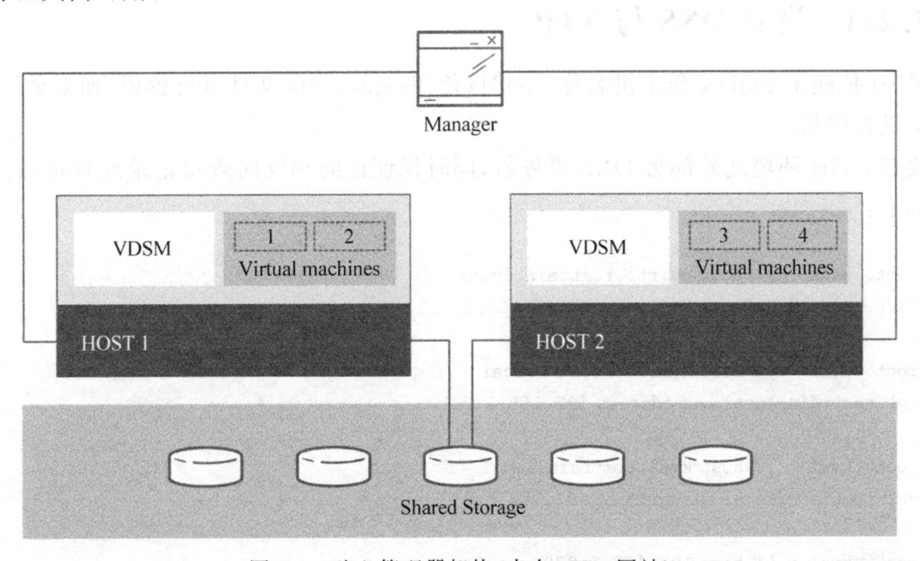


图 7-2 独立管理器架构(来自 oVirt 网站)

# 7.2 oVirt 安装

本章将构建一个自托管引擎架构的实验环境，拓扑结构与图 7-1 类似。主要包括以下步骤：

（1）准备 DNS 与 NTP。

（2）准备 NFS 存储。

（3）安装 Cockpit 的 oVirt 插件。

（4）安装 oVirt 引擎的映像文件。

（5）使用 Cockpit 部署 oVirt 引擎。

（6）访问管理门户。

（7）为 oVirt 安装、添加宿主机。

IP 地址及角色如表 7-1 所示。

表 7-1 IP 地址及角色

| FQDN | IP 地址 | 角 色 | 备 注 |
|---|---|---|---|
| ovirt1. tomtrain. local | 192.168.1.230 | oVirt 引擎 | 自托管引擎 |
| kvm1. tomtrain. local | 192.168.1.231 | 虚拟化宿主机 | CentOS 8.3.2011 |
| kvm2. tomtrain. local | 192.168.1.232 | 虚拟化宿主机 | CentOS 8.3.2011 |
| kvm3. tomtrain. local | 192.168.1.233 | 虚拟化宿主机 | oVirt node4.4.4-2021020810 |
| stor1. tomtrain. local | 192.168.1.235 | 存储服务器 | CentOS 8.3.2011 |

## 7.2.1 准备 DNS 与 NTP

oVirt 依赖于 FQDN 和主机名称，不建议使用/etc/hosts 文件进行解析，而要使用 DNS 服务器进行解析。

要在 oVirt 环境之外部署 DNS 服务器，同时保证正向和反向查找记录配置正确。示例命令如下：

```
[root@kvm1 ~]# host ovirt1.tomtrain.local
ovirt1.tomtrain.local has address 192.168.1.230

[root@kvm1 ~]# host kvm1.tomtrain.local
kvm1.tomtrain.local has address 192.168.1.231

[root@kvm1 ~]# host kvm2.tomtrain.local
kvm2.tomtrain.local has address 192.168.1.232

[root@kvm1 ~]# host 192.168.1.230
230.1.168.192.in-addr.arpa domain name pointer ovirt1.tomtrain.local.
```

```
[root@kvm1 ~]# host 192.168.1.231
231.1.168.192.in-addr.arpa domain name pointer kvm1.tomtrain.local.

[root@kvm1 ~]# host 192.168.1.232
232.1.168.192.in-addr.arpa domain name pointer kvm2.tomtrain.local.
```

由于宿主机要组成群集，所以要保证时钟的准确。可以使用外部 NTP 服务器进行校时。

### 7.2.2 准备 NFS 存储

数据中心(Data Center)是 oVirt 环境中所有物理和逻辑资源的最高级别的容器，它是群集、虚拟机、存储域和网络的集合。一个数据中心至少有一个数据存储域(Data Storage Domain)，它包含独立映像存储库的逻辑实体。每个存储域用于存储虚拟磁盘或 ISO 映像，以及用于导入和导出虚拟机映像。

目前 oVirt 支持以下存储类型：

(1) NFS 存储。

(2) iSCSI 存储。

(3) 光纤通道(FCP)存储。

(4) Gluster 储存。

自托管引擎必须具有额外的数据域，至少要有 74GB 空间用于引擎虚拟机。安装自托管引擎之前，需要准备好存储。

本次实验将使用 NFS 存储，其配置文件如下：

```
[root@stor1 ~]# cat /etc/exports
/vmdata *(rw,no_root_squash,sync)
```

oVirt 引擎中有多个用户和组，它需要以账号 vdsm(UID 是 36)和组 kvm(GID 是 36)的身份来对数据存储进行文件操作，所以需要在 NFS 服务器中创建相同名称与 ID 的用户及组，示例命令如下：

```
[root@stor1 ~]# groupadd kvm -g 36

[root@stor1 ~]# useradd vdsm -u 36 -g 36
```

还需要修改导出目录的所有权及权限，示例命令如下：

```
[root@stor1 ~]# chown -R 36:36 /vmdata/

[root@stor1 ~]# chmod 0755 /vmdata/

[root@stor1 ~]# ll -d /vmdata/
drwxr-xr-x. 2 vdsm kvm 6 Feb 20 09:08 /vmdata/
```

### 7.2.3 安装 Cockpit 的 oVirt 插件

既可以在 RHEL/CentOS 宿主机上部署自托管引擎，也可以在 oVirt Node 上安装。本次实验将在 CentOS 8.3 上进行部署。安装操作系统时采用最小化安装，安装后升级到最新的版本。查看版本信息，示例命令如下：

```
[root@kvm1 ~]# cat /etc/redhat-release
CentOS Linux release 8.3.2011

[root@kvm1 ~]# uname -a
Linux kvm1 4.18.0-240.10.1.el8_3.x86_64 #1 SMP Mon Jan 18 17:05:51 UTC 2021 x86_64 x86_64
x86_64 GNU/Linux
```

除了基本软件仓库和 oVirt 引擎所需的软件仓库外，不需要启用其他软件仓库。查看当前的软件仓库，示例命令如下：

```
[root@kvm1 ~]# dnf repolist
repo id                                    repo name
appstream                                  CentOS Linux 8 - AppStream
baseos                                     CentOS Linux 8 - BaseOS
extras                                     CentOS Linux 8 - Extras
```

oVirt 的组件比较多，最简单的方法是通过安装 cockpit-ovirt-dashboard 软件包进行安装。它是一个 oVirt 的 Cockpit 插件，通过 dnf 或 yum 安装时会自动安装所依赖的软件包。

当前 CentOS 软件仓库中没有这个软件包，所以需要先安装官方版本的 oVirt 4.4 的软件仓库配置文件，示例命令如下：

```
[root@kvm1 ~]# dnf -y install \
    https://resources.ovirt.org/pub/yum-repo/ovirt-release44.rpm
```

这将新增加多个 oVirt 的软件仓库。查看当前软件仓库的配置，示例命令如下：

```
[root@kvm1 ~]# dnf repolist
repo id                                    repo name
appstream                                  CentOS Linux 8 - AppStream
baseos                                     CentOS Linux 8 - BaseOS
extras                                     CentOS Linux 8 - Extras
ovirt-4.4                                  Latest oVirt 4.4 Release
ovirt-4.4-advanced-virtualization          Advanced Virtualization packages for x86_64
ovirt-4.4-CentOS-gluster7                  CentOS-8 - Gluster 7
ovirt-4.4-CentOS-nfv-openvswitch           CentOS-8 - NFV OpenvSwitch
```

```
ovirt-4.4-CentOS-opstools             CentOS-8 - OpsTools - collectd
ovirt-4.4-CentOS-ovirt44              CentOS-8 - oVirt 4.4
ovirt-4.4-copr:copr.fedorainfracloud.org:mdbarroso:ovsdbapp
Copr repo for ovsdbapp owned by mdbarroso
ovirt-4.4-copr:copr.fedorainfracloud.org:networkmanager:NetworkManager-1.26 Copr repo
for NetworkManager-1.26 owned by networkmanager
ovirt-4.4-copr:copr.fedorainfracloud.org:nmstate:nmstate-0.3
Copr repo for nmstate-stable owned by nmstate
ovirt-4.4-copr:copr.fedorainfracloud.org:sac:gluster-ansible
Copr repo for gluster-ansible owned by sac
ovirt-4.4-copr:copr.fedorainfracloud.org:sbonazzo:EL8_collection
Copr repo for EL8_collection owned by sbonazzo
ovirt-4.4-epel                        Extra Packages for Enterprise Linux 8 - x86_64
ovirt-4.4-virtio-win-latest
virtio-win builds roughly matching what will be shipped in upcoming RHEL
powertools                            CentOS Linux 8 - PowerTools
```

查看cockpit-ovirt-dashboard软件的信息，示例命令如下：

```
[root@kvm1 ~]# dnf list | grep  cockpit-ovirt-dashboard
cockpit-ovirt-dashboard.noarch     0.14.17-1.el8     @ovirt-4.4

[root@kvm1 ~]# dnf info cockpit-ovirt-dashboard
Last metadata expiration check: 0:10:50 ago on Fri 19 Feb 2021 05:09:59 PM CST.
Available Packages
Name         : cockpit-ovirt-dashboard
Version      : 0.14.17
Release      : 1.el8
Architecture : noarch
Size         : 3.5 M
Source       : cockpit-ovirt-0.14.17-1.el8.src.rpm
Repository   : ovirt-4.4
Summary      : Dashboard for Cockpit based on oVirt
URL          : https://gerrit.ovirt.org/gitweb?p=cockpit-ovirt.git;a=summary
License      : ASL 2.0
Description  : This package provides a Cockpit dashboard for use with oVirt.
```

cockpit-ovirt-dashboard是一个Cockpit的插件，本身并不大，但是由于它依赖Ansible、oVirt、pacemaker、fence、virt-v2v等组件，所以会安装很多软件包(大约230个)。示例命令如下：

```
[root@kvm1 ~]# dnf -y install cockpit-ovirt-dashboard
```

安装完成后，会在Cockpit的左边导航中看到一个新的功能菜单，如图7-3所示。

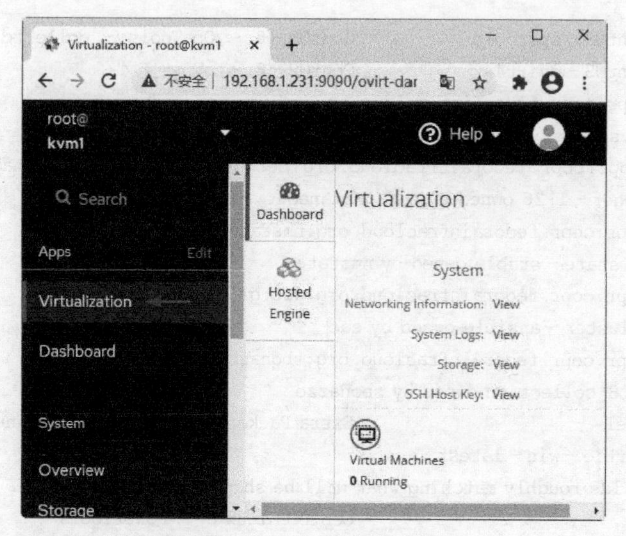


图 7-3 oVirt 的 Cockpit 插件

## 7.2.4 安装 oVirt 引擎的映像文件

Cockpit 需要通过映像文件来部署 oVirt 引擎虚拟机，这个映像包含在 ovirt-engine-appliance 软件包中，可以通过 dnf 或 yum 命令进行安装，示例命令如下：

```
[root@kvm1 ~]# dnf info ovirt-engine-appliance
Last metadata expiration check: 1:09:08 ago on Fri 19 Feb 2021 12:36:55 PM CST.
Available Packages
Name         : ovirt-engine-appliance
Version      : 4.4
Release      : 20201221110111.1.el8
Architecture : x86_64
Size         : 2.5 G
Source       : ovirt-engine-appliance-4.4-20201221110111.1.el8.src.rpm
Repository   : ovirt-4.4
Summary      : The oVirt Engine Appliance image (OVA)
URL          : https://www.ovirt.org/
License      : GPLv2
Description  : This package contains the prebuild oVirt Engine appliance image. It is intended
to be used with hosted-engine setup.

[root@kvm1 ~]# dnf -y install ovirt-engine-appliance
```

这个文件比较大(2.5GB 左右)，联机安装对网络带宽要求比较高，还可以采用手工下载 RPM 文件进行安装。

首先从 oVirt 软件仓库的配置文件中获得仓库的 URL，示例命令如下：

```
[root@kvm1 ~]# cat /etc/yum.repos.d/ovirt-4.4.repo
[ovirt-4.4]
name=Latest oVirt 4.4 Release
#baseURL=https://resources.ovirt.org/pub/ovirt-4.4/rpm/el$releasever/
mirrorlist=https://mirrorlist.ovirt.org/mirrorlist-ovirt-4.4-el$releasever
enabled=1
gpgcheck=1
gpgkey=file:///etc/pki/rpm-gpg/RPM-GPG-ovirt-4.4
```

访问 https://resources.ovirt.org/pub/ovirt-4.4/rpm/，找到文件的 URL https://resources.ovirt.org/pub/ovirt-4.4/rpm/el8/x86_64/ovirt-engine-appliance-4.4-20201221110111.1.el8.x86_64.rpm。

下载之后就可以手工进行安装了，示例命令如下：

```
[root@kvm1 ~]# ls -l ovirt-engine-appliance-4.4-20201221110111.1.el8.x86_64.rpm
-rw-r--r--. 1 root root 2679784778 Feb 19 14:10 ovirt-engine-appliance-4.4-
20201221110111.1.el8.x86_64.rpm

[root@kvm1 ~]# rpm -ivh \
    ovirt-engine-appliance-4.4-20201221110111.1.el8.x86_64.rpm
Verifying...                          ################################# [100%]
Preparing...                          ################################# [100%]
Updating / installing...
   1:ovirt-engine-appliance-4.4-202012################################# [100%]
```

查看 ovirt-engine-appliance 软件包中的文件，示例命令如下：

```
[root@kvm1 ~]# rpm -ql ovirt-engine-appliance
/etc/ovirt-hosted-engine/10-appliance.conf
/usr/share/ovirt-engine-appliance
/usr/share/ovirt-engine-appliance/ovirt-engine-appliance-4.4-20201221110111.1.
el8.ova
```

软件包中包含了 oVirt 引擎的 OVA 格式的映像文件。

### 7.2.5 使用 Cockpit 部署 oVirt 引擎

有了 oVirt 引擎的映像文件，就可以进行部署了。以 root 用户的身份登录到 Cockpit，单击 Virtualization 下的 Hosted Engine，单击托管引擎选项下的 Start 按钮，如图 7-4 所示。

自托管引擎部署共分为 5 个步骤，如图 7-5 所示。

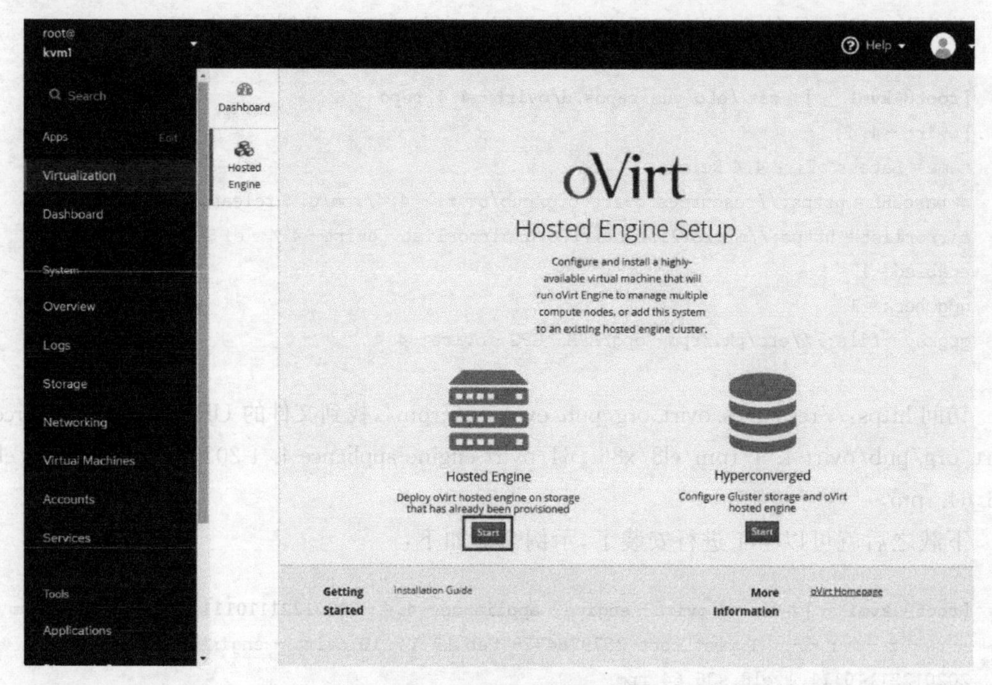


图 7-4 自托管引擎管理界面

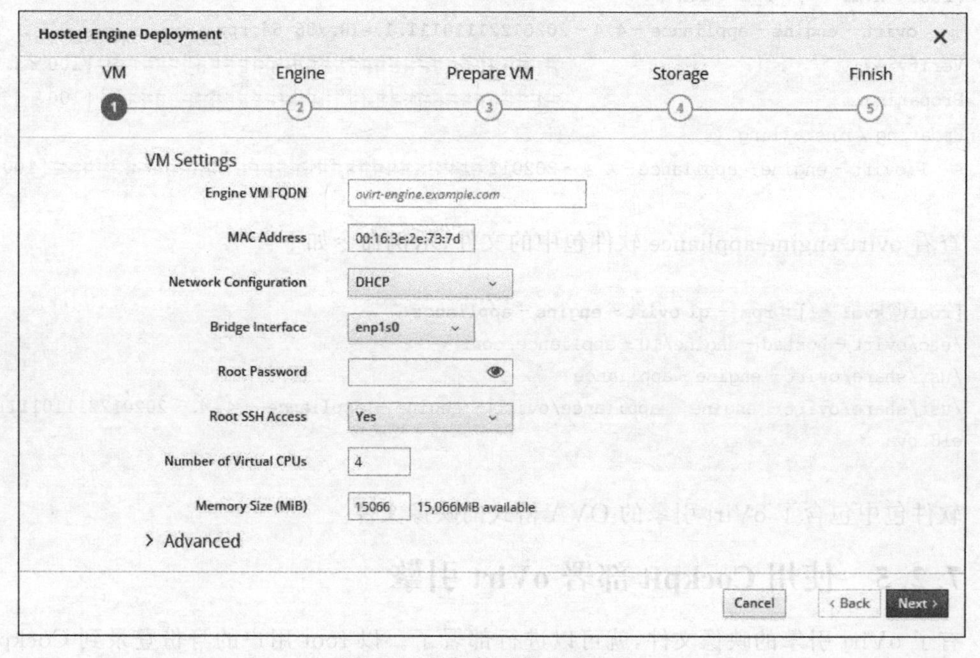


图 7-5 自托管引擎部署向导

## 1. 虚拟机设置

首先需要设置运行 oVirt 引擎虚拟机的信息，如图 7-6 所示。

VM Settings

| | |
|---|---|
| Engine VM FQDN | ovirt1.tomtrain.local |
| MAC Address | 00:16:3e:4e:e5:71 |
| Network Configuration | Static |
| VM IP Address | 192.168.1.230 / 24 |
| Gateway Address | 192.168.1.254 |
| DNS Servers | 192.168.1.11 |
| Bridge Interface | enp1s0 |
| Root Password | •••••• |
| Root SSH Access | Yes |
| Number of Virtual CPUs | 1 |
| Memory Size (MiB) | 5120　14,874MiB available |

图 7-6　自托管引擎部署向导-虚拟机设置

(1) 输入引擎虚拟机的 FQDN。输入后，向导会进行验证，验证通过后会出现绿色的对号。

(2) 使用随机生成的虚拟机网卡的 MAC 地址，也可以输入自定义 MAC 地址。

(3) 从下拉列表中选择 DHCP 或静态 IP 地址。如果选中的是静态 IP 地址，则还需要手工输入相应的配置。

(4) 从下拉列表中选择网桥接口。

(5) 输入并确认虚拟机 root 用户的密码。

(6) 允许 root 用户通过 SSH 访问。

(7) 输入虚拟机的 vCPU 数量。

(8) 输入内存大小(MiB)，在输入字段旁边显示可用内存。虚拟机最小内存为 4096MiB。

单击 Advanced 展开高级字段，这些是可选项，如图 7-7 所示。

(1) 输入 root 用户的 SSH 公共密钥，这将用于对引擎虚拟机的 root 访问。

(2) 如果选中 Edit Hosts File 复选框，则会在/etc/hosts 文件添加引擎虚拟机和宿主机的条目。

(3) 自定义管理网桥名称，默认名称为 ovirtmgmt。

(4) 输入管理网桥的网关地址。

(5) 输入引擎虚拟机的主机 FQDN。

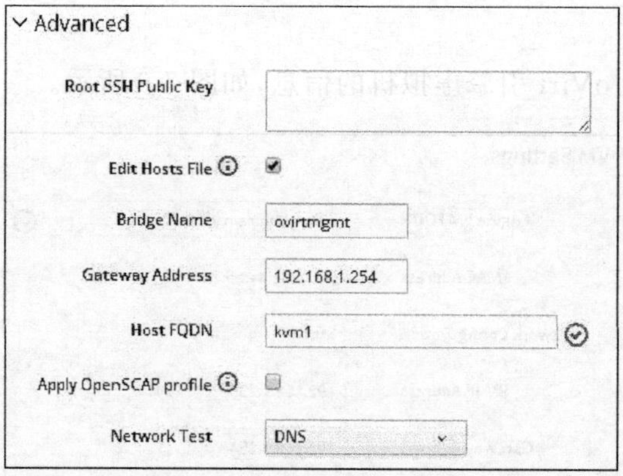

Advanced
Root SSH Public Key
Edit Hosts File
Bridge Name ovirtmgmt
Gateway Address 192.168.1.254
Host FQDN kvm1
Apply OpenSCAP profile
Network Test DNS

图 7-7 自托管引擎部署向导-虚拟机高级设置

(6) 单击 Next 按钮继续。

### 2. oVirt 引擎设置

设置 oVirt 管理员 admin@internal 的密码。

配置事件通知,包括以下几个配置:

(1) 输入 SMTP 服务器的服务器名称和端口号。

(2) 输入发件人的电子邮件地址。

(3) 输入收件人的电子邮件地址。

单击 Next 按钮继续。

### 3. 虚拟机准备

查看引擎及其虚拟机的配置,示例信息如下:

```
Please review the configuration. Once you click the 'Prepare VM' button, a local virtual machine
will be started and used to prepare the management services and their data. This operation may
take some time depending on your hardware.
VM
Engine FQDN:ovirt1.tomtrain.local
MAC Address:00:16:3e:4d:57:03
Network Configuration:Static
VM IP Address:192.168.1.230/24
Gateway Address:192.168.1.254
DNS Servers:192.168.1.11
Root User SSH Access:yes
Number of Virtual CPUs:1
Memory Size (MiB):5120
Root User SSH Public Key:(None)
Add Lines to /etc/hosts:yes
Bridge Name:ovirtmgmt
Apply OpenSCAP profile:no
Engine
```

```
SMTP Server Name:localhost
SMTP Server Port Number:25
Sender E-Mail Address:root@localhost
Recipient E-Mail Addresses:root@localhost
```

如果需要调整，则可单击 Back 按钮。如果信息正确，则可单击 Prepare VM 按钮，单击此按钮后将开始部署虚拟机、安装 oVirt 引擎。由于步骤多，而且还需要从 oVirt 的软件仓库下载软件包，所以时间会比较长，如图 7-8 所示。

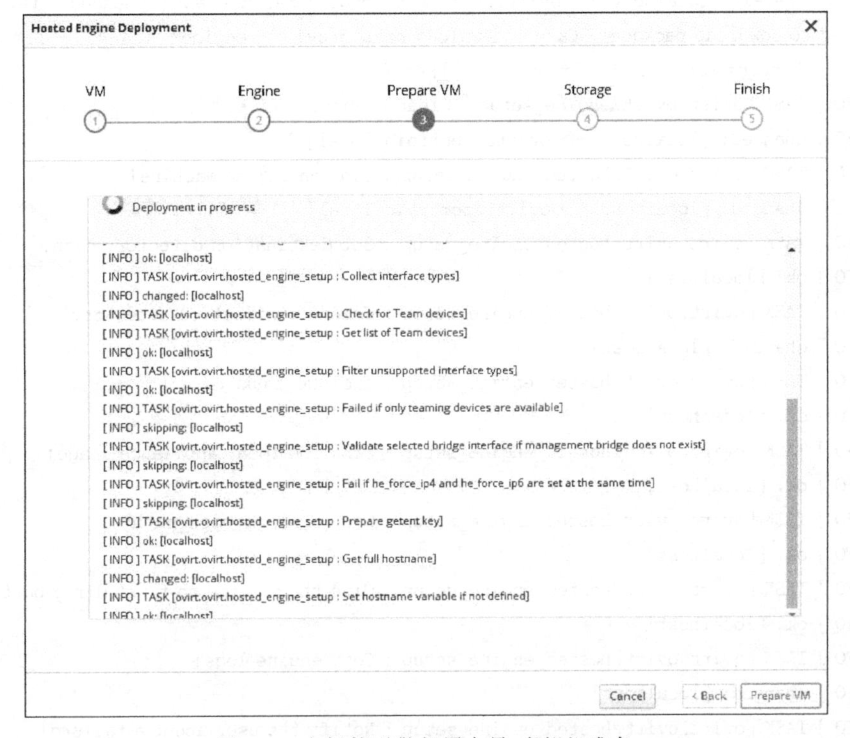

图 7-8 自托管引擎部署向导-虚拟机准备

在部署虚拟机时，如果出现错误，则会终止部署。可以根据提示信息进行排错，然后单击 Prepare VM 按钮重新部署。

常见错误 1：主机名与管理网桥接口不唯一匹配。

```
...
[ INFO ] TASK [ovirt.ovirt.hosted_engine_setup : Ensure the resolved address resolves only on the selected interface]
[ ERROR ] fatal: [localhost]: FAILED! => {"changed": false, "msg": "hostname 'kvm1' doesn't uniquely match the interface 'enp1s0' selected for the management bridge; it matches also interface with IP ['fe80::82fa:1693:cca9:5208', '192.168.122.1']. Please make sure that the hostname got from the interface for the management network resolves only there.\n"}
```

解决方法是在/etc/hosts 中添加主机名的记录，示例命令如下：

```
[root@kvm1 ~]# cat /etc/hosts
127.0.0.1   localhost localhost.localdomain localhost4 localhost4.localdomain4
::1         localhost localhost.localdomain localhost6 localhost6.localdomain6
192.168.1.231 kvm1
```

常见错误 2：从 oVirt 软件仓库下载软件出错，从而导致部署失败。

```
[ ERROR ] fatal: [localhost -> ovirt1.tomtrain.local]: FAILED! => {"changed": false, "msg":
"Failed to download packages: Cannot download noarch/ovirt-engine-4.4.4.7-1.el8.noarch.
rpm: All mirrors were tried", "results": []}
[ INFO ] TASK [ovirt.ovirt.engine_setup : Clean temporary files]
[ INFO ] changed: [localhost -> ovirt1.tomtrain.local]
[ INFO ] TASK [ovirt.ovirt.hosted_engine_setup : Sync on engine machine]
[ INFO ] changed: [localhost -> ovirt1.tomtrain.local]
[ INFO ] TASK [ovirt.ovirt.hosted_engine_setup : Set destination directory path]
[ INFO ] ok: [localhost]
[ INFO ] TASK [ovirt.ovirt.hosted_engine_setup : Create destination directory]
[ INFO ] changed: [localhost]
[ INFO ] TASK [ovirt.ovirt.hosted_engine_setup : include_tasks]
[ INFO ] ok: [localhost]
[ INFO ] TASK [ovirt.ovirt.hosted_engine_setup : Find the local appliance image]
[ INFO ] ok: [localhost]
[ INFO ] TASK [ovirt.ovirt.hosted_engine_setup : Set local_vm_disk_path]
[ INFO ] ok: [localhost]
[ INFO ] TASK [ovirt.ovirt.hosted_engine_setup : Give the vm time to flush dirty buffers]
[ INFO ] ok: [localhost]
[ INFO ] TASK [ovirt.ovirt.hosted_engine_setup : Copy engine logs]
[ INFO ] changed: [localhost]
[ INFO ] TASK [ovirt.ovirt.hosted_engine_setup : Notify the user about a failure]
[ ERROR ] fatal: [localhost]: FAILED! => {"changed": false, "msg": "There was a failure
deploying the engine on the local engine VM. The system may not be provisioned according to the
playbook results: please check the logs for the issue, fix accordingly or re-deploy from
scratch.\n"}
```

解决方法是重试或切换互联网链路。

部署成功后，单击 Next 按钮继续，如图 7-9 所示。

#### 4. 存储域设置

从下拉列表中选择存储类型，然后输入自托管引擎存储域的详细信息，如图 7-10 所示。

本实验选择 NFS。在 Storage Connection 字段中输入完整的地址和存储路径。

如果需要，则可输入其他的挂载选项。单击 Next 按钮继续。

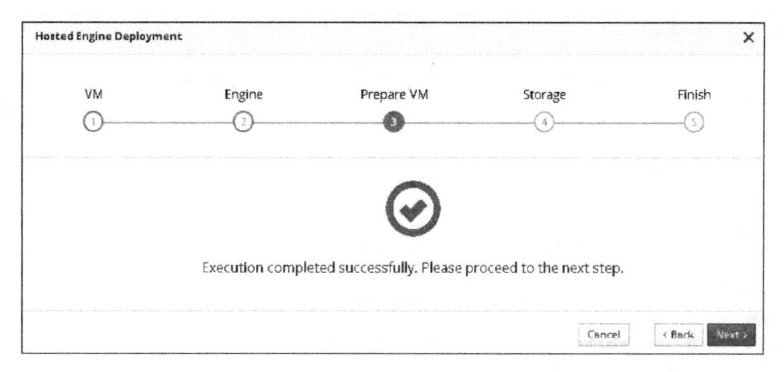


图 7-9 自托管引擎部署向导-虚拟机准备完毕

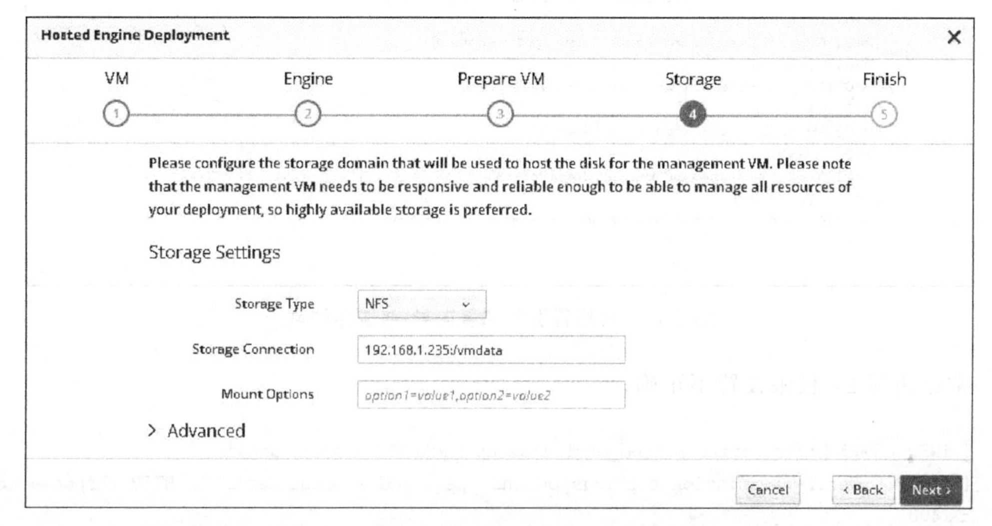


图 7-10 自托管引擎部署向导-存储域设置

## 5. 完成

查看存储配置。如果详细信息正确,则可单击 Finish Deployment 按钮开始部署,如图 7-11 所示。

在部署存储域时,如果出现错误,则会终止部署。可以根据提示信息进行排错,然后重新部署。

常见错误 1:存储空间不足。

```
[ ERROR ] fatal: [localhost]: FAILED! => {"changed": false, "msg": "Error: the target storage domain contains only 26.0GiB of available space while a minimum of 61.0GiB is required If you wish to use the current target storage domain by extending it, make sure it contains nothing before adding it."}
```

解决方法是增加存储空间。

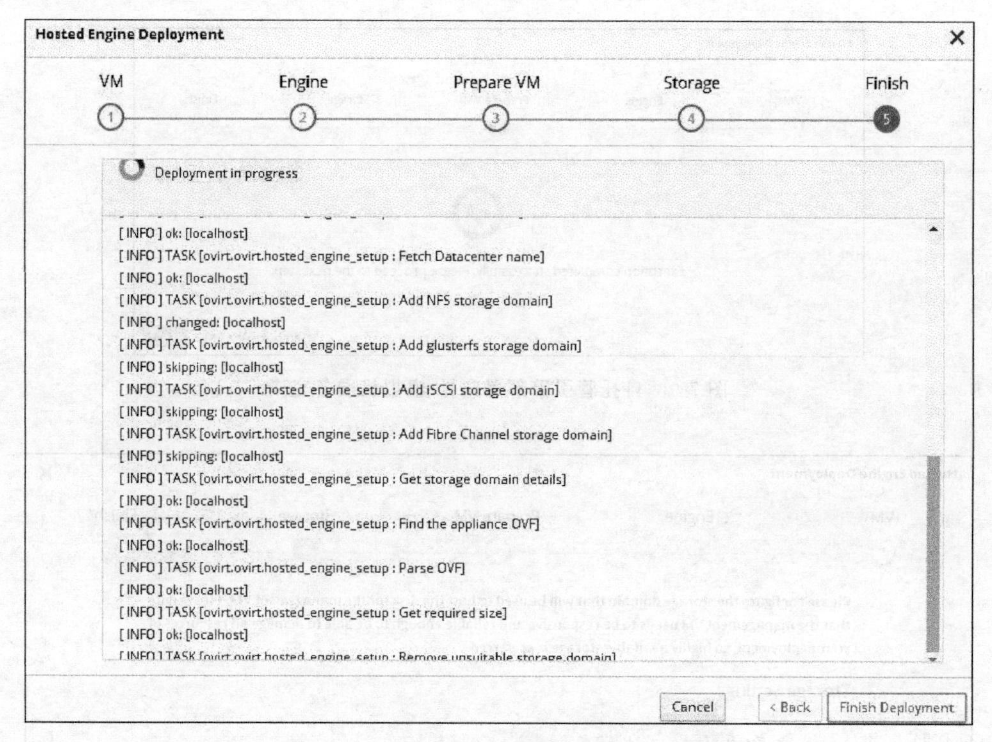


图 7-11　自托管引擎部署向导-部署存储域

常见错误 2：权限设置不正确。

```
[ INFO ] TASK [ovirt.ovirt.hosted_engine_setup : Add NFS storage domain]
[ ERROR ] Verify permission settings on the specified storage path.]". HTTP response code
is 400.
[ ERROR ] fatal: [localhost]: FAILED! => {"changed": false, "msg": "Fault reason is \"
Operation Failed\". Fault detail is \"[Permission settings on the specified path do not allow
access to the storage.\nVerify permission settings on the specified storage path.]\". HTTP
response code is 400."}
```

解决方法是在 NFS 服务器上检查以下配置：

（1）账号 vdsm 和组 kvm。

（2）导出目录的宿主及权限。

（3）NFS 导出目录的配置。

部署完成后，单击 Close 按钮。

## 7.2.6　访问管理门户

oVirt 提供多种管理方式，其中最常用的是通过 Web 浏览器访问管理门户。建议使用对 HTML5 支持比较好的浏览器，例如：Chrome、Firefox、国产浏览器的极速模式（Blink、

Webkit)等。

在浏览器网址栏中输入安装过程中所设置的 FQDN 名称并按 Enter 键，会出现欢迎页面，如图 7-12 所示。

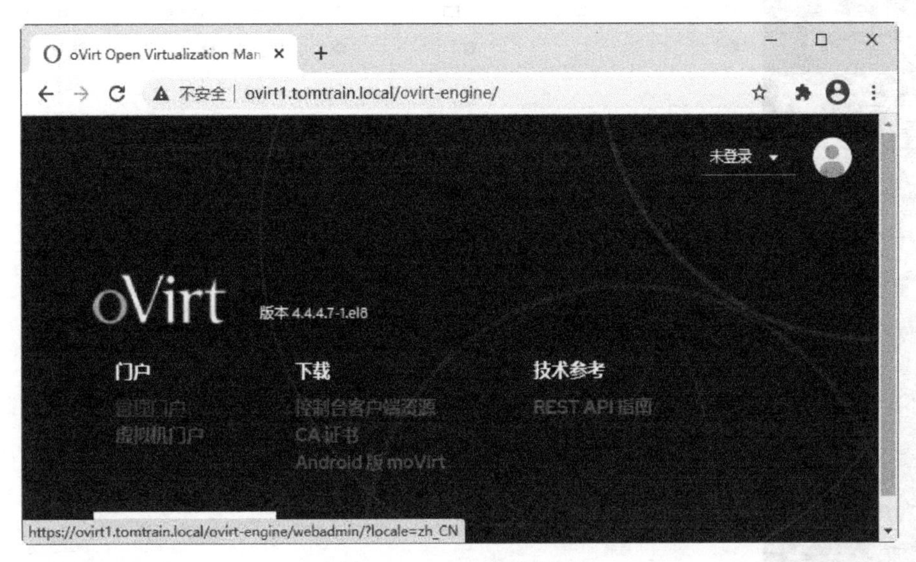


图 7-12　oVirt 门户站点

单击“管理门户”按钮，会出现管理门户的 SSO 登录页面，如图 7-13 所示。输入用户名和密码。如果是首次登录，则需要使用用户名 admin 及在安装过程中指定的密码。选择要验证的域。如果使用内部管理员用户名登录，则应选择 internal 域。单击“登录”按钮。登录成功后，oVirt 管理门户页面如图 7-14 所示。

图 7-13　oVirt 管理门户登录

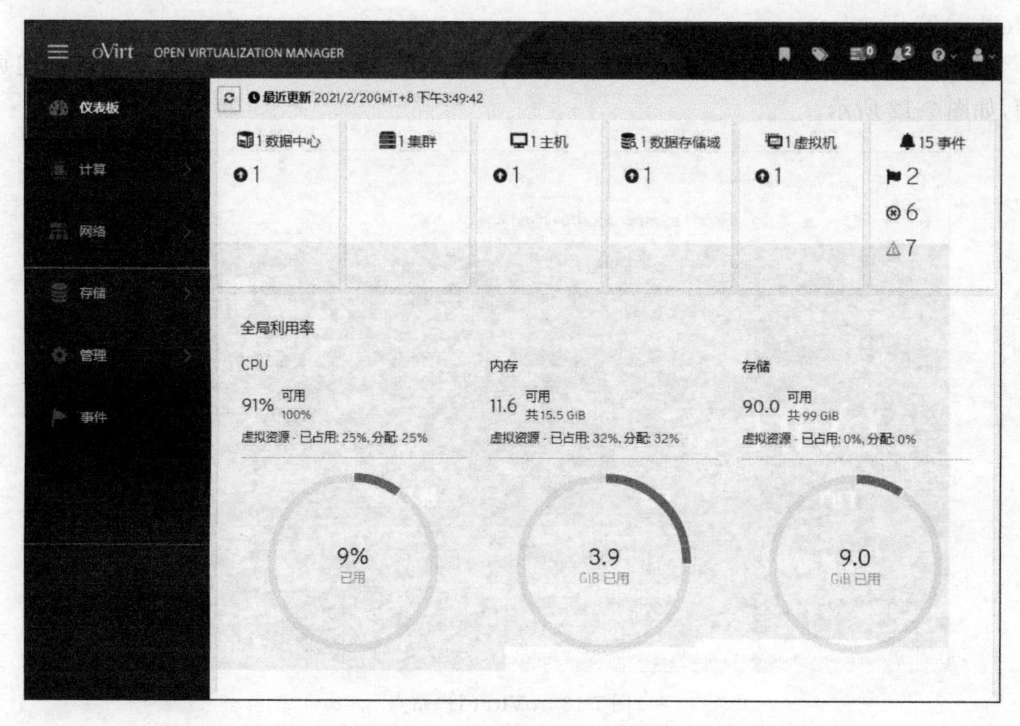


图 7-14 oVirt 管理门户页面

oVirt 管理门户页面的左边是功能导航窗格，右边是细节窗格，上面是标题栏。

要注销 oVirt 管理门户，可以在标题栏中单击用户名，然后单击“注销”按钮，就会返回欢迎页面。

**提示**：管理门户支持多种语言，默认设置与 Web 浏览器的语言环境设置一致。如果要使用默认语言以外的其他语言，则可以从欢迎页面的下拉列表中选择首选语言。

## 7.2.7 查看引擎安装结果

oVirt 对虚拟化平台进行了安全加固，不再允许 virsh 的无密码访问了，示例命令如下：

```
[root@kvm1 ~]# virsh list
Please enter your authentication name: 回车
Please enter your password: 回车
error: failed to connect to the hypervisor
error: authentication failed: Failed to start SASL negotiation: -1 (SASL(-1): generic
failure: All-whitespace username.)
```

查看配置文件/etc/ovirt-hosted-engine/virsh_auth.conf，可以获得账号信息，示例命令如下：

```
[root@kvm1 ~]# cat /etc/ovirt-hosted-engine/virsh_auth.conf
[credentials-vdsm]
authname=vdsm@ovirt
password=shibboleth

[auth-libvirt-localhost]
credentials=vdsm
```

通过-c 选项将这个配置文件的凭证信息提供给 virsh,示例命令如下：

```
[root@kvm1 ~]# virsh -c \
    qemu:///system?authfile=/etc/ovirt-hosted-engine/virsh_auth.conf
Welcome to virsh, the virtualization interactive terminal.

Type:  'help' for help with commands
       'quit' to quit

virsh # pool-list --all
 Name                                       State      Autostart
------------------------------------------------------------------
 4127dd88-bcd1-41f7-9314-84275028fd2c-1    active     yes
 4127dd88-bcd1-41f7-9314-84275028fd2c-2    inactive   yes
 default                                    active     yes
 localvma8ag8a2r                            active     yes
 localvminxsjdpl                            active     yes

virsh # net-list --all
 Name              State     Autostart   Persistent
----------------------------------------------------
 ;vdsmdummy;       active    no          no
 vdsm-ovirtmgmt    active    yes         yes

virsh # list --all
 Id     Name           State
--------------------------------
 1      HostedEngine   running

virsh # domblklist HostedEngine
 Target     Source
------------------------------------------------------------------
 sdc        -
 vda        /run/vdsm/storage/35264c95-fb5b-4a4c-a607-c13c7ebb6355/5ca09cd3-04a8-
4c0b-9ae5-d3e135ad17c8/2bdf4b18-42bd-4664-a461-45c73c318426
```

自托管引擎是一个名为 HostedEngine 的虚拟机。在部署时,会创建新的存储池和网络。

### 7.2.8 为 oVirt 安装、添加宿主机

oVirt 的宿主机有两种：RHEL/CentOS 8 主机和 oVirt 节点(oVirt Node)。

如果采用的是 RHEL/CentOS 8 主机，则建议安装时采用最小化的软件包，然后安装 oVirt 软件仓库配置文件及 cockpit-ovirt-dashboard 软件包，示例命令如下：

```
[root@kvm2 ~]# dnf -y install \
    https://resources.ovirt.org/pub/yum-repo/ovirt-release44.rpm

[root@kvm2 ~]# dnf -y install cockpit-ovirt-dashboard
```

oVirt 节点是基于 CentOS 的再发行版本，它提供了一种将物理计算机配置成 oVirt 宿主机的最简单的方法。oVirt 节点仅包含充当 Hypervisor 所需的软件包和 Cockpit 软件包。

从 https://www.ovirt.org/download/node.html 下载 oVirt 节点的 ISO 文件，然后写入安装介质 USB、CD 或 DVD。

通过 oVirt 安装介质启动的计算机，从启动菜单中选择 Install oVirt Node 4.4.4，如图 7-15 所示。

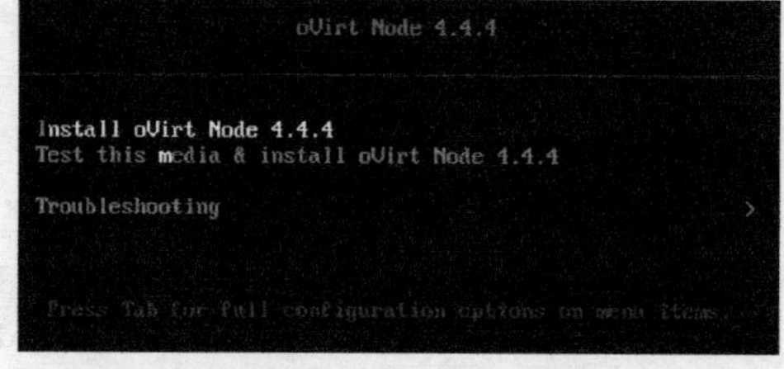


图 7-15 oVirt Node 安装介质启动菜单

oVirt 的安装过程与安装 CentOS 基本相同：选择安装语言及键盘布局，配置安装目标、选择时区并设置时间和日期、设置 root 密码，配置网络及主机名。配置完成后单击 Begin Installation 按钮进行安装，如图 7-16 所示。

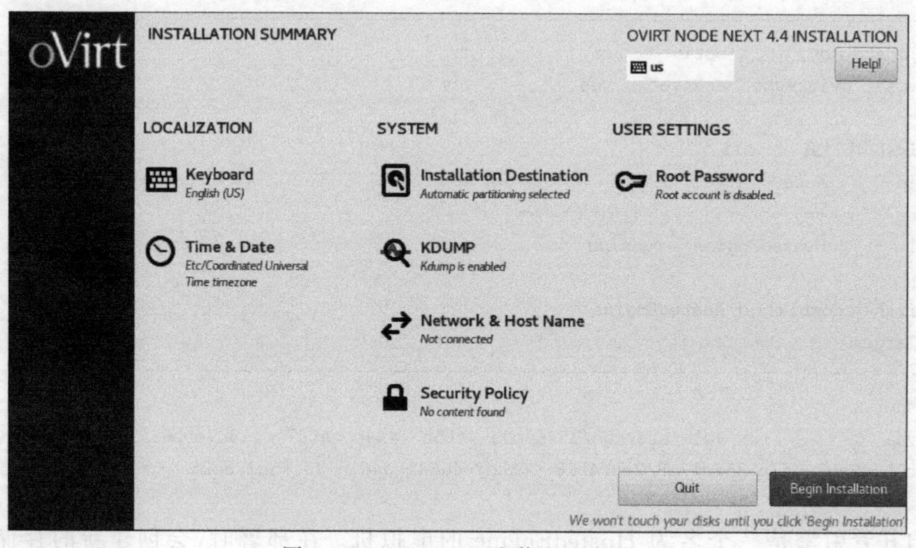


图 7-16 oVirt Node 安装配置选项

将 RHEL/CentOS 8 主机和 oVirt 节点添加到 oVirt 环境中的操作步骤是一样的。在管理门户中,单击“计算”中的“主机”,会显示当前主机列表,如图 7-17 所示。

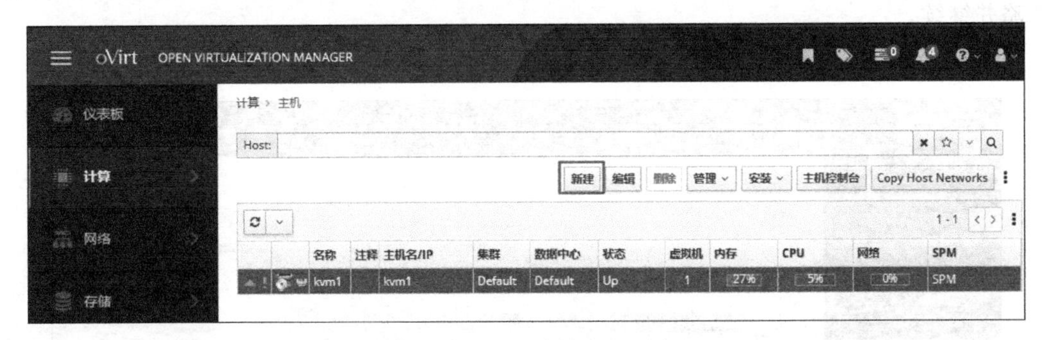


图 7-17　oVirt 中的主机列表

单击“新建”按钮,在新窗口中创建新的主机,如图 7-18 所示。

新建主机
常规
电源管理
SPM
控制台和 GPU
内核
承载的引擎
关联
主机集群 Default
数据中心: Default
使用 Foreman/Satellite
名称 kvm2
注释 CentOS 8.3
主机名/IP 192.168.1.232
SSH 端口 22
安装后激活主机
验证
用户名 root
密码 ••••••
SSH 公共密钥
高级参数
确定 取消

图 7-18　新建主机

使用下拉列表为新主机选择数据中心和主机群集。

输入新主机的名称和地址。标准 SSH 端口(端口 22)将自动填充在“SSH 端口”字段中。

输入 root 用户的密码以使用密码验证。或者将 SSH PublicKey 字段中显示的密钥复制到主机的/root/. ssh/authorized_keys 文件中，这样可以使用公共密钥身份验证。

单击“确定”按钮。由于仅仅提供了基本信息，所以会有一些警告信息，单击“确定”按钮忽略并继续。

新主机将显示在主机列表中，状态为 Installing，如图 7-19 所示。

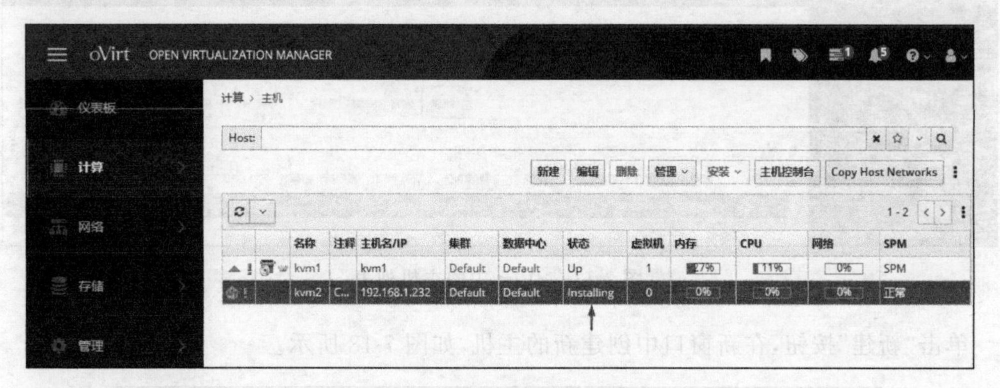


图 7-19　正在安装新主机

将主机添加到 oVirt 环境需要进行虚拟化检查、软件包安装和创建网桥等多个操作，这些需要一些时间，可以在“事件”中查看安装进度，如图 7-20 所示。

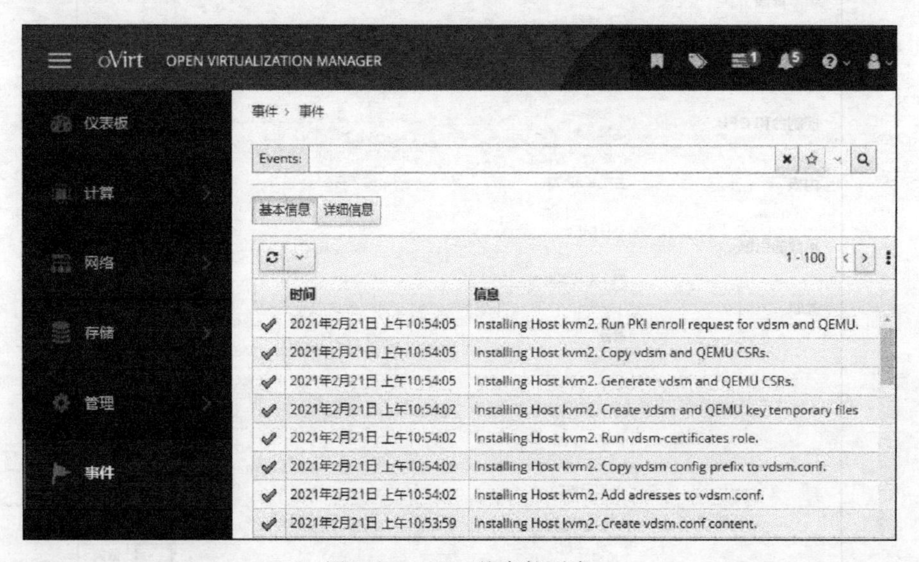


图 7-20　oVirt 的事件列表

最终主机的状态变为 Up，如图 7-21 所示。

向 oVirt 中添加宿主机有时会失败，需要根据事件日志进行排错。一个常见的错误原因是宿主机从 oVirt 软件仓库中下载软件超时。解决方法是修改宿主机的 oVirt 软件仓库的配置，通过 mirrorlist 参数使用镜像 URL，示例命令如下：

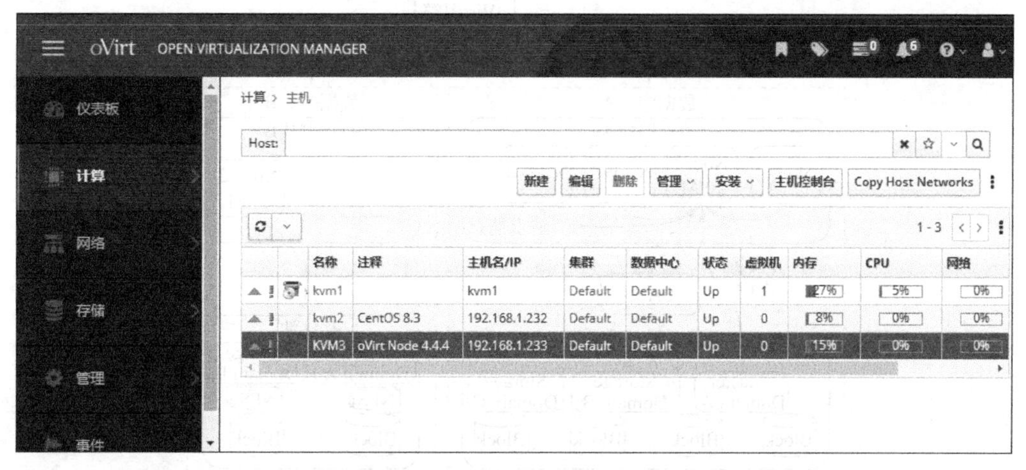


图 7-21　新主机添加成功

```
#vi /etc/yum.repos.d/ovirt-4.4.repo
[ovirt-4.4]
name=Latest oVirt 4.4 Release
#baseURL=https://resources.ovirt.org/pub/ovirt-4.4/rpm/el$releasever/
mirrorlist=https://mirrorlist.ovirt.org/mirrorlist-ovirt-4.4-el$releasever
enabled=1
gpgcheck=1
gpgkey=file:///etc/pki/rpm-gpg/RPM-GPG-ovirt-4.4
```

## 7.3　数据中心管理

oVirt 的数据中心是定义特定环境中所使用的资源的逻辑集合。数据中心是资源的容器,资源包括群集、宿主机、网络资源(逻辑网络和物理网卡)和存储资源。一个 oVirt 环境可以包含多个数据中心,所有数据中心均通过单个管理门户进行管理。通过数据中心可以分隔基础架构,如图 7-22 所示。

一个数据中心可以包含多个群集,而这些群集又可以包含多台宿主机,每台宿主机上可以支持多个虚拟机。数据中心可以有多个关联的存储域。

### 7.3.1　查看默认的数据中心

oVirt 在安装过程中会创建一个名为 Default 的默认数据中心。在管理门户中,选择"计算"→"数据中心",会显示当前数据中心列表,如图 7-23 所示。

选中 Default 数据中心,单击"编辑"按钮,可以查看、编辑数据中心,如图 7-24 所示。

数据中心的属性和说明如表 7-2 所示。

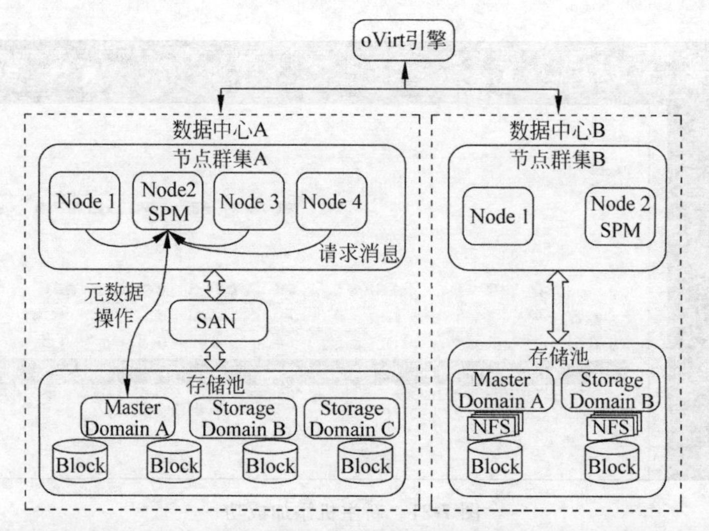


图 7-22 数据中心示例

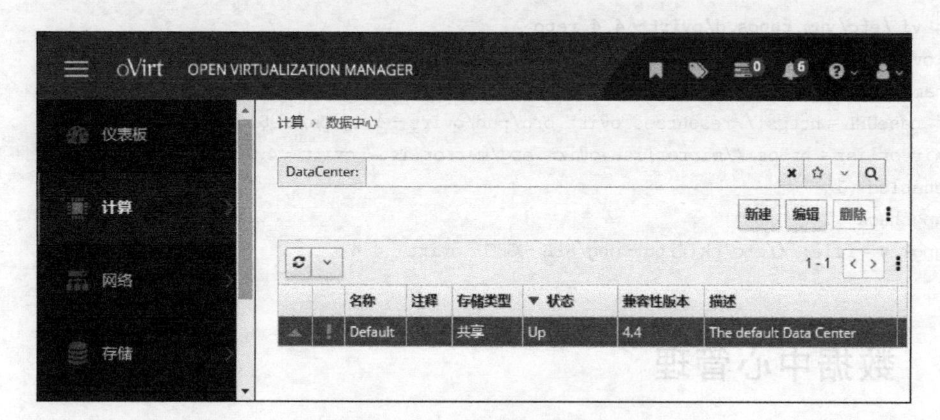


图 7-23 oVirt 中的数据中心列表

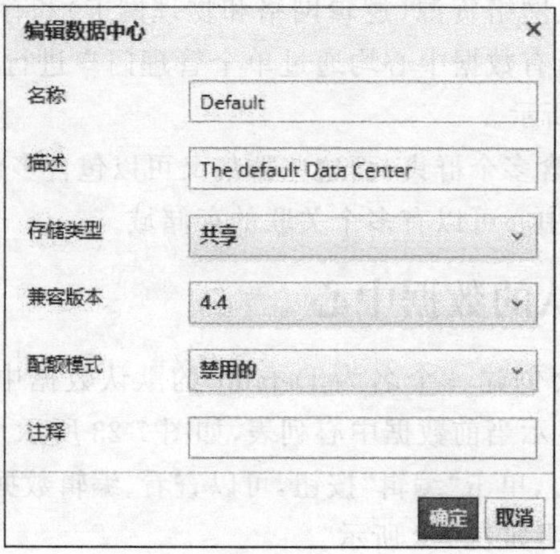

编辑数据中心 ×

名称 Default

描述 The default Data Center

存储类型 共享

兼容版本 4.4

配额模式 禁用的

注释

确定 取消

图 7-24 编辑数据中心的属性

表 7-2　数据中心的属性和说明

| 属　　性 | 说　　明 |
|---|---|
| 名称 | 数据中心的名称。长度不超过 40 个字符，并且必须保证唯一性，只能包含大小写字母、数字、连字符和下画线 |
| 描述 | 数据中心的描述 |
| 存储类型 | 有“共享”或“本地”两种存储类型。可以将不同类型的存储域(iSCSI、NFS、FC、POSIX 和 Gluster)添加到同一数据中心。本地存储和共享存储不能混合使用。数据中心初始化后，可以更改存储类型 |
| 兼容版本 | oVirt 的版本。升级 oVirt 引擎之后，主机、群集和数据中心可能仍处于早期版本。在升级数据中心的兼容性级别之前，需要确保先升级所有主机，然后升级集群 |
| 配额模式 | 配额是 oVirt 随附的资源限制工具，共有 3 个可选项。<br>(1) 禁用的：不实施配额。<br>(2) 审核：希望编辑配额设置时，可启用此选项。<br>(3) 强制的：实施配额 |
| 注释 | (可选) 有关数据中心的纯文本注释 |

存储池管理器(Storage Pool Manager，SPM)用于赋予数据中心中某台宿主机的角色，只有它可以管理数据中心的存储域，图 7-22 中数据中心 A 的 Node2 就是这个数据中心的 SPM。SPM 角色可以由数据中心的任何宿主机来承担，它并不妨碍宿主机进行其他操作，例如运行虚拟机。

SPM 通过在存储域之间协调元数据来控制对存储的访问，这包括创建、删除和操作虚拟磁盘(映像)、快照和模板等操作。

为确保元数据的完整性，在同一时刻只能由一台宿主机担任数据中心中的 SPM。oVirt 引擎可确保 SPM 始终可用并且唯一。如果 SPM 主机在访问存储时遇到问题，则引擎会将 SPM 角色移至其他主机，此过程可能需要一些时间。

宿主机的 SPM 优先级会控制着分配 SPM 角色的可能性。可以在 SPM 选项卡中更改优先级，如图 7-25 所示。

图 7-25　宿主机的 SPM 优先级

还可以手工更改 SPM 角色，如图 7-26 所示。

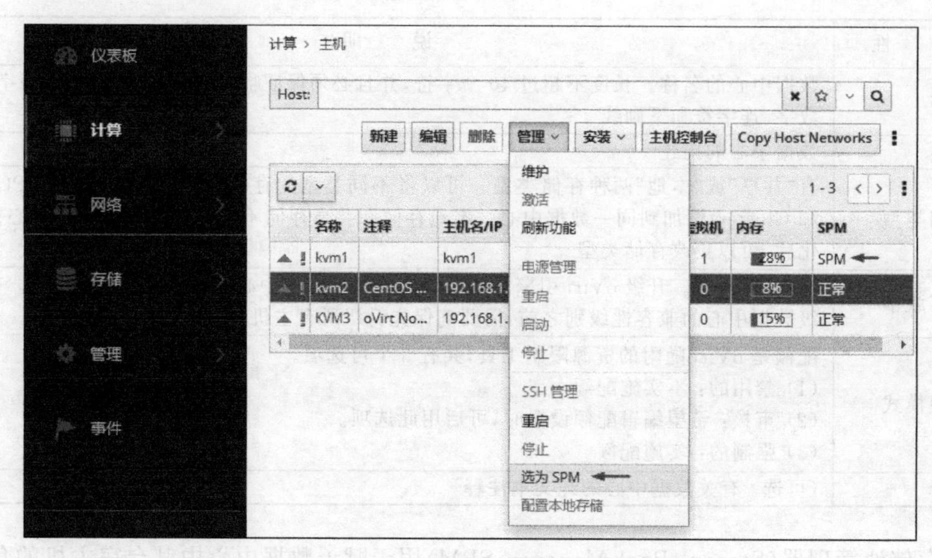


图 7-26　更改数据中心的 SPM 角色

## 7.3.2　创建新的数据中心

可以根据需要创建新的数据中心。在管理门户中，选择“计算”→“数据中心”，会显示当前数据中心列表。单击“新建”按钮，创建新的数据中心，如图 7-27 所示。

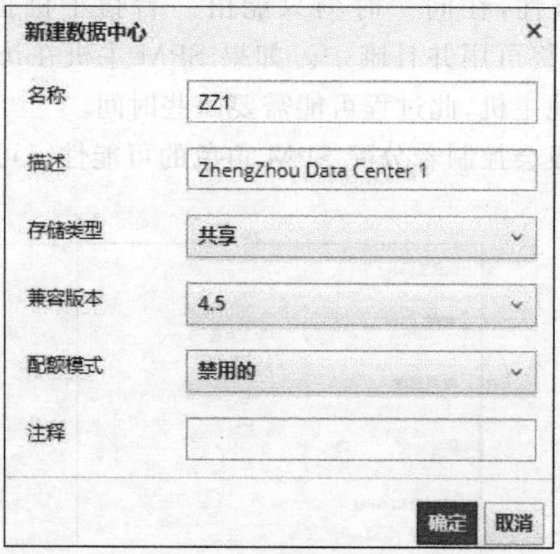

新建数据中心

名称　ZZ1

描述　ZhengZhou Data Center 1

存储类型　共享

兼容版本　4.5

配额模式　禁用的

注释

确定　取消

图 7-27　新建数据中心(一)

输入新数据中心的名称、描述，从下拉菜单中选择数据中心的“存储类型”“兼容版本”和“配额模式”，单击“确定”按钮后便可创建数据中心，然后会出现“数据中心-引导操作”窗口，如图 7-28 所示。

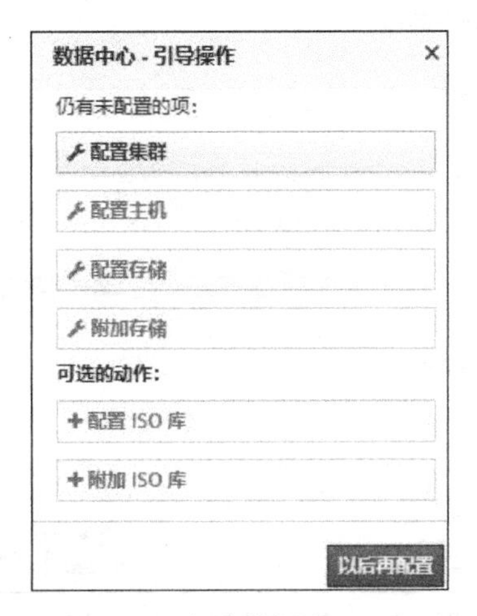


图 7-28　新建数据中心(二)

**提示**：窗口的英文原标题是 Data Center-Guide Me，被翻译成"数据中心-引导操作"，翻译得不准确，常常让人误解。

数据中心需要有群集、宿主机和存储域才能运行。设置兼容版本后，无法降低版本号。

通过单击"以后再配置"按钮推迟配置。以后还可以在数据中心中单击更多操作标识符"⋮"，然后选择"引导操作"进行配置，如图 7-29 所示。

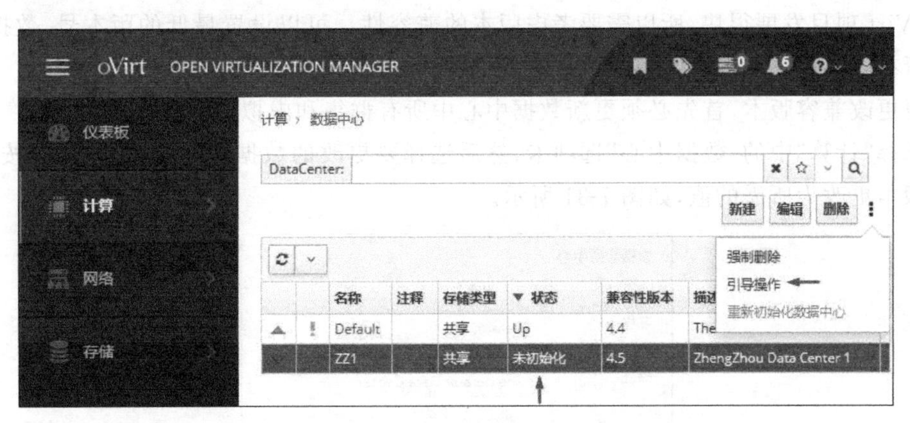


图 7-29　未初始化的数据中心

新数据中心将保持未初始化状态，直到为其配置了集群、宿主机和存储域为止。

## 7.3.3　更改数据中心存储类型

可以在数据中心初始化后更改其存储类型。这对于用于移动虚拟机或模板的数据域很有用。

单击"计算"中的"数据中心"，然后选择要更改的数据中心，单击"编辑"按钮，将存储类型更改为所需的值，如图 7-30 所示。

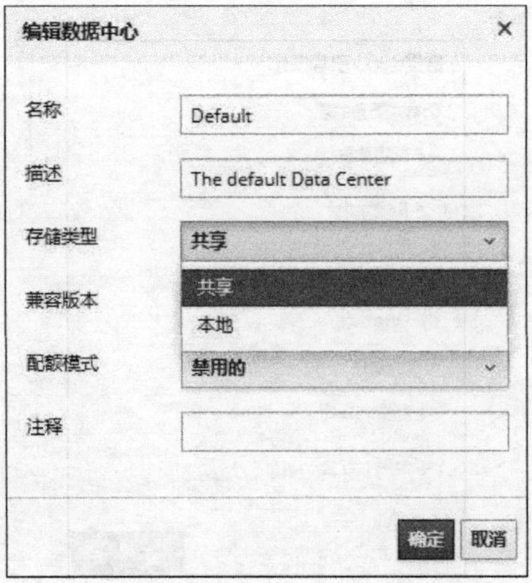

图 7-30　修改数据中心的存储类型

更改存储类型,有以下限制。

(1) 共享到本地:数据中心只能有一个集群和一台宿主机。

(2) 本地到共享:数据中心不能包含本地存储域。

### 7.3.4 更改数据中心兼容版本

oVirt 项目发展很快,所以需要考虑版本的兼容性。可以设置最低的版本号,数据中心中的所有群集都必须支持所需的兼容性级别。

要更改兼容版本,首先必须更新数据中心中所有群集和虚拟机的兼容性版本。

选择"计算"中的"数据中心"选项卡,然后选择要更改的数据中心,单击"编辑"按钮,将兼容版本更改为所需的值,如图 7-31 所示。

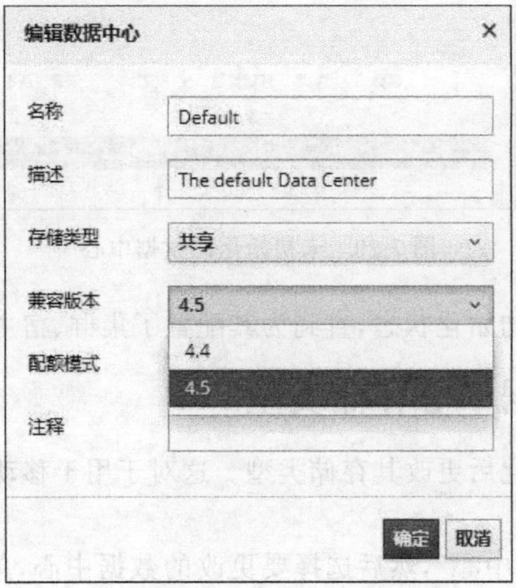

图 7-31　修改数据中心的兼容版本

### 7.3.5 重新初始化数据中心

重新初始化数据中心是一个恢复操作。如果主数据域的数据发生损坏且无法修复，则必须执行重新初始化操作，用新的主数据域替换坏的主数据域。重新初始化数据中心可以还原与数据中心相关的所有资源，包括群集、宿主机和存储域。可以将任何备份或导出的虚拟机或模板导入新的主数据域。

在进行初始化操作之前，要确保连接到数据中心的所有存储域都处于维护模式。

单击"计算"中的"数据中心"，然后选择要更改的数据中心，单击更多操作标识符"⋮"，然后单击"重新初始化数据中心"。

"数据中心重新初始化"窗口列出了所有可用(分离的、处于维护模式中的)存储域。单击要添加到数据中心的存储域的单选按钮，选择"批准操作"复选框，单击"确定"按钮。

存储域作为主数据域连接到数据中心并被激活，然后可以将任何备份、导出的虚拟机或模板导入新的主数据域中。

### 7.3.6 删除数据中心

删除数据中心有两种方法：正常删除和强制删除。

**1. 正常删除**

正常地删除数据中心需要有活动的宿主机。这种方法不会删除关联的资源。

删除之前，要确保连接到数据中心的存储域处于维护模式。

单击"计算"中的"数据中心"，然后选择要删除的数据中心，单击"删除"按钮，最后单击"确定"按钮。

**2. 强制删除**

如果存储域或宿主机损坏，则数据中心状态会变为 Non Responsive。在这种情况下可能无法正常地删除数据中心，只能进行强制删除。

强制删除不需要有活动的宿主机，它还会永久删除附加的存储域。在强制删除数据中心之前，可能还需要销毁损坏的存储域。

要确保连接到数据中心的存储域处于维护模式。单击"计算"中的"数据中心"，然后选择要删除的数据中心，单击更多操作标识符"⋮"，然后单击"强制删除"按钮，选择"批准操作"复选框，单击"确定"按钮，如图 7-32 所示。

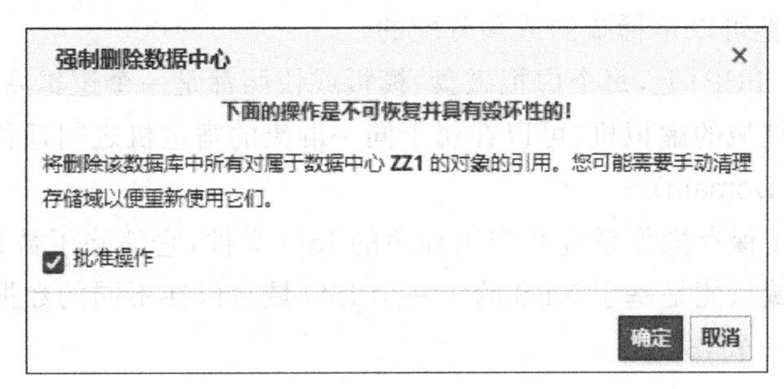


图 7-32　强制删除数据中心

## 7.4 存储管理

oVirt 使用集中式存储系统保存虚拟磁盘、ISO 文件和快照。可以使用以下方式实现存储：

(1) 网络文件系统(NFS)。

(2) iSCSI。

(3) 光纤通道协议(FCP)。

(4) 宿主机的本地存储。

(5) 并行 NFS(pNFS)。

(6) GlusterFS。

(7) 其他符合 POSIX 的文件系统。

设置存储是初始化新数据中心的先决条件，因为除非连接并激活了存储域，否则将无法初始化数据中心。

### 7.4.1 存储域概述

确定数据中心的存储需求后，就可以配置环境和附加存储了。oVirt 具有 3 种类型的存储域。

1. 数据域(Data Domain)

数据域包含数据中心中所有虚拟机和模板的磁盘文件、OVF 文件和快照。

数据域不能在数据中心之间共享。可以将多种类型的数据域(iSCSI、NFS、FC、POSIX 和 Gluster)添加到同一数据中心，前提是它们都是共享的而不是本地的域。

必须先将数据域附加到数据中心，然后才能将其他类型的域附加到该数据中心。

存储域可以由块设备(iSCSI 或 FCP)或文件系统(NFS、GlusterFS 或其他 POSIX 兼容文件系统)组成。

默认情况下，GlusterFS 域和本地存储域支持 4KB 大小的块。4KB 块大小可以提供更好的性能，尤其是在使用大文件时。

在 NFS 上，所有虚拟磁盘、模板和快照都是文件。虚拟磁盘文件可以是 QCOW2 或 RAW 格式，存储类型可以是稀疏的或预分配的。

在 SAN(iSCSI/FCP)上，每个虚拟磁盘、模板或快照都是一个逻辑卷。

保存在共享存储域的虚拟机，可以在属于同一群集的宿主机之间迁移。

2. ISO 域(ISO Domain)

ISO 域存储用于保存操作系统和应用程序的 ISO 文件，它消除了数据中心对物理光盘介质的需求。ISO 域只能是基于 NFS 的。一个 ISO 域可以在不同的数据中心之间共享，一个数据中心只能有一个 ISO 域。

### 3. 导出域(Export Domain)

这是早期版本的 oVirt 使用的域,它是临时存储库,用于在数据中心和 oVirt 环境之间复制和移动映像,可用于备份虚拟机。导出域可以在数据中心之间移动,但是一次只能在一个数据中心中处于活动状态。导出域只能是基于 NFS 的,并且只能将一个导出域添加到数据中心。

现在版本的 oVirt 已经不再使用导出域了,因为可以将数据域从现有数据中心断开,然后导入另一个数据中心,这样就可以在新的数据中心中访问这些虚拟机、虚拟磁盘和模板了。

## 7.4.2 管理 NFS 存储

### 1. 创建 NFS 存储域

在 NFS 服务器上为 oVirt 创建新的 NFS 共享。如前所述,oVirt 需要特定的系统用户账号 vdsm 和系统用户组 kvm 访问 NFS 共享。

在 NFS 服务器创建新的目录,设置目录的权限、所有者及所有者组,示例命令如下:

```
[root@stor1 ~]# mkdir /nfs1

[root@stor1 ~]# chown 36:36 /nfs1

[root@stor1 ~]# chmod 755 /nfs1

[root@stor1 ~]# ls - ld /nfs1
drwxr - xr - x. 2 vdsm kvm 6 Feb 23 08:18 /nfs1

[root@stor1 ~]# echo "/nfs1 * (rw,no_root_squash,sync)" >> /etc/exports
```

重新加载配置文件以便生效,示例命令如下:

```
[root@stor1 ~]# exportfs - r

[root@stor1 ~]# showmount - e localhost
Export list for localhost:
/nfs1    *
/vmdata *
```

下面将此 NFS 存储作为数据域附加到 oVirt 环境。在管理门户中,单击“存储”中的“域”,会显示当前存储的列表,如图 7-33 所示。

单击“新建域”按钮,输入存储域的名称。采用“数据中心”“域功能”“存储类型”和“主机”列表的默认值,如图 7-34 所示。

输入用于存储域的导出路径,导出路径的格式为“IP 地址或域名:/路径”。

单击“确定”按钮开始创建新的存储域。

还可以根据要求配置自定义连接参数及高级参数,如图 7-35 所示。

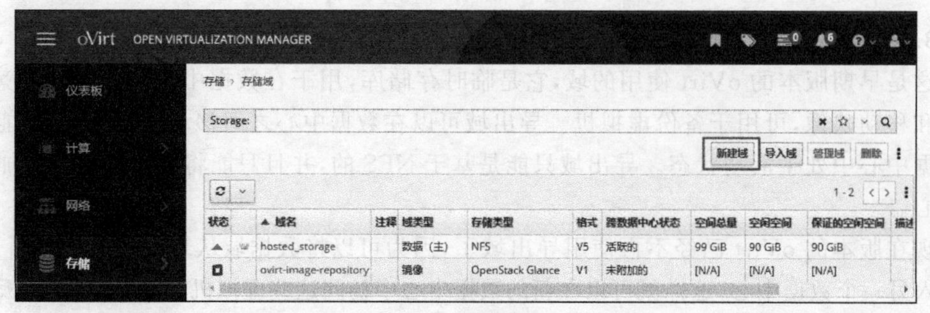


图 7-33　存储域列表

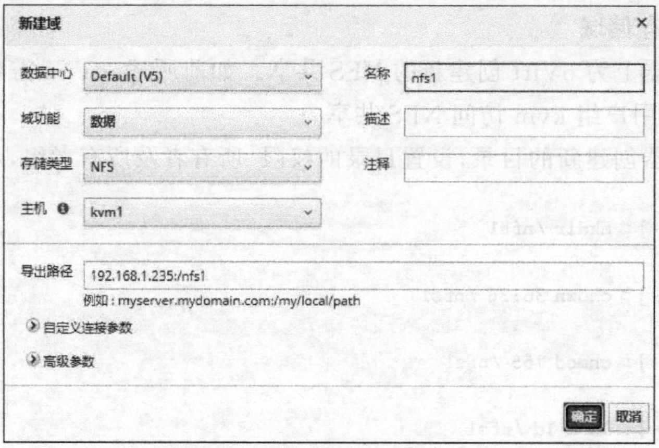

新建域 ×

数据中心 Default (V5)　　名称 nfs1

域功能 数据　　描述

存储类型 NFS　　注释

主机 kvm1

导出路径 192.168.1.235:/nfs1

例如：myserver.mydomain.com:/my/local/path

自定义连接参数

高级参数

确定　取消

图 7-34　新建 NFS 存储域(一)

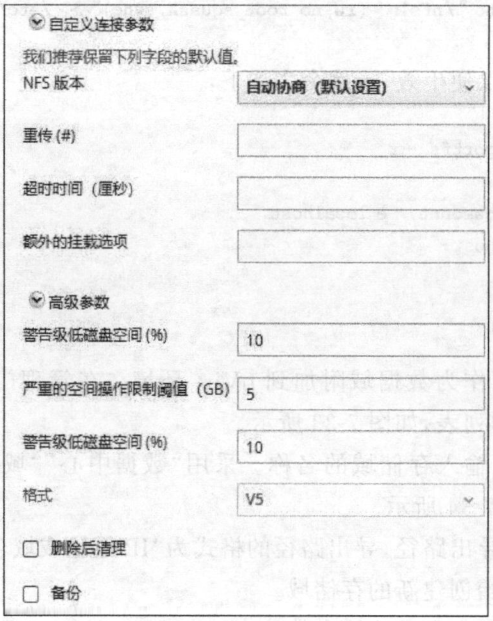

自定义连接参数

我们推荐保留下列字段的默认值。

NFS 版本　自动协商（默认设置）

重传 (#)

超时时间（厘秒）

额外的挂载选项

高级参数

警告级低磁盘空间 (%)　10

严重的空间操作限制阈值 (GB)　5

警告级低磁盘空间 (%)　10

格式　V5

☐ 删除后清理

☐ 备份

图 7-35　新建 NFS 存储域(二)

新数据域的初始状态为“已锁定(Locked)”，等磁盘准备好后才会变成“活跃”，这时数据域将会被附加到数据中心，如图 7-36 所示。

图 7-36　新域创建成功

oVirt 在创建 NFS 存储域时，会在 NFS 目录中创建目录结构及写入元数据，示例命令如下：

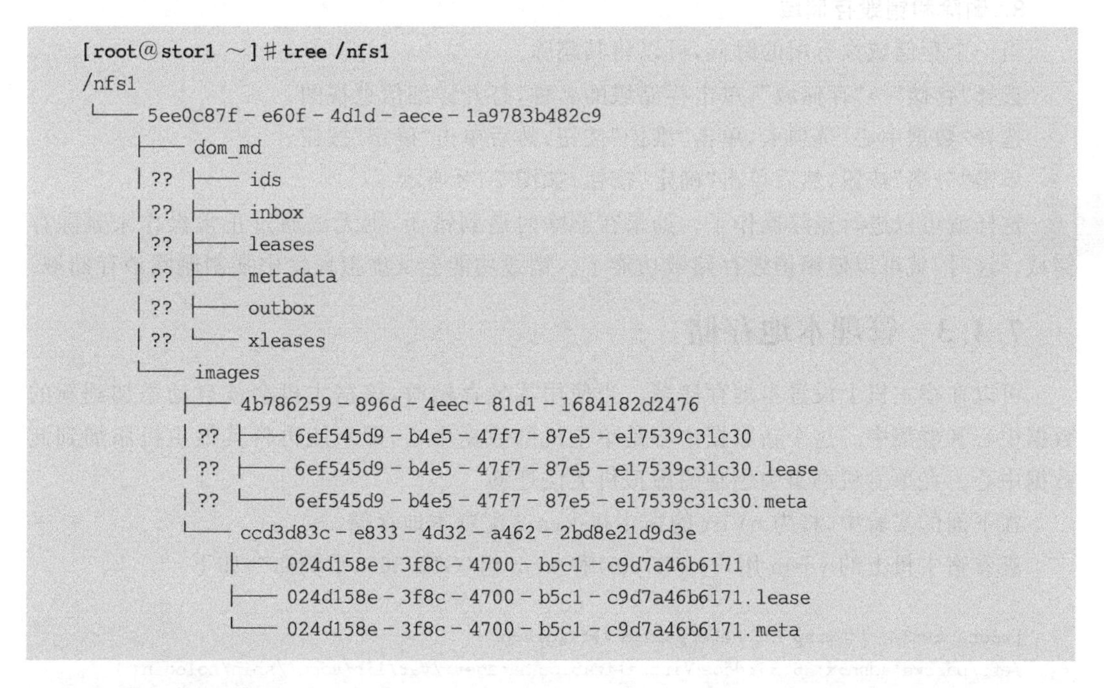

```
[root@stor1 ~]# tree /nfs1
/nfs1
└── 5ee0c87f-e60f-4d1d-aece-1a9783b482c9
    ├── dom_md
    | ?? ├── ids
    | ?? ├── inbox
    | ?? ├── leases
    | ?? ├── metadata
    | ?? ├── outbox
    | ?? └── xleases
    └── images
        ├── 4b786259-896d-4eec-81d1-1684182d2476
        | ?? ├── 6ef545d9-b4e5-47f7-87e5-e17539c31c30
        | ?? ├── 6ef545d9-b4e5-47f7-87e5-e17539c31c30.lease
        | ?? └── 6ef545d9-b4e5-47f7-87e5-e17539c31c30.meta
        └── ccd3d83c-e833-4d32-a462-2bd8e21d9d3e
            ├── 024d158e-3f8c-4700-b5c1-c9d7a46b6171
            ├── 024d158e-3f8c-4700-b5c1-c9d7a46b6171.lease
            └── 024d158e-3f8c-4700-b5c1-c9d7a46b6171.meta
```

**2. 增加 NFS 存储**

要增加 NFS 存储的容量，既可以创建一个新的存储域并将其添加到现有的数据中心，又可以采用增加 NFS 服务器上的可用空间的方法。

选择“存储”→“存储域”，选择 NFS 存储域的名称，将打开详细信息视图。

选择“数据中心”选项卡，然后单击“维护”按钮，会将存储域置于维护模式，如图 7-37 所示。

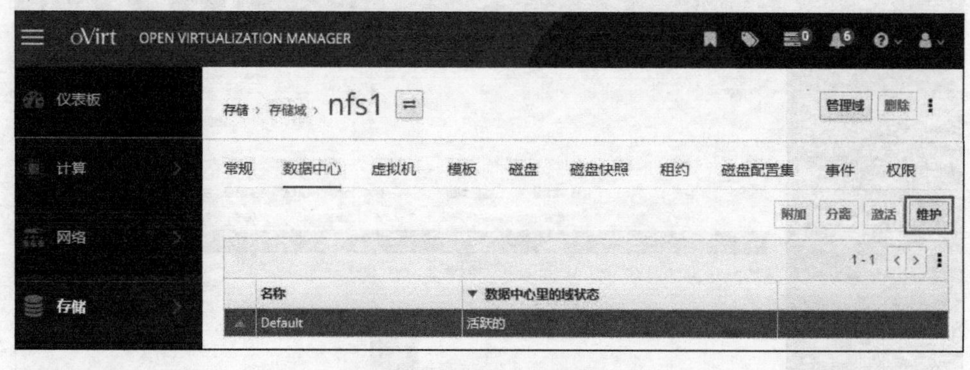


图 7-37　将域置于维护模式

当存储域置于维护模式时将卸载现有共享，这样就可以调整存储域的大小了。

调整完毕后，在存储域的详细信息视图中，选择“数据中心”选项卡，然后单击“激活”按钮以挂载存储域。

**3. 删除和销毁存储域**

当一个存储域没有用的时候，可以将其删除。

选择“存储”→“存储域”，单击存储域的名称，打开详细信息视图。

选择“数据中心”选项卡，单击“维护”按钮，然后单击“确定”按钮。

单击“分离”按钮，然后单击“确定”按钮，如图 7-38 所示。

这样就可以进行删除操作了。如果在删除时遇到错误，则无法通过正常操作来删除存储域。这时，就可以使用销毁存储域功能了。销毁功能会从虚拟环境中强制删除该存储域。

## 7.4.3　管理本地存储

可以在宿主机上设置本地存储域。当使用本地存储时，该宿主机会被自动添加到新的数据中心和群集中。这个新数据中心是单主机的数据中心，即不能再将其他主机添加到此数据中心。在单主机群集中创建的虚拟机无法迁移。

在下面的实验中，将为 oVirt 的宿主机 kvm2 配置本地存储。

查看宿主机上的 vdsm 用户(UID 36)和 kvm 组(GID 36)，示例命令如下：

```
[root@kvm2 ~]# grep :36 /etc/passwd /etc/group
/etc/passwd:vdsm:x:36:36:Node Virtualization Manager:/var/lib/vdsm:/sbin/nologin
/etc/group:kvm:x:36:qemu,ovirtimg,sanlock
```

在准备本地存储时，要保证 vdsm、kvm 有适合的读写权限，示例命令如下：

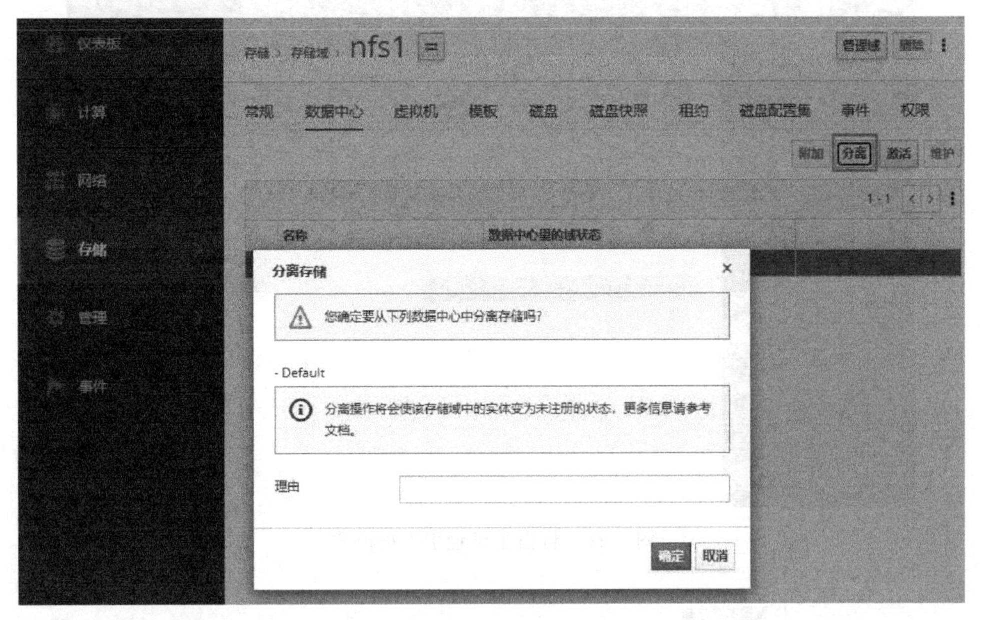


图 7-38 分离存储域置

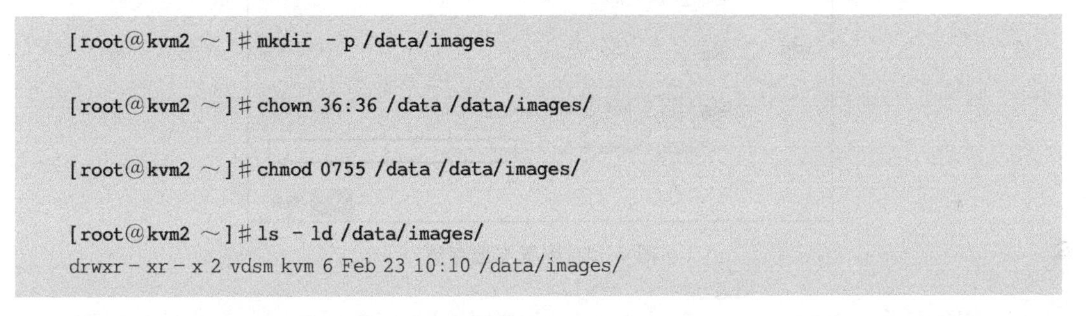

```
[root@kvm2 ~]# mkdir -p /data/images

[root@kvm2 ~]# chown 36:36 /data /data/images/

[root@kvm2 ~]# chmod 0755 /data /data/images/

[root@kvm2 ~]# ls -ld /data/images/
drwxr-xr-x 2 vdsm kvm 6 Feb 23 10:10 /data/images/
```

**提示**：建议本地存储使用单独的逻辑卷或磁盘，特别是 oVirt 节点，这样可以防止升级时丢失数据。

在管理门户中单击“计算”中的“主机”，然后选择主机。选择“管理”下拉菜单中的“维护”单选项，如图 7-39 所示。

当置于维护模式后，再次单击“管理”下拉菜单，单击“配置本地存储”选项。

可以根据需要单击“数据中心”“集群”和“存储”字段旁边的“编辑”按钮修改默认的名称，如图 7-40 所示。

在文本框中输入本地存储的路径/data/images，单击“确定”按钮。

此宿主机会移动到自己的数据中心中，在主机管理中单击“激活”按钮以退出维护状态。

与其他类似的存储域一样，oVirt 也会创建相应的目录结构及元数据文件，示例命令如下：

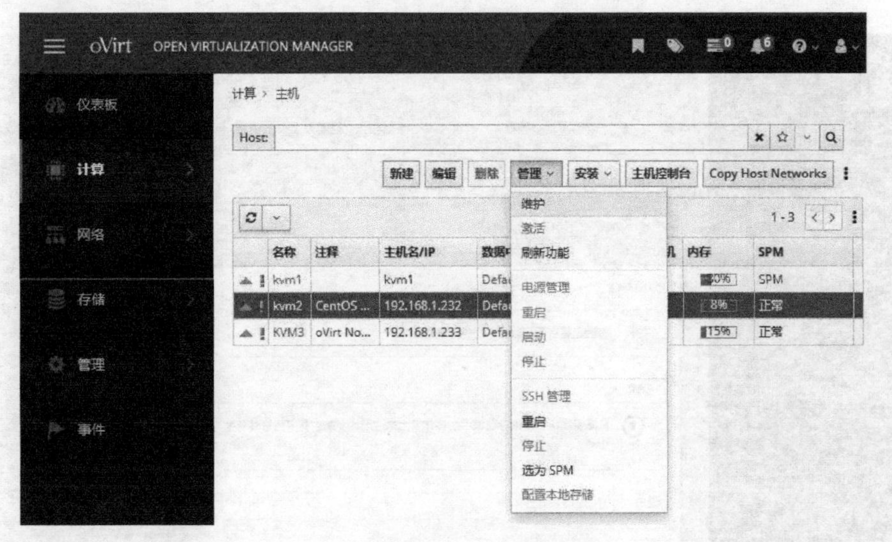


图 7-39　将宿主机置于维护模式

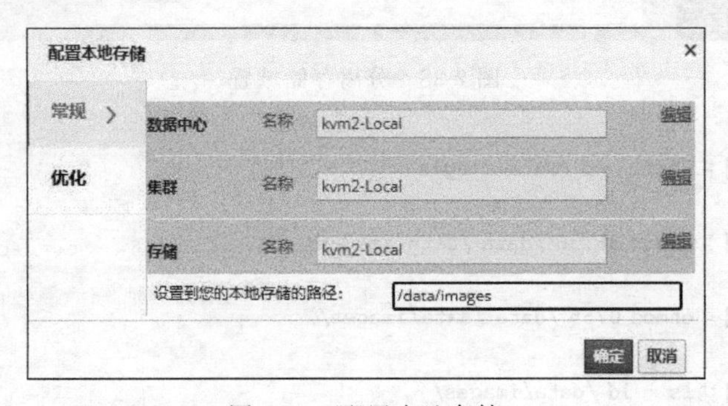


图 7-40　配置本地存储

```
[root@kvm2 ~]# tree /data/
/data/
└── images
    └── a715bcb3-6f3d-4c60-aac6-542ffec55e08
        ├── dom_md
        │   ├── ids
        │   ├── inbox
        │   ├── leases
        │   ├── metadata
        │   └── outbox
        ├── images
        └── master
            ├── tasks
            └── vms

7 directories, 5 files
```

### 7.4.4 管理 iSCSI 存储

oVirt 支持的块存储包括 iSCSI 存储和 FCP 存储，它们是由 LUN 组成的卷组所创建的存储域。这些卷组和 LUN 一次只能附加给一个存储域。

在配置时，需要为每台宿主机配置 iSCSI 的启动器。建议为启动器配置有意义、易辨识的启动器名称，示例命令如下：

```
[root@kvm1 ~]# cat /etc/iscsi/initiatorname.iscsi
InitiatorName=iqn.1994-05.com.redhat:kvm1
[root@kvm2 ~]# cat /etc/iscsi/initiatorname.iscsi
InitiatorName=iqn.1994-05.com.redhat:kvm2
[root@kvm3 ~]# cat /etc/iscsi/initiatorname.iscsi
InitiatorName=iqn.1994-05.com.redhat:kvm3
```

在本实验中，由 CentOS 8.3 来构建 iSCSI 存储，同时给 3 个 iSCSI 启动器分配一个 LUN。查看配置好的存储，示例命令如下：

```
[root@stor1 ~]# targetcli ls
o- / ............................................................... [...]
  o- backstores .................................................... [...]
  | o- block ........................................ [Storage Objects: 0]
  | o- fileio ....................................... [Storage Objects: 1]
  | | o- disk0 ....... [/iscsifiles/disk0.img (10.0GiB) write-back activated]
  | |   o- alua ......................................... [ALUA Groups: 1]
  | |     o- default_tg_pt_gp ............... [ALUA state: Active/optimized]
  | o- pscsi ........................................ [Storage Objects: 0]
  | o- ramdisk ...................................... [Storage Objects: 0]
  o- iscsi .................................................. [Targets: 1]
  | o- iqn.2003-01.org.Linux-iscsi.stor1.x8664:sn.e246346cc56a .... [TPGs: 1]
  |   o- tpg1 .................................... [no-gen-acls, no-auth]
  |     o- acls ................................................ [ACLs: 3]
  |     | o- iqn.1994-05.com.redhat:kvm1 ................. [Mapped LUNs: 1]
  |     | | o- mapped_lun0 ....................... [lun0 fileio/disk0 (rw)]
  |     | o- iqn.1994-05.com.redhat:kvm2 ................. [Mapped LUNs: 1]
  |     | | o- mapped_lun0 ....................... [lun0 fileio/disk0 (rw)]
  |     | o- iqn.1994-05.com.redhat:kvm3 .................. [Mapped LUNs: 1]
  |     |   o- mapped_lun0 ....................... [lun0 fileio/disk0 (rw)]
  |     o- luns ................................................ [LUNs: 1]
  |     | o- lun0 . [fileio/disk0 (/iscsifiles/disk0.img) (default_tg_pt_gp)]
  |     o- portals .......................................... [Portals: 1]
  |       o- 0.0.0.0:3260 ........................................... [OK]
  o- loopback ............................................... [Targets: 0]
```

在管理门户中，单击“存储”中的“域”，单击“新建域”按钮，输入存储域的名称。从下拉列表中选择一个数据中心。在“域功能”中选择“数据”，然后选择“iSCSI”作为“存储类型”。选择一个活动主机作为“主机”，如图 7-41 所示。

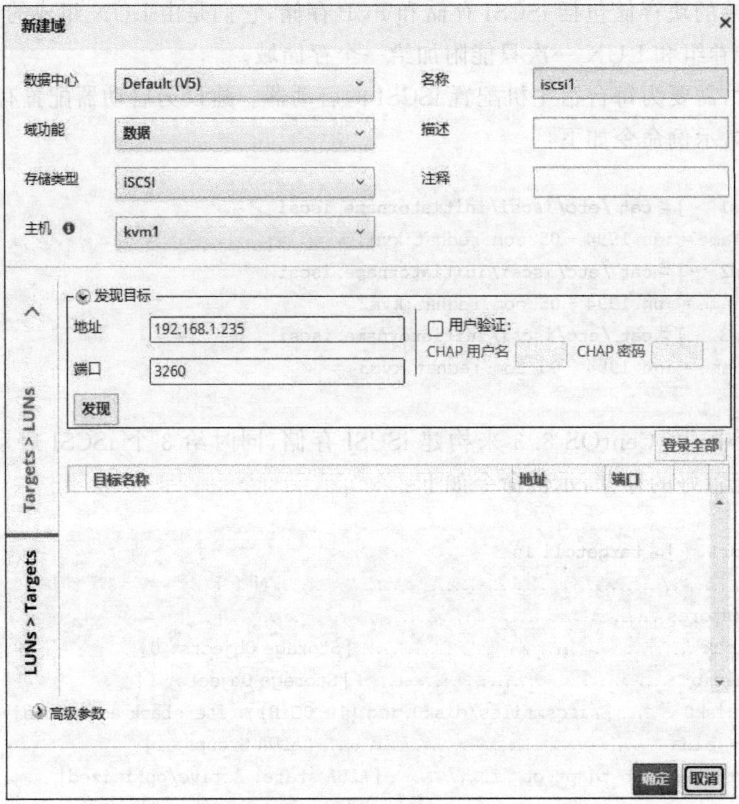

图 7-41 新建 iSCSI 存储域(一)

与存储域的通信是从所选主机(本示例中是 kvm1)而不是直接从引擎进行的，因此，在配置存储域之前，所有宿主机都必须有权访问存储设备。

在“地址”字段中输入 iSCSI 服务器的 FQDN 或 IP 地址。

在“端口”字段中输入端口号，默认值为 3260。

如果使用 CHAP 来保护存储，则可选中“用户身份验证”复选框。输入 CHAP 用户名和密码。

单击“发现”按钮进行目标发现操作。

从发现结果中选择一个或多个目标，单击“全部登录”按钮，如图 7-42 所示。

单击所需目标旁边的“+”按钮，这将展开条目并显示附加到目标的所有未使用的 LUN。

选中用于创建存储域的每个 LUN 的复选框，如图 7-43 所示。

单击“高级参数”，可以配置高级参数(可选)：

(1) 在“警告级低磁盘空间(%)”字段中输入百分比值。如果存储域上的可用空间低于

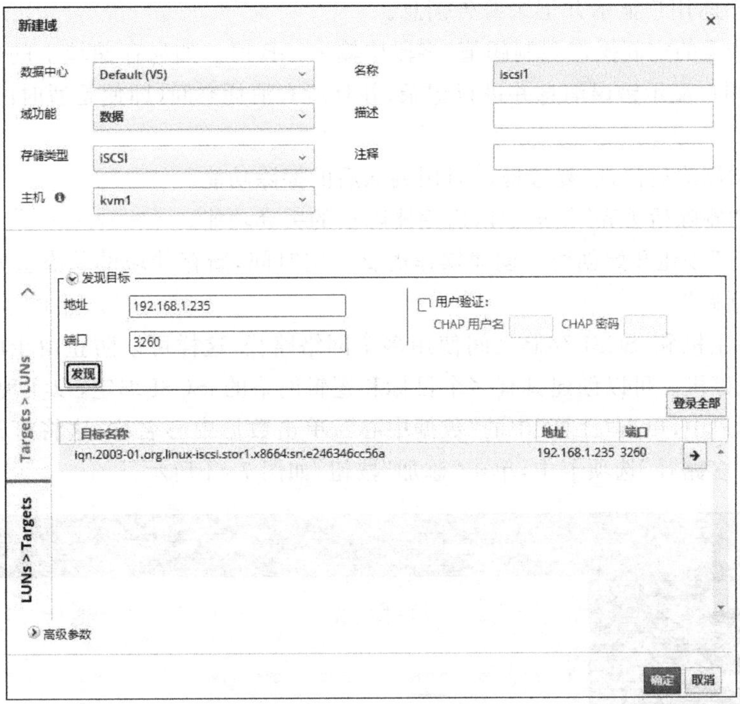

新建域

数据中心 Default (V5)　名称 iscsi1
域功能 数据　描述
存储类型 iSCSI　注释
主机 kvm1

Targets > LUNs　LUNs > Targets

发现目标
地址 192.168.1.235
端口 3260
发现
用户验证：
CHAP 用户名　CHAP 密码
登录全部

| 目标名称 | 地址 | 端口 |
| --- | --- | --- |
| iqn.2003-01.org.linux-iscsi.stor1.x8664:sn.e246346cc56a | 192.168.1.235 | 3260 |

高级参数
确定 取消

图 7-42 新建 iSCSI 存储域(二)

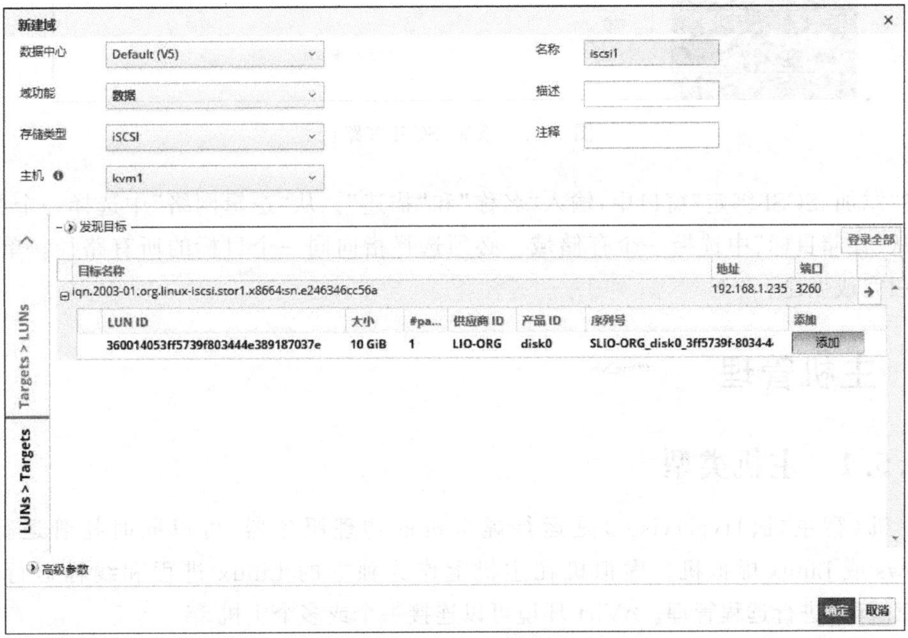

新建域

数据中心 Default (V5)　名称 iscsi1
域功能 数据　描述
存储类型 iSCSI　注释
主机 kvm1

Targets > LUNs　LUNs > Targets

发现目标
登录全部

| 目标名称 | 地址 | 端口 |
| --- | --- | --- |
| iqn.2003-01.org.linux-iscsi.stor1.x8664:sn.e246346cc56a | 192.168.1.235 | 3260 |

| LUN ID | 大小 | #pa... | 供应商 ID | 产品 ID | 序列号 | 添加 |
| --- | --- | --- | --- | --- | --- | --- |
| 360014053ff5739f803444e389187037e | 10 GiB | 1 | LIO-ORG | disk0 | SLIO-ORG_disk0_3ff5739f-8034-4 | 添加 |

高级参数
确定 取消

图 7-43 新建 iSCSI 存储域(三)

此百分比，则会向用户显示并记录警告消息。

(2) 在“严重的空间操作限制阈值”字段中输入 GB 值。如果存储域上的可用空间低于此值，则会向用户显示错误消息并进行记录，并且所有消耗空间(即使是暂时的)的新操作都将被阻止。

(3) 选中“删除后清理”复选框以启用删除后的擦除功能。

(4) 选中“删除后丢弃”复选框以启用删除后的丢弃功能。

单击“确定”按钮开始创建。创建操作需要一定时间，新存储域的状态会从“已锁定”变成最终状态“活跃”。

建议在宿主机和 iSCSI 存储之间使用多个网络路径，这样可以防止由于网络路径故障而导致的主机停机。可以创建具有多个目标和逻辑网络的 iSCSI 绑定，以实现冗余。

在管理门户中，单击“计算”中的“数据中心”，单击数据中心名称，这将打开详细信息视图。在“iSCSI 多路径”选项卡中，单击“添加”按钮，如图 7-44 所示。

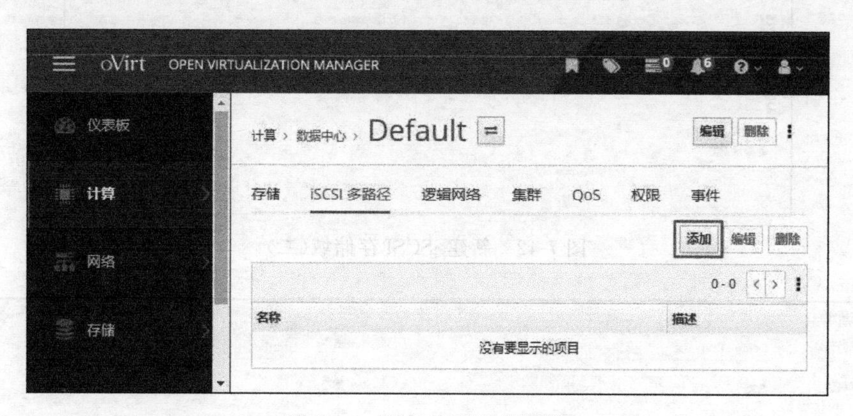


图 7-44　添加 iSCSI 多路径

在“添加 iSCSI 绑定”窗口中，输入“名称”和“描述”。从“逻辑网络”中选择一个逻辑网络，并从“存储目标”中选择一个存储域。必须选择指向同一个目标的所有路径。单击“确定”按钮完成创建。

# 7.5 主机管理

## 7.5.1 主机类型

主机(宿主机、Hypervisor)是运行虚拟机的物理服务器，可以同时托管运行多个 Windows 或 Linux 虚拟机。虚拟机在主机上作为独立的 Linux 进程和线程运行，并由 oVirt Engine 进行远程管理。oVirt 环境可以连接一个或多个主机。

oVirt 支持 2 种类型的主机：oVirt 节点、RHEL/CentOS。

**提示：**在主机的常规信息视图中“软件”用于显示主机的类型，例如：oVirt Node 4.4.4、

CentOS Linux 8 等。

主机默认会使用提供虚拟化优化的 tuned 配置文件。查看当前活动的配置文件，示例命令如下：

```
[root@kvm3 ~]# tuned-adm active
Current active profile: virtual-host
```

主机默认会启用安全功能。安全增强型 Linux(SELinux)和防火墙已完全配置，并且默认情况下处于启用状态。

查看 SELinux 的状态，示例命令如下：

```
[root@kvm3 ~]# getenforce
Enforcing

[root@kvm3 ~]# getsebool -a  | grep virt
staff_use_svirt --> off
unprivuser_use_svirt --> off
use_virtualbox --> off
virt_lockd_blk_devs --> off
virt_qemu_ga_read_nonsecurity_files --> off
virt_read_qemu_ga_data --> off
virt_rw_qemu_ga_data --> off
virt_sandbox_share_apache_content --> off
virt_sandbox_use_all_caps --> on
virt_sandbox_use_audit --> on
virt_sandbox_use_fusefs --> off
virt_sandbox_use_mknod --> off
virt_sandbox_use_netlink --> off
virt_sandbox_use_sys_admin --> off
virt_transition_userdomain --> off
virt_use_comm --> off
virt_use_execmem --> off
virt_use_fusefs --> on
virt_use_glusterd --> off
virt_use_nfs --> on
virt_use_pcscd --> off
virt_use_rawip --> off
virt_use_samba --> on
virt_use_sanlock --> on
virt_use_usb --> on
virt_use_xserver --> off
```

查看防火墙的状态，示例命令如下：

```
[root@kvm3 ~]# firewall-cmd --list-all
public (active)
  target: default
  icmp-block-inversion: no
  interfaces: enp1s0 ovirtmgmt
  sources:
  services: cockpit dhcpv6-client glusterfs libvirt libvirt-tls ovirt-imageio ovirt-vmconsole snmp ssh vdsm
  ports: 22/tcp 6081/udp
  protocols:
  masquerade: no
  forward-ports:
  source-ports:
  icmp-blocks:
  rich rules:
```

推荐在生产环境中使用 oVirt 节点，它是特殊版本的 Enterprise Linux，仅带有托管虚拟机所需的软件包。推荐使用 Cockpit 来管理 oVirt 节点，而不是通过 SSH 或控制台直接访问。

不管是哪种类型的节点，都不建议创建额外的用户账号及安装虚拟化之外的软件，包括第三方监控程序。

## 7.5.2 编辑主机配置

主机添加成功之后，还可以修改部分配置。

单击“计算”中的“主机”，然后选择要更改的主机，单击“编辑”按钮，会显示“编辑主机”窗口，如图 7-45 所示。

### 1. 常规设置

可以修改名称和注释。名称不能超过 40 个字符，而且必须是唯一的，可以包含大小写字母、数字、连字符和下画线。注释用于帮助管理员了解主机。

### 2. 电源管理设置

如果主机上安装了受支持的电源管理卡，就可以配置电源管理了，如图 7-46 所示。

电源管理由主机群集的调度策略控制。如果启用了电源管理并且达到了定义的低利用率值，则引擎将关闭主机电源以节省能源，并可在需要进行负载平衡或群集中没有足够的可用主机时再次启动主机。如果选中“禁用电源管理的策略控制”复选框，则会禁用策略控制。

电源管理中的代理是指“隔离代理(Fence Agent)”，在本书第 2 章“实现虚拟机高可用”中详细介绍过隔离的机制及隔离代理的配置，在此处可以通过图形化的界面来配置隔离代理。

### 3. SPM 优先级设置

定义主机被授予存储池管理器(Storage Pool Manager，SPM)角色的可能性。低优先级

编辑主机 ×

常规 >

电源管理

SPM

控制台和 GPU

内核

承载的引擎

关联

主机集群 Default

数据中心：Default

使用 Foreman/Satellite

名称 kvm2

注释

主机名/IP 192.168.1.232

SSH 端口 22

安装后激活主机

验证

用户名 root

SSH 公共密钥 ssh-rsa AAAAB3NzaC1yc2EAAAADAQABAAABAQCIYK

高级参数

SSH 指印

SHA256:m0an9atouZiL7aY0dqP2/XYmj0sAwo6Aa7L6WZbqVY4

输入主机指印或手动从主机获取

确定 取消

图 7-45　编辑主机-常规

编辑主机 ×

常规

电源管理 >

SPM

控制台和 GPU

内核

承载的引擎

关联

启用电源管理

kdump 集成

禁用电源管理的策略控制

按先后顺序排序的代理

添加隔离代理 +

高级参数

确定 取消

图 7-46　编辑主机-电源管理

意味着将主机授予SPM角色的可能性低,而高优先级则意味着增加主机的可能性。默认设置为“正常”,如图7-47所示。

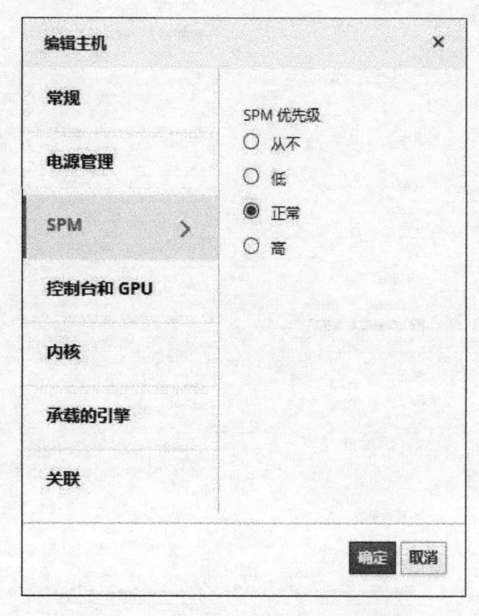


图7-47　编辑主机-SPM优先级

### 4. 控制台和GPU设置

使用指定的显示地址来覆盖此主机上运行的所有虚拟机的设置,可以是FQDN或IP地址,如图7-48所示。

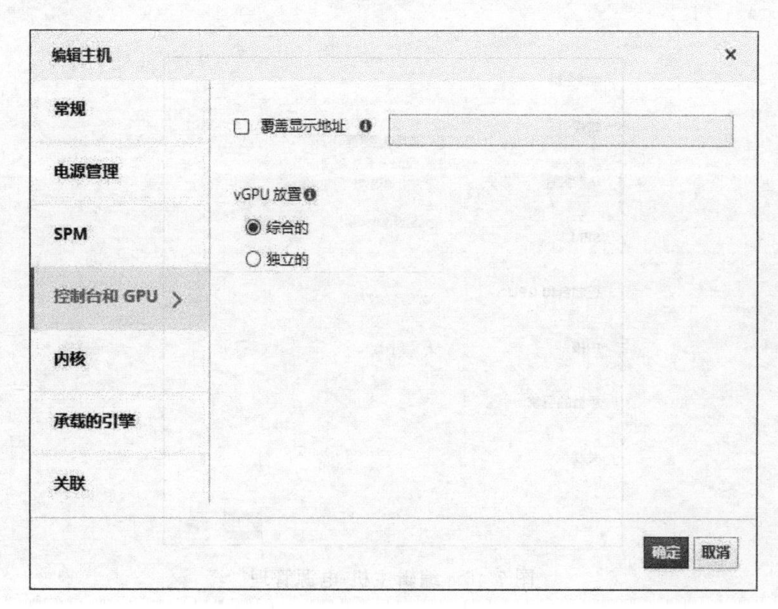


图7-48　编辑主机-控制台与GPU

如果主机位于 NAT 防火墙后面，当用户从外部连接到虚拟机时，可以通过此功能返回公共 IP 或 FQDN，而不是返回运行虚拟机的主机的私有地址。

oVirt 4.1.4 新增了 vGPU 透传(passthrough)的功能。vGPU 技术可以将物理 GPU(具有 GRID 的功能，例如 NVIDIA Tesla M60)"分片(shard)"为多个较小的 vGPU，然后可以将每个 vGPU 分配给虚拟机。这样便可以在虚拟机中运行需要 GPU 加速的工作负载。vGPU 的放置有两个选项：综合的(Consolidated)和独立的(Separated)。

**5. 内核设置**

可以在"内核"选项卡中通过复选框轻松设置宿主机的内核参数，如图 7-49 所示。

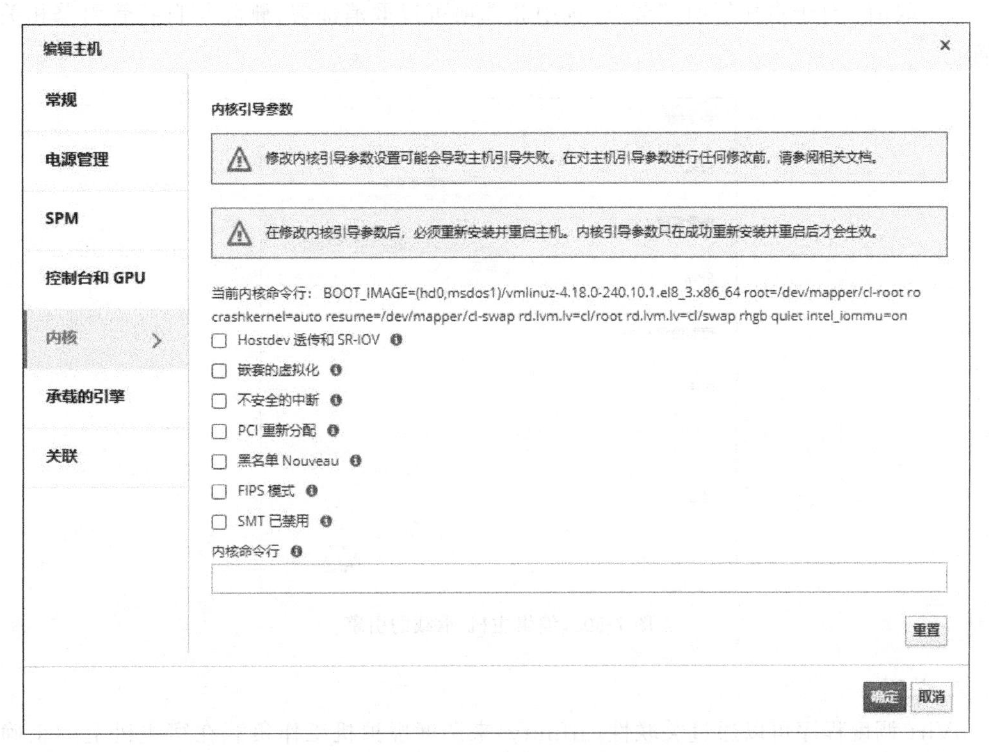


图 7-49　编辑主机-内核

当前 oVirt 版本有以下可配置参数：

(1) Hostdev 透传和 SR-IOV。

(2) 嵌套的虚拟化。

(3) 不安全的中断。

(4) PCI 重新分配。

(5) 黑名单 Nouveau。

(6) FIPS 模式。

(7) SMT 已禁用。

对于更复杂的更改，可以在"内核命令行"旁边的文本框中输入所需的任何其他参数。

在更改之前必须将主机置于维护模式。更改任何内核命令行参数都需要重新启动主机才能生效，重新启用之后还需要重新安装主机。

**6. 承载引擎**

设置主机是否可以托管 oVirt 引擎，即是否可以运行 oVirt 引擎的虚拟机，如图 7-50 所示。

提供的 3 个选项如下。

(1) 无：无须任何操作。

(2) 部署：选择此选项可将主机部署为自托管引擎节点。

(3) 取消：对于自托管引擎节点，选中此选项可以取消部署，删除与自托管引擎相关的配置。

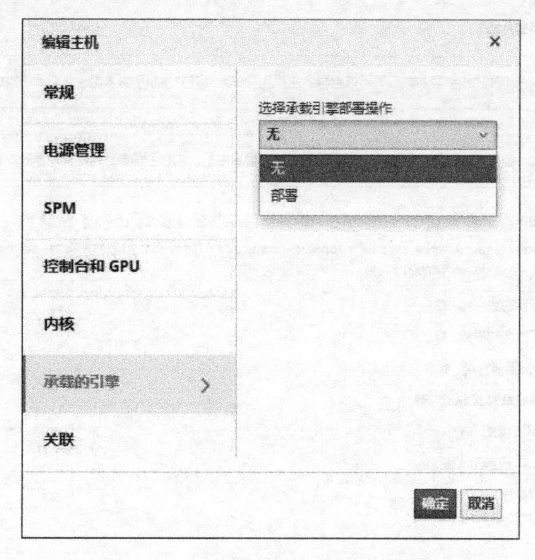


图 7-50　编辑主机-承载的引擎

**7. 关联**

oVirt 调度程序可以通过关联性(affinity)来调度虚拟机工作负载在宿主机上的正确分布。在本选项卡中，可以将宿主机分配到关联组(Affinity Group)，也可以分配关联标签(Affinity Label)，如图 7-51 所示。

## 7.5.3　主机维护模式

在执行任何可能导致 VDSM 停止工作的操作(例如重新启动、配置网络或存储)前，都需要将主机置于维护模式(Maintenance Mode)。

当主机进入维护模式时，oVirt 引擎会尝试将所有正在运行的虚拟机迁移到其他主机。要满足实时迁移的先决条件，特别是群集中必须至少有一个活动主机，且该主机具有运行要迁移虚拟机的能力。

固定(pinned)到某台主机或无法迁移的虚拟机将被关闭。

编辑主机
常规
电源管理
SPM
控制台和 GPU
内核
承载的引擎
关联
选择一个关联组　添加
选择的关联组　没有选择关联组
选择一个关联标签　添加
选择的关联标签　没有选择关联标签
确定　取消

图 7-51　编辑主机-关联

如果主机是存储池管理器(SPM),则 SPM 角色将迁移到另一台主机。

当维护操作完成后,可以对处于维护模式中的主机进行激活操作。换句话说,维护与激活是一对逆操作。

单击“计算”中的“主机”,然后选择要操作的主机,单击“管理”下拉菜单中的“维护”选项,如图 7-52 所示。确认后就可以进行虚拟机的迁移操作了。

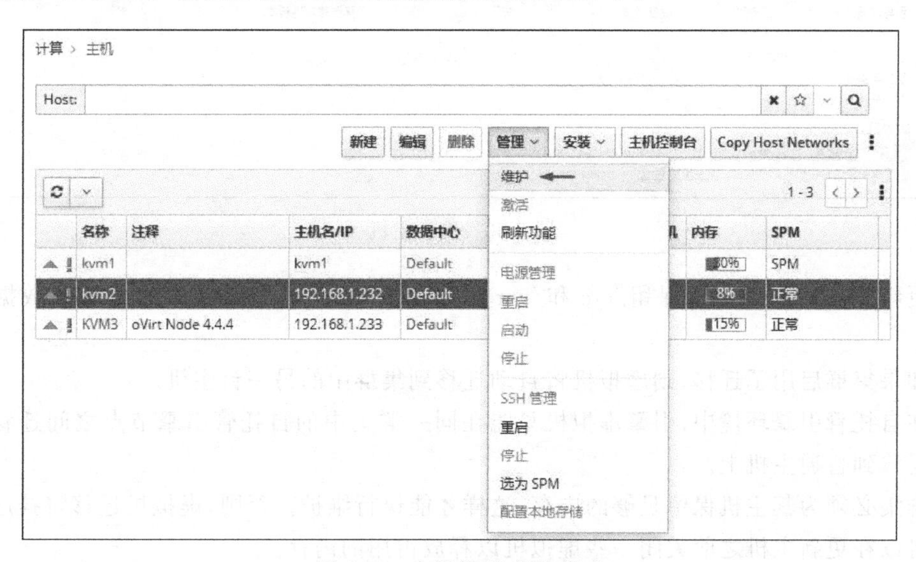


图 7-52　主机管理-维护模式

如果在任何虚拟机上迁移失败,则可单击“管理”下拉菜单中的“激活”以停止将其置于维护模式,然后在虚拟机上单击“取消迁移”按钮。排除迁移失败故障之后,再进行操作。

## 7.5.4 更新主机

可以更新群集中的所有主机，也可以更新单个主机。在生产环境中，更新操作要谨慎。建议先更新一台或几台主机，确认没有问题之后再更新群集中的其他所有主机。

**1. 更新个别主机**

在管理门户中，单击“计算”中的“主机”，然后选择要更新的主机，单击“安装”下拉菜单中的“检查升级”，然后单击“确定”按钮。

如果有更新，则可单击“安装”下拉菜单中的“升级”，或者在主机详细信息中“动作项目”下单击“升级”，如图 7-53 所示。

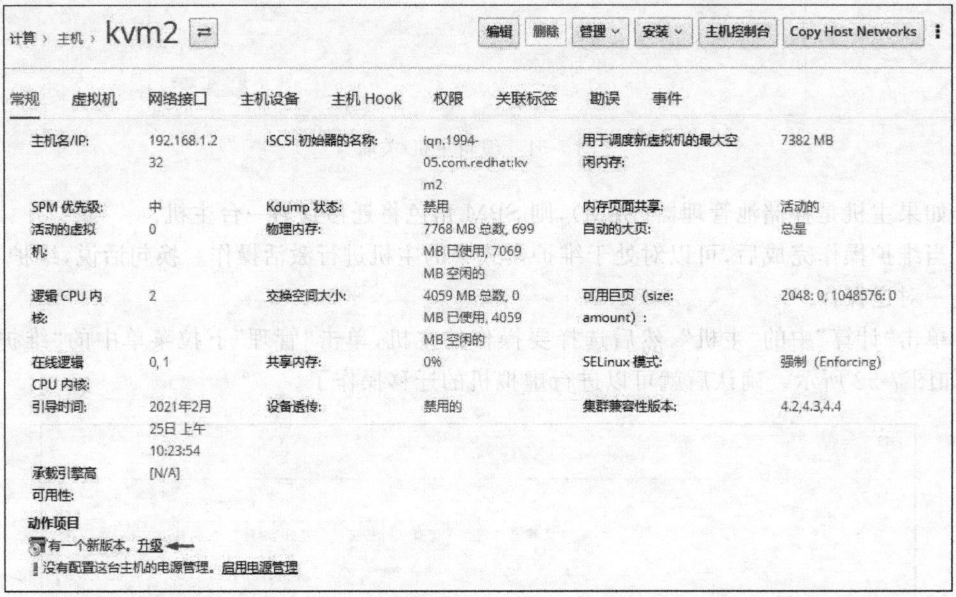

图 7-53 更新主机

更新 oVirt 节点，仅会保留/etc 和/var 目录中的内容，其他路径中的已修改数据将被覆盖。

如果集群启用了迁移，则虚拟机将自动迁移到集群中的另一台主机。

在自托管引擎环境中，引擎虚拟机只能在同一群集中的自托管引擎节点之间迁移。它不能迁移到普通主机上。

群集必须为其主机保留足够的内存，这样才能执行维护。否则，虚拟机迁移将挂起并失败。可以在更新主机之前关闭一些虚拟机以释放占用的内存。

不要同时更新所有主机，因为至少有一台主机必须保持可用状态，这样才能执行 Storage Pool Manager(SPM)任务。

无法将固定的虚拟机(例如使用 vGPU 的虚拟机)迁移到另一台主机。在更新主机之前，必须关闭固定的虚拟机。

可以在事件中看到更新过程中的信息，如图 7-54 所示。

图 7-54　更新主机的日志信息

**2. 更新群集中的所有主机**

可以更新群集中的所有主机，这在升级到新版本的 oVirt 时特别有用。

**提示：**也可以通过 dnf 或 yum 命令手工更新主机，不过通过 oVirt 进行更新会获得更多日志信息。

## 7.5.5 重新安装主机

从管理门户重新安装 oVirt 节点和 RHEL/CentOS 主机。该过程包括停止并重新启动主机。

如果集群启用了迁移，则虚拟机会自动迁移到集群中的另一台主机。

确保群集有足够的内存供其主机执行维护。如果群集缺少内存，则虚拟机的迁移将挂起，然后会失败。为了减少内存使用，在将主机置于维护模式之前，应关闭部分或全部虚拟机。

执行重新安装之前，需要确保群集包含多个主机。不要尝试同时重新安装所有主机。一台主机必须保持可用状态才能执行 Storage Pool Manager(SPM)任务。

需要先将要重新安装的主机置于维修模式，然后单击“安装”下拉菜单中的“重新安装”，单击“确定”按钮以重新安装主机。

重新安装主机并将其状态恢复为 Up 后，可以将虚拟机迁移回该主机。

如果重新安装后，管理门户显示主机状态为 Install Failed，则可以单击“管理”下拉菜单中的“激活”，等主机变为 Up 状态后就可以继续使用了。

**提示**：先将主机从 oVirt 中删除，然后通过“新建”的方式再将主机添加到 oVirt 中，其效果相当于重新安装主机。

## 7.6 虚拟机管理

可以在管理门户和虚拟机门户中完成 oVirt 中的大多数虚拟机的管理任务，而有些管理任务只能在管理门户中完成。2 个门户之间的用户界面有些差异，本书只讲解在管理门户中执行的任务。

### 7.6.1 在客户端计算机上安装支持组件

需要在客户端计算机上安装 Remote Viewer。如果是 Windows 客户端，则需要安装 usbdk 驱动程序。

1. Remote Viewer

在 oVirt 中，打开虚拟机控制台的默认应用程序是 Remote Viewer，所以必须将其安装在客户端计算机上。安装后，在尝试于虚拟机打开 SPICE 会话时会自动调用它，或者作为独立应用程序来运行。

在 RHEL/CentOS 安装 Remote Viewer，示例命令如下：

```
[root@tompc1 ~]# dnf -y install virt-viewer
```

在 Windows 上安装 Remote Viewer，可以从 https://virt-manager.org/download/上下载 32 位或 64 位的 MSI 格式安装包进行安装。

2. Windows 上的 usbdk 驱动程序

usbdk 是一个驱动程序，用于 Windows 操作系统上 remote-viewer 对 USB 设备进行独占访问。

先从 https://www.spice-space.org/download.html 上下载 32 位或 64 位的 MSI 格式安装包，然后进行安装。

两个组件的有些版本在 Windows 平台上安装结束时，可能不会出现提示信息。我们可通过事件查看器中的应用程序日志来判断是否安装成功，如图 7-55 所示。

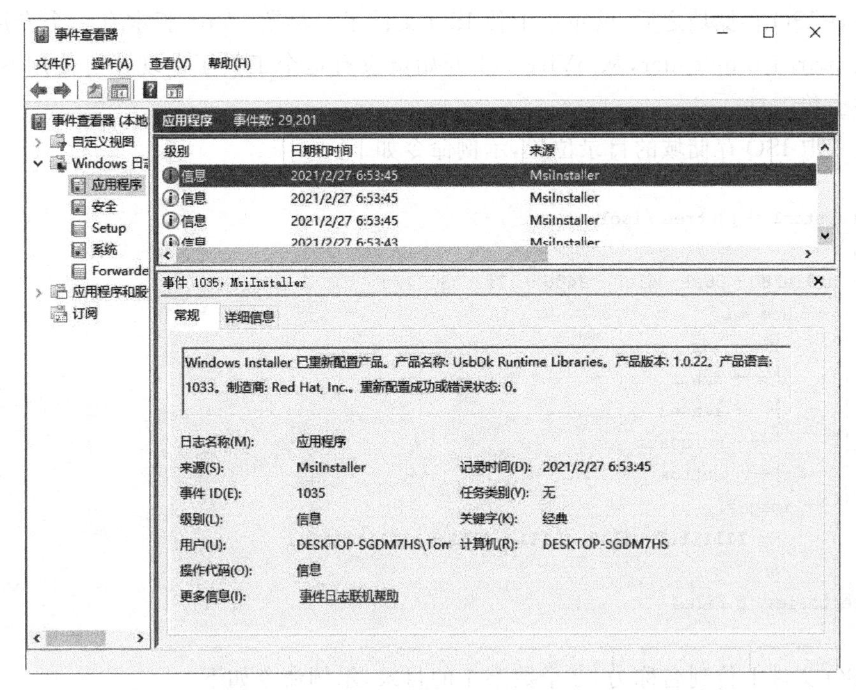


图 7-55　应用程序日志中的软件安装信息

## 7.6.2　准备 ISO 存储域及 ISO 文件

需要准备好 ISO 存储和 ISO 文件，这样才能安装虚拟机操作系统。

在管理门户中，选择“存储”→“存储域”，单击“新建域”按钮。输入存储域的名称，将域功能设置为 ISO，并将存储类型设置为 NFS，输入“导出路径”，单击“确定”按钮开始创建，如图 7-56 所示。

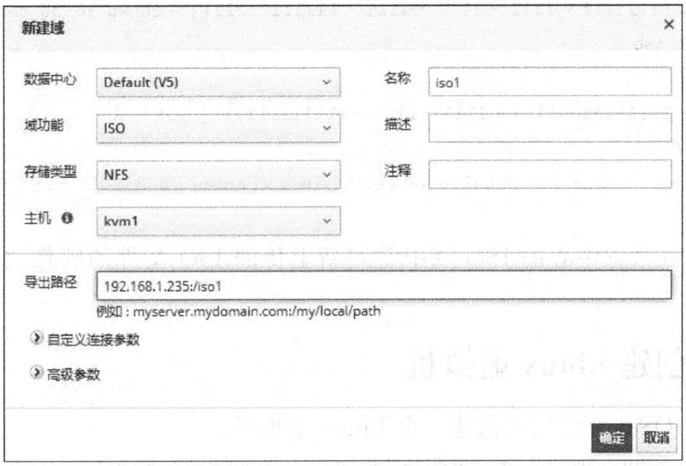

新建域
数据中心 Default (V5)
名称 iso1
域功能 ISO
描述
存储类型 NFS
注释
主机 kvm1
导出路径 192.168.1.235:/iso1
例如：myserver.mydomain.com:/my/local/path
自定义连接参数
高级参数
确定 取消

图 7-56　创建 ISO 存储域

准备好ISO存储域之后，就可以上传ISO文件了。早期oVirt版本有一个专用的ISO上传工具ovirt-iso-uploader，从oVirt 4.4开始就没有这个工具了。现在可使用SCP等普通的文件上传工具。

首先获取ISO存储域的目录位置，示例命令如下：

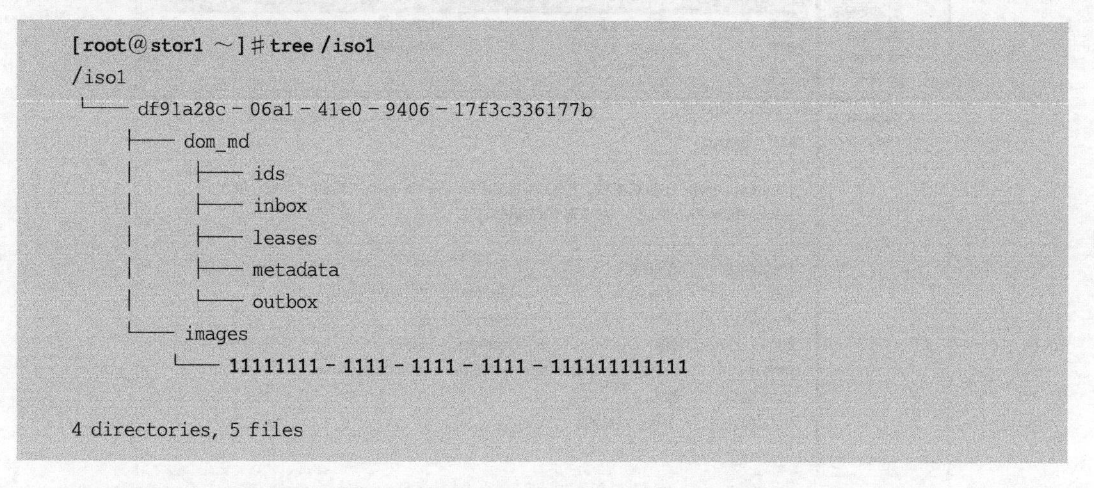

```
[root@stor1 ~]# tree /iso1
/iso1
└── df91a28c-06a1-41e0-9406-17f3c336177b
    ├── dom_md
    │   ├── ids
    │   ├── inbox
    │   ├── leases
    │   ├── metadata
    │   └── outbox
    └── images
        └── 11111111-1111-1111-1111-111111111111

4 directories, 5 files
```

将ISO文件上传到名称为32个数字1的目录，示例命令如下：

```
[root@tomkvm1 iso]# scp centos-8.3.2011-x86_64-dvd1.iso 192.168.1.235:/iso1/df91a28c
-06a1-41e0-9406-17f3c336177b/images/11111111-1111-1111-1111-111111111111/
```

将新ISO文件的用户和组所有权更改为36：36(vdsm的用户和组)，示例命令如下：

```
[root@stor1 ~]# cd /iso1/df91a28c-06a1-41e0-9406-17f3c336177b/images/11111111-
1111-1111-1111-111111111111/

[root@stor1 11111111-1111-1111-1111-111111111111]# chown 36:36 centos-8.3.2011-
x86_64-dvd1.iso

[root@stor1 11111111-1111-1111-1111-111111111111]# ls -l
total 9047040
-rw-r--r--. 1 vdsm kvm 9264168960 Feb 27 10:13 centos-8.3.2011-x86_64-dvd1.iso
```

现在就可以在数据中心的ISO域中看到新上传的ISO文件的映像文件了，如图7-57所示。

## 7.6.3 创建Linux虚拟机

下面以CentOS 8.3为例来创建一个Linux虚拟机。

单击"计算"中的"虚拟机"，单击"新建"按钮打开"新建虚拟机"窗口。从下拉列表中选择一个操作系统，由于 oVirt 中没有 CentOS，所以选择类似的 Red Hat Enterprise Linux 8. x x64，如图 7-58 所示

图 7-57 ISO 存储域中的文件列表

图 7-58 创建 Linux 虚拟机-常规

实例类型选择 Small(后期可以调整),优化目标选择“服务器”。输入虚拟机名称、描述、注释等信息。在实例镜像中单击“创建”按钮,以便创建虚拟磁盘,如图 7-59 所示。

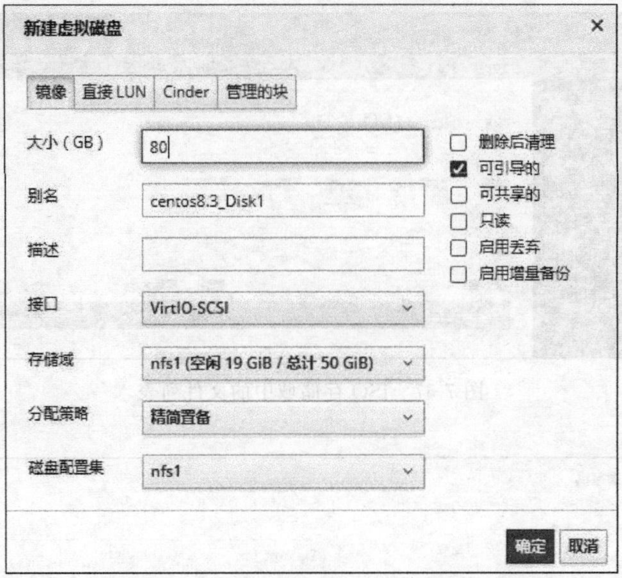

图 7-59 创建 Linux 虚拟机-新建虚拟磁盘

为新的虚拟磁盘设置大小(GB)和别名。既可以接受所有其他字段的默认设置,也可以根据需要更改它们,本实验将存储域设置为 nfs1。

“普通”选项卡底部有一个 nic1 下拉列表,从中选择一个 vNIC 配置文件,这样就可以将虚拟机连接到网络,如图 7-58 所示。

单击左下角的“显示高级选项”按钮,可以根据需要在“系统”选项卡上修改虚拟机的内存、vCPU 等配置,在“启动选项”选项卡中的“附加 CD”中选择 CentOS 8.3 的 ISO 文件,将引导启动的第 1 个设备设置为 CD-ROM,将第 2 个设备设置为硬盘,如图 7-60 所示。

单击“确定”按钮将创建新的虚拟机,它会显示在虚拟机列表中,此时虚拟机的状态为 Down。单击“运行”按钮启动虚拟机。

单击“控制台”按钮,会下载一个名为 console.vv 的文件。打开这个文件,会自动打开虚拟机的控制台窗口。

**提示:** 如果没有打开控制台窗口,则需检查文件关联设置。

可以在控制台中进行操作系统的安装。在 oVirt 中安装 Linux 与普通安装步骤相同,如图 7-61 所示。

**提示:** 在控制台中,按下 Shift+F12 组合键可以从虚拟机中释放鼠标。

安装完成后可以通过 virsh 查看新虚拟机的属性,示例命令如下:

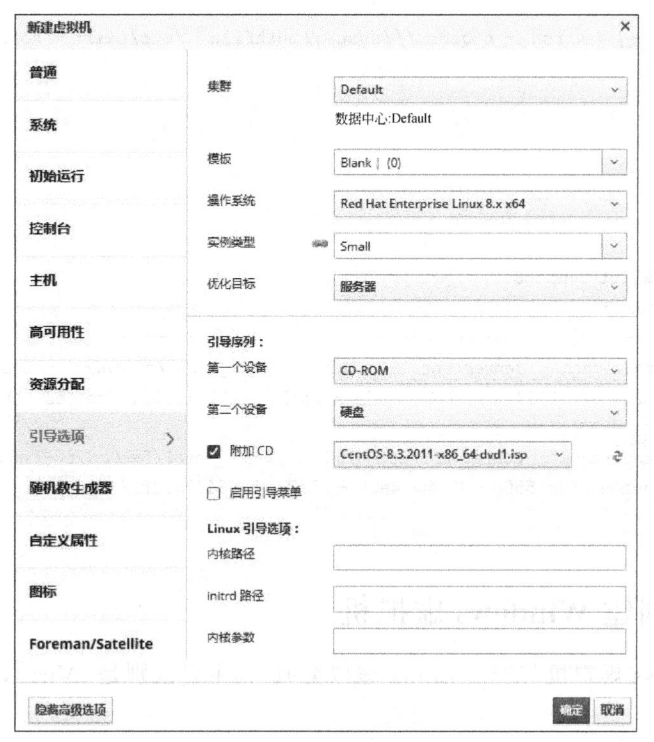


图 7-60　创建 Linux 虚拟机-设置引导选项

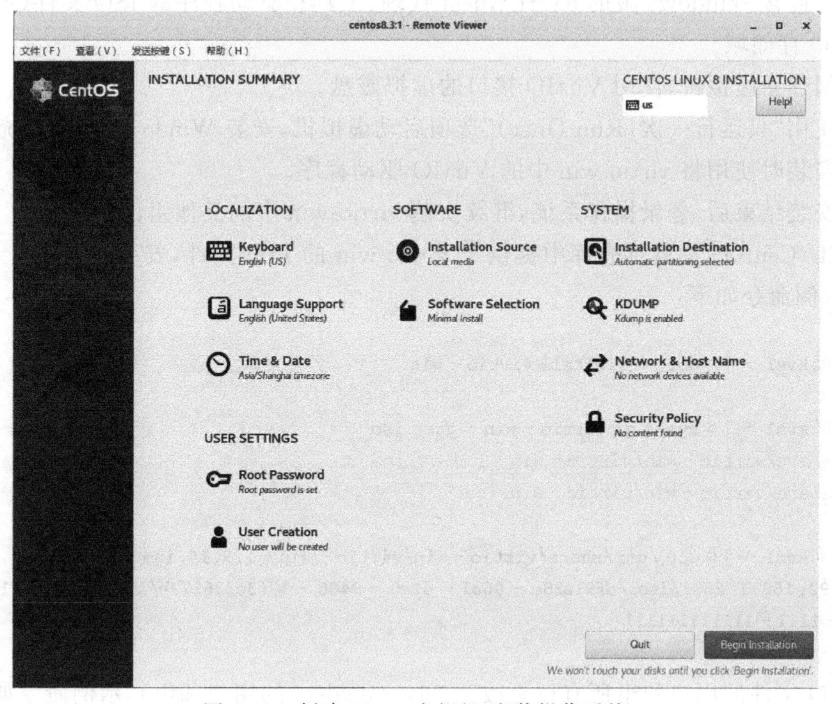


图 7-61　创建 Linux 虚拟机-安装操作系统

```
[root@kvm2 ~]# virsh -c qemu:///system?authfile=/etc/ovirt-hosted-engine/virsh_
auth.conf

virsh # list
 Id   Name        State
----------------------------
 1    centos8.3   running

virsh # domblklist centos8.3
 Target   Source
-------------------------------------------------------------------------------
 sdc      /rhev/data-center/mnt/192.168.1.235:_iso1/df91a28c-06a1-41e0-9406-
17f3c336177b/images/11111111-1111-1111-1111-111111111111/centos-8.3.2011-x86_64-
dvd1.iso
 sda      /rhev/data-center/mnt/192.168.1.235:_nfs1/5ee0c87f-e60f-4d1d-aece-
1a9783b482c9/images/58c1550d-92d4-4a00-b3a0-f466447d92f1/4c8d2330-b82f-4901-923b-
ade5667839ab
```

### 7.6.4 创建 Windows 虚拟机

安装 Windows 虚拟机与安装 Linux 虚拟类似，其主要区别是 oVirt Guest 代理、工具和驱动程序的安装。

安装过程包括以下步骤：

(1) 将包含 Windows 版的 oVirt Guest 代理、工具和驱动程序的 ISO 文件(virtio-win)上传到 ISO 存储域。

(2) 创建新虚拟机，使用 VirtIO 接口的虚拟磁盘。

(3) 使用“只运行一次(Run Once)”选项启动虚拟机，安装 Windows 操作系统。

(4) 安装时使用将 virtio-win 中的 VirtIO 驱动程序。

(5) 安装结束后，登录操作系统，继续安装 virtio-win 中的其他组件。

RHEL/CentOS 的软件仓库中提供了 virtio-win 的 ISO 文件，安装后将其上传到 ISO 存储中，示例命令如下：

```
[root@kvm1 ~]# dnf -y install virtio-win

[root@kvm1 ~]# rpm -ql virtio-win | grep iso
/usr/share/virtio-win/virtio-win-1.9.14.iso
/usr/share/virtio-win/virtio-win.iso

[root@kvm1 ~]# scp /usr/share/virtio-win/virtio-win-1.9.14.iso \
    192.168.1.235:/iso1/df91a28c-06a1-41e0-9406-17f3c336177b/images/11111111-1111-
1111-1111-111111111111
```

将 ISO 文件的用户和组所有权更改为 36：36(vdsm 的用户和组)，示例命令如下：

```
[root@stor1 11111111-1111-1111-1111-111111111111]# chown 36:36 \
    virtio-win-1.9.14.iso
```

创建新的虚拟机，根据需要设置 vCPU 的数量、内存大小、网络接口、磁盘大小及接口等参数。

因为在安装 Windows 操作系统的过程中需要使用 VirtIO 驱动程序，所以需要使用“只运行一次”选项来启动虚拟机。单击“计算”中的“虚拟机”，然后选择刚创建的新虚拟机。单击“运行”中的“只运行一次”，如图 7-62 所示。

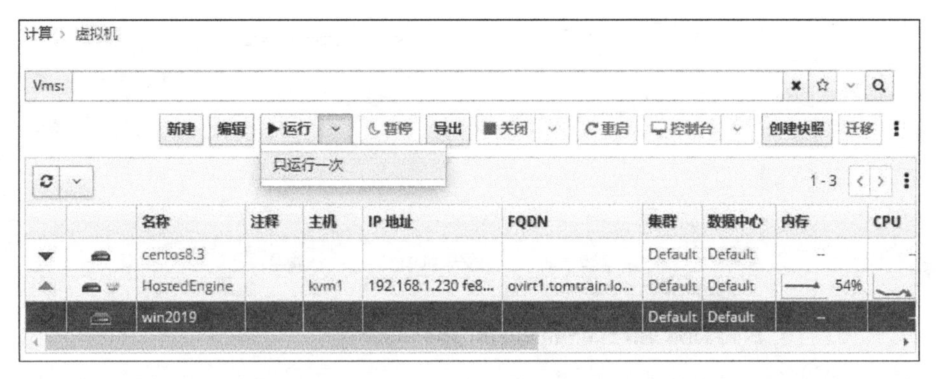


图 7-62 使用“只运行一次”选项来启动虚拟机

在“运行虚拟机”窗口中配置引导选项。选择“附加 CD”复选框，然后从下拉列表中选择 Windows 的安装介质 ISO。选择“附加 Windows 客户端工具 CD”复选框，这会将 virt-win 的 ISO 文件也附加给虚拟机，如图 7-63 所示。

图 7-63 运行虚拟机设置窗口

通过“上移”“下移”按钮将 CD-ROM 移到“引导序列”字段的顶部。

单击“确定”按钮，将虚拟机的状态更改为 Up，并开始操作系统安装。打开虚拟机的控制台进行安装。

当 Windows 安装程序提示找不到任何驱动器时，单击“加载驱动程序”，如图 7-64 所示。

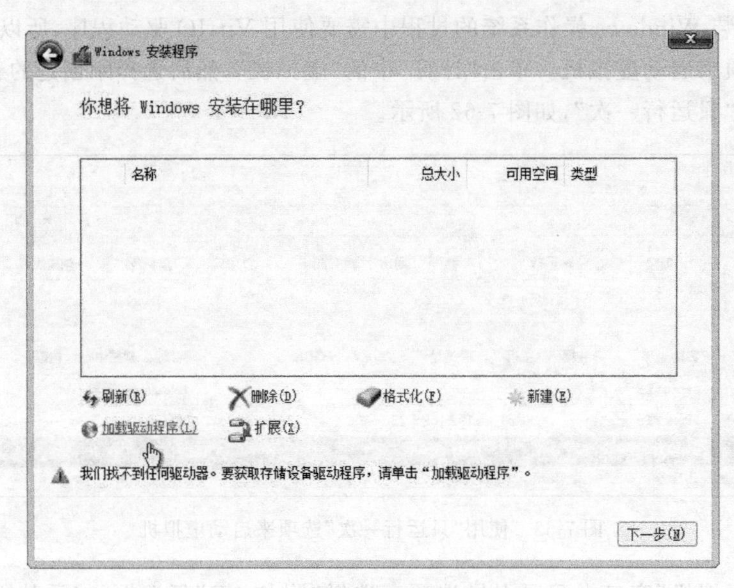


图 7-64　无法找到任何驱动器

在“选择要安装的驱动程序”列表中，根据 Windows 版本选择适合的驱动程序。例如，对于 Windows Server 2019，选择 Red Hat VirtIO SCSI 控制器(E:\amd64\2k19\vioscsi.inf)，单击“下一步”按钮继续，如图 7-65 所示。

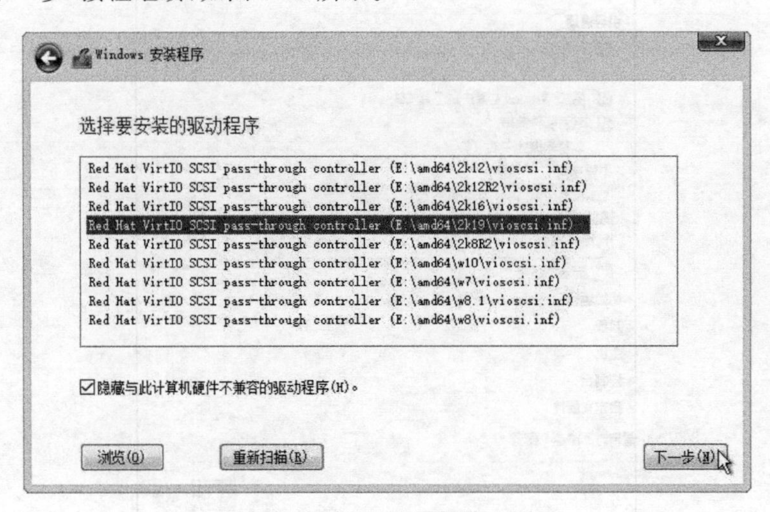


图 7-65　选择要安装的驱动程序

加载驱动程序之后，就可以识别 VirtIO 接口的磁盘了，如图 7-66 所示。其余安装操作按常规进行即可。

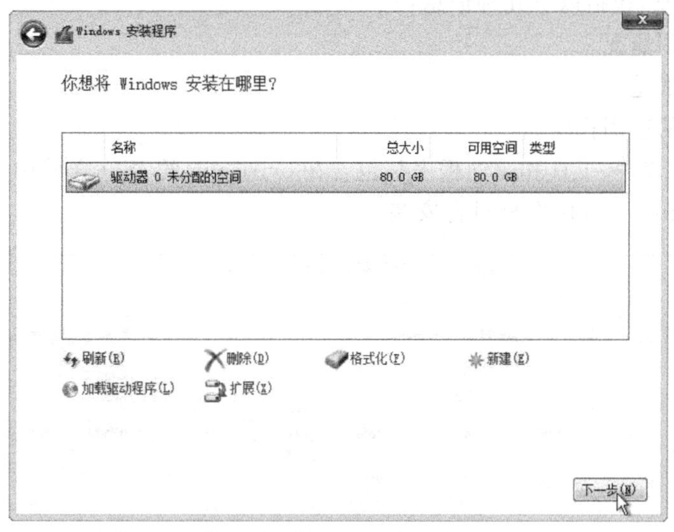


图 7-66 选择要安装的磁盘

在操作系统安装完成之后，打开设备管理器，会发现有多个没有驱动程序的设备，如图 7-67 所示。

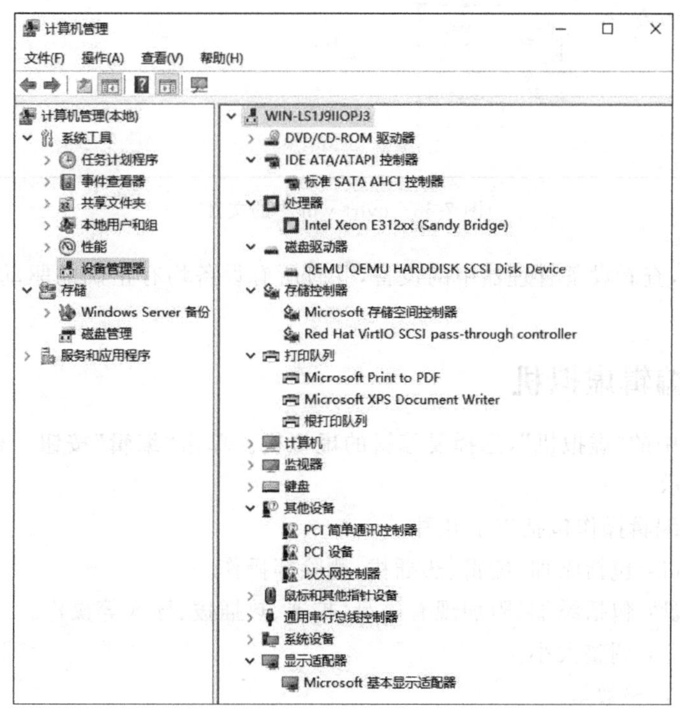


图 7-67 存在没有驱动程序的设备

除了为这些设备安装驱动程序之外，还需要安装 oVirt Guest 代理和工具，这样才能在管理门户和虚拟机门户中正常关闭或重新启动虚拟机。同时这些代理和工具还可以提供与虚拟机有关的信息，包括以下几种信息：

(1) 资源使用情况。

(2) IP 地址信息。

(3) 已安装的应用程序。

oVirt Guest 代理、工具和驱动程序都包含在 ovirt-win 的 ISO 文件中，如图 7-68 所示。双击 virtio-win-guest-tools 图标进行安装。

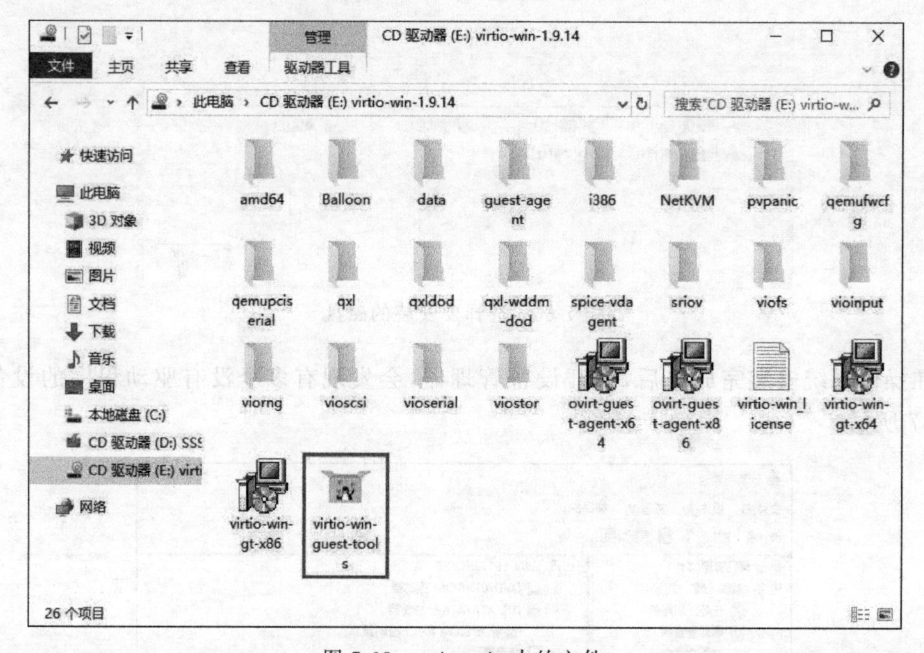

图 7-68 ovirt-win 中的文件

安装完成后，查看设备管理器中的设备，此时所有设备均有正确的驱动程序，如图 7-69 所示。

## 7.6.5 编辑虚拟机

单击“计算”中的“虚拟机”，选择要编辑的虚拟机。单击“编辑”按钮。根据需要更改设置，如图 7-70 所示。

比较常见的编辑操作包括以下几种。

(1) 网络接口：包括添加、编辑、热插拔、删除等操作。

(2) 虚拟磁盘：包括添加、附加现有磁盘、扩展、热插拔、导入等操作。

(3) 虚拟内存：调整大小。

(4) VCPU：调整数量。

图 7-69 所有设备均有正确的驱动程序

图 7-70 编辑虚拟机

(5) 将虚拟机固定到多个主机。

(6) 更改虚拟机的 CD。

在进行任何更改之前,应确保了解正确的参数。对虚拟机的某些更改会立即生效,而有些更改则必须关闭并重新启动虚拟机才会生效。

## 7.6.6 虚拟机常规操作

可以在 oVirt 中对虚拟机进行全生命周期管理。日常的操作包括以下几种。

### 1. 关闭虚拟机

关闭(Shutdown)是正常关闭虚拟机,而断电(Power Off)则执行强制关机操作。正常关闭通常比断电更可取。

### 2. 暂停虚拟机

暂停虚拟机的效果类似于物理机的休眠,虚拟机的状态将被更改为 Suspended。

### 3. 重新启动虚拟机

这与关闭虚拟机后再启动的效果相同。

### 4. 删除虚拟机

虚拟机正在运行时,“删除”按钮是被禁用的。必须先关闭虚拟机,然后才能将其删除。

选中“删除磁盘”复选框,可以将附加到虚拟机的虚拟磁盘与虚拟机一起删除。如果清除了“删除磁盘”复选框,则仅仅删除虚拟机配置,而虚拟磁盘将作为浮动磁盘(无主磁盘)保留在环境中。

### 5. 克隆虚拟机

可以用克隆虚拟机的方式快速创建一台新的虚拟机。

单击“计算”中的“虚拟机”,然后选择要克隆的虚拟机。单击更多操作标识符“⋮”,然后单击“克隆 VM”。输入新虚拟机的克隆名称,设置其他的参数,如图 7-71 所示。单击“确定”按钮,这样就完成了克隆操作。

## 7.6.7 快照管理

快照是虚拟机的操作系统和应用程序在指定时间点上的视图。建议在对虚拟机进行重大更改之前,先创建快照,如果出现意外情况,则可以使用快照将虚拟机恢复到以前的状态。

### 1. 创建快照

单击“计算”中的“虚拟机”,单击虚拟机的名称以转到详细信息视图。单击“快照”选项卡,然后单击“创建”按钮。输入快照的描述。使用复选框选择要包括的磁盘,如图 7-72 所示。

如果未选择任何磁盘,则将创建不包含磁盘的虚拟机的部分快照。

如果选中了“保存内存”选项,则会将正在运行的虚拟机的内存也包括在快照中。

单击“确定”按钮开始创建,创建完毕后快照会出现在快照列表中。

图 7-71　克隆虚拟机

图 7-72　创建快照

### 2. 还原快照

快照可用于将虚拟机还原到之前的状态。在虚拟机详细信息视图中单击“快照”选项卡以列出可用的快照。

选择要还原的快照。单击“预览(Preview)”下拉按钮，然后选择“自定义”选项，如图 7-73 所示。

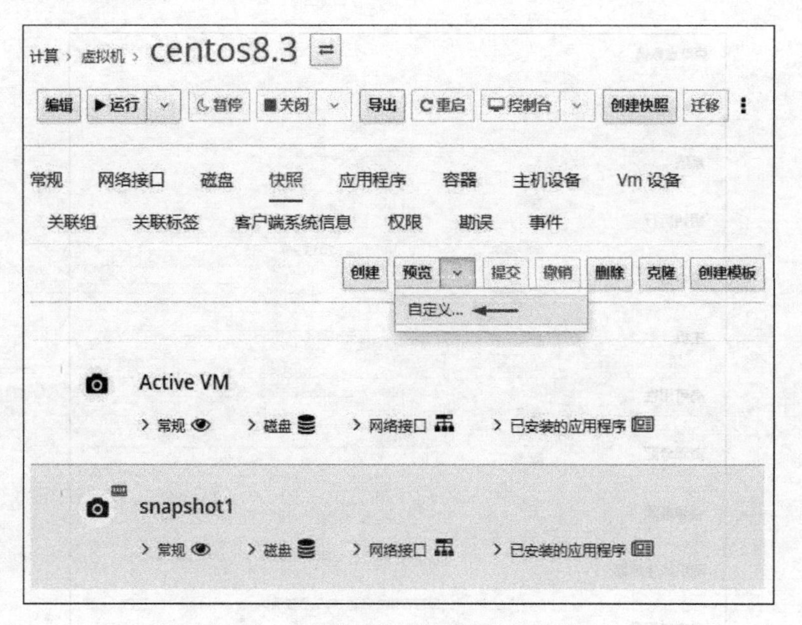


图 7-73 使用快照还原虚拟机-快照列表

根据需要使用复选框选择要还原的虚拟机配置、内存和磁盘，然后单击“确定”按钮，如图 7-74 所示。

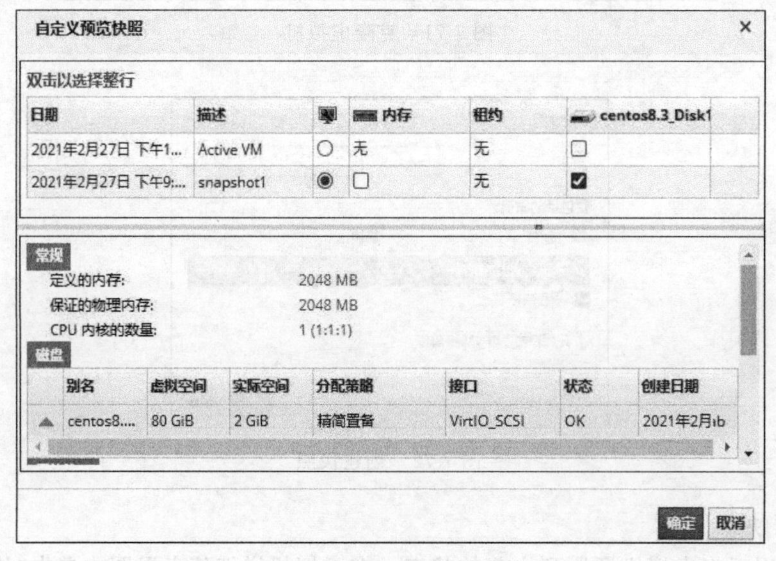


图 7-74 使用快照还原虚拟机-自定义预览快照

单击“提交”按钮将虚拟机永久还原到快照的状态，任何后续快照都将被删除。或者单击“撤销”按钮以停用快照，并使虚拟机返回其先前状态。

本示例中单击“提交”按钮进行还原，如图 7-75 所示。

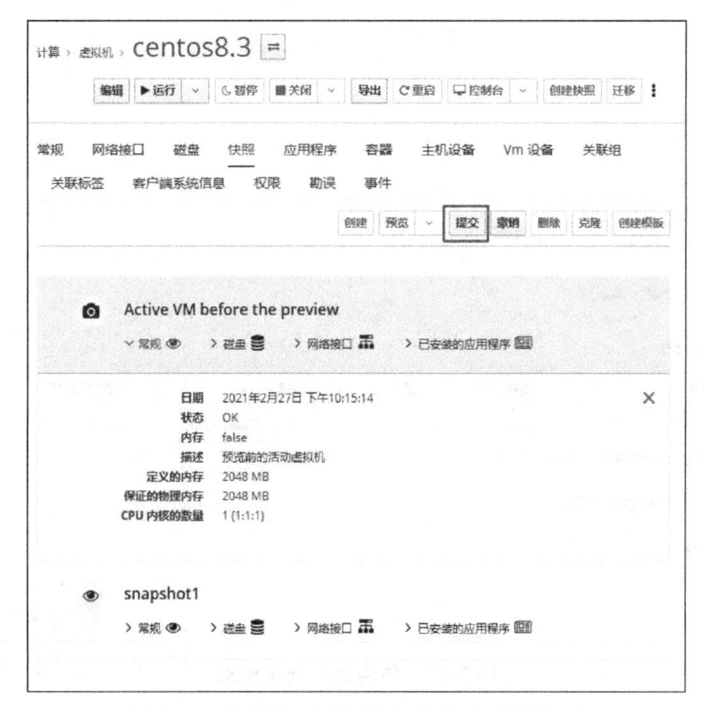


图 7-75　使用快照还原虚拟机-提交预览快照

可以在 oVirt 的事件日志中看到操作的详细信息，如图 7-76 所示。

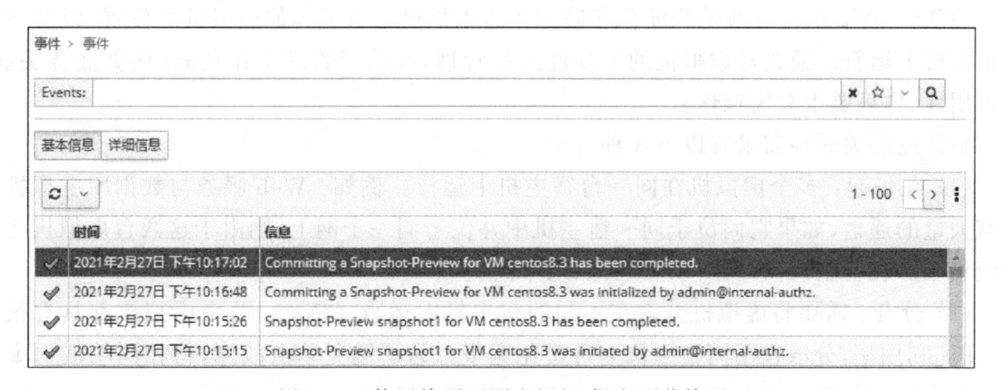


图 7-76　使用快照还原虚拟机-提交预览快照

### 3. 使用快照创建新虚拟机

有两种使用快照创建虚拟机的方法：

(1) 直接克隆快照，这与克隆虚拟机类似。

(2) 先通过快照来创建模板，如图 7-77 所示，然后由模板创建新的虚拟机。

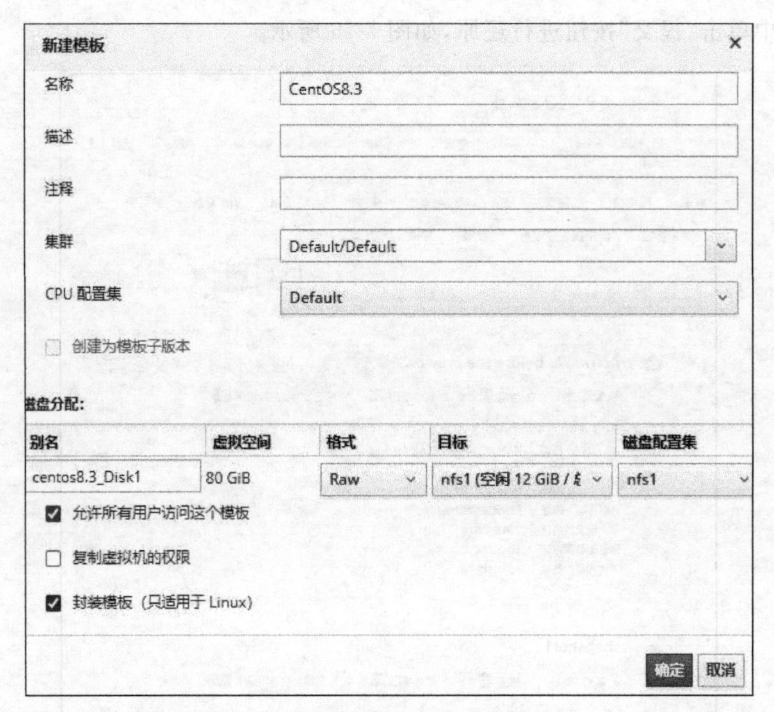


图 7-77　使用快照创建模板

## 7.6.8　关联性管理

关联性(Affinity)可帮助调度程序确定一台虚拟机相对于其他虚拟机的位置,以及在哪台宿主机上运行。通过对虚拟机的关联性进行管理,有助于管理工作负载,例如业务关联、高可用性、许可要求和灾难恢复。

最常见的关联性需求有以下 3 种。

(1) 在一起：多个虚拟机在同一台宿主机上运行。例如：Web 服务与数据库虚拟机之间有大量的通信,如果它们位于同一宿主机中并位于同一子网上,则由于这两台虚拟机之间的流量全部在宿主机内部,所以延迟低、效率高。

(2) 分开：需要将虚拟机分散在不同的宿主机上运行。例如：两个虚拟机都具有很高的 CPU 利用率,分散在不同宿主机上可避免竞争。为了避免单点故障,将相同功能的虚拟机运行在不同的宿主机上。

(3) 远离：某个虚拟机不要运行在特定的宿主机上。

在 oVirt 中实现关联性的技术元素有以下几个：

(1) 虚拟机关联性规则。

(2) 宿主机关联性规则。

(3) 默认权重模块。

管理员可以通过关联组和标签来管理虚拟机的关联性，如图 7-78 和图 7-79 所示。

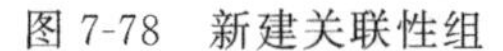


图 7-78 新建关联性组

图 7-79 新建关联性标签

## 7.6.9 实时迁移

实时迁移(Live Migration)提供了在物理主机之间移动正在运行的虚拟机的功能。在将虚拟机重定位到新的物理主机时，虚拟机仍保持开机状态，虚拟机的内存数据从源主机复制到目标宿主机，存储和网络连接不变，所以应用程序会继续运行、服务不会中断。

oVirt 环境必须正确配置以支持实时迁移，需要满足以下几个条件：

(1) 源主机和目标主机是同一个群集的成员，要确保它们之间的 CPU 的兼容性。

(2) 源主机和目标主机的状态为 Up。

(3) 源主机和目标主机可以访问相同的虚拟网络和 VLAN。

(4) 源主机和目标主机可以访问虚拟机所在的数据存储域。

(5) 目标主机具有足够的 CPU、内存来支持虚拟机的要求。

(6) 虚拟机没有 cache！=none 的自定义属性。

(7) 虚拟机没有使用 vGPU。

**提示**：在不同群集之间进行虚拟机的实时迁移，有时可以成功，但不建议这么做。

虚拟机实时迁移是一项资源密集型操作。为了优化实时迁移，可以为群集和虚拟机设置迁移策略。

单击“计算”中的“群集”，然后选择一个群集，单击“编辑”按钮，单击“迁移策略”选项卡。可以为整个群集设置迁移策略，如图 7-80 所示。

还可以在虚拟机的级别配置迁移选项。单击“计算”中的“虚拟机”，然后选择一个虚拟机。单击“编辑”按钮，单击“主机”选项卡，可以为此虚拟机配置个性化的迁移策略，如图 7-81 所示。

当主机进入维护模式时，oVirt 会自动将主机上运行的所有虚拟机实时迁移到其他主机上。在迁移虚拟机时，会评估每个虚拟机所适合的目标主机，从而将负载分散到整个群集中。

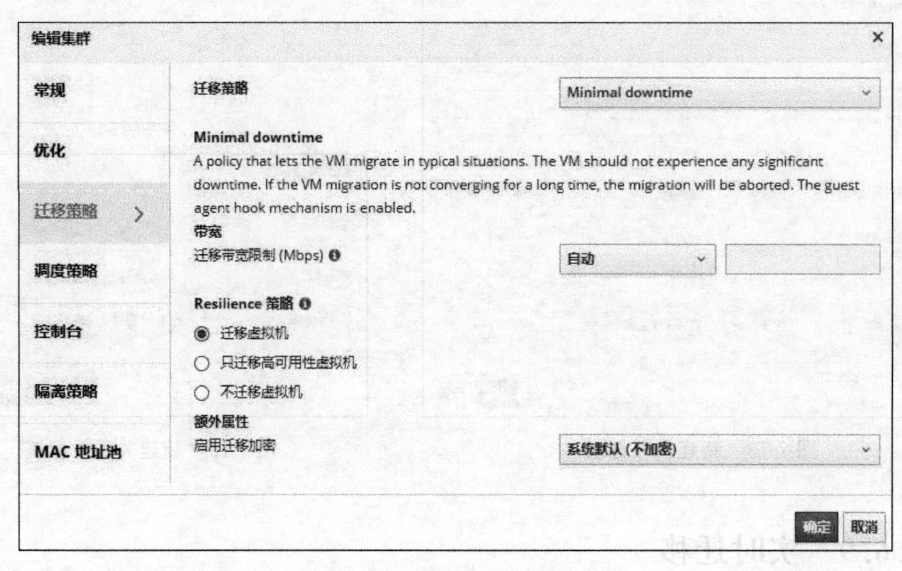


图 7-80　群集的迁移策略

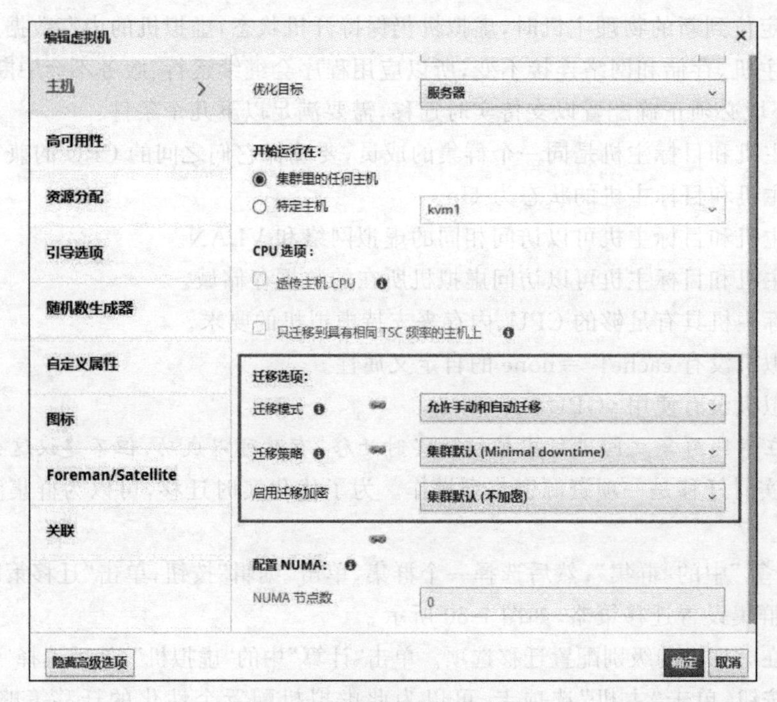


图 7-81　虚拟机的迁移策略

还可以进行手动实时迁移。单击“计算”中的“虚拟机”，然后选择正在运行的虚拟机，单击“迁移”按钮。

可以设置自动选择或者指定目标主机，如图 7-82 所示。当采用自动选择主机时，系统会根据调度策略中的负载平衡规则来确定目标主机。

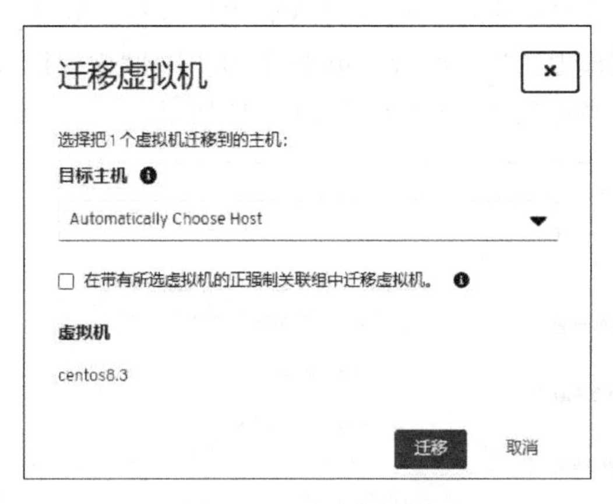


图 7-82 迁移虚拟机

迁移期间，在迁移进度栏中会有图标显示。同时在事件日志中也会看到相应的日志信息，如图 7-83 所示。

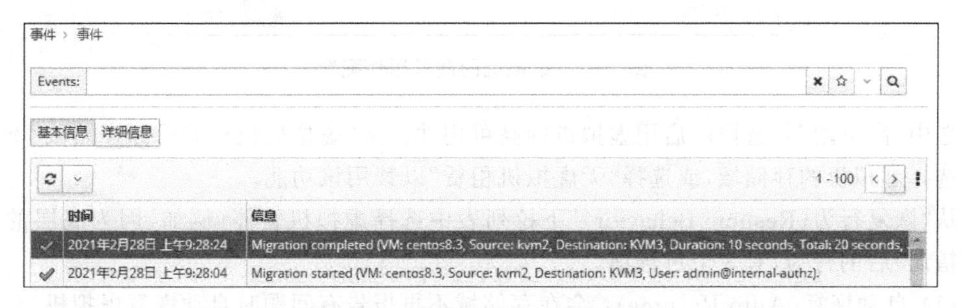


图 7-83 虚拟机迁移的日志信息

## 7.6.10 虚拟机高可用

对于运行关键工作负载的虚拟机，一定要使用高可用性。凭借高可用性，虚拟机可以在很短的时间内重新启动，而无须用户干预，因此可以将业务中断的时间降至最低。

使用存储域 V4 或更高版本，虚拟机具有在存储上的特定卷上获取租约的附加功能，即使原始主机断电，虚拟机也可以在另一台主机上启动。该功能还可以防止在两台不同的主机上启动同一个虚拟机，从而避免虚拟机磁盘损坏。

如果发生存储 I/O 错误，则虚拟机将暂停。可以定义在重新建立与存储域的连接后主机如何处理高可用性虚拟机，例如恢复、强制关闭或继续暂停。

实现虚拟机的高可用性，需要满足以下几个条件：

(1) 主机需要有电源管理设备和隔离(fencing)参数。

(2) 主机必须是具有其他可用主机的群集的一部分。

(3) 目标主机必须正在运行。

(4) 源主机和目标主机必须有权访问虚拟机所在的数据域、相同的虚拟网络和 VLAN。

(5) 目标主机上必须有足够的 CPU 和 RAM 来运行新虚拟机。

需要为每个高可用虚拟机进行配置。单击"计算"中的"虚拟机",然后选择一个虚拟机。单击"编辑"按钮,单击"高可用性"选项卡,如图 7-84 所示。

图 7-84 虚拟机的高可用性配置

选中"高可用"复选框以启用虚拟机的高可用性。从"虚拟机租赁的目标存储域"下拉列表中选择虚拟机的存储域,或选择"无虚拟机租赁"以禁用该功能。

从"恢复行为(Resume Behavior)"下拉列表中选择虚拟机暂停(例如,因为底层的存储访问错误)后的行为,有 3 个可选项。

(1) 自动恢复(Auto Resume):会在存储域不再报告有问题时自动恢复虚拟机。

(2) 保持暂停(Leave Paused):会使虚拟机处于暂停状态以等待用户来解决问题。

(3) 终止(Kill):在另外主机上重启虚拟机之前,需要将虚拟机关闭。如果定义了虚拟机租约,则终止是唯一可用的选项。

从"优先级"下拉列表中选择"低""中"或"高"。触发迁移后,将创建一个队列,在该队列中首先迁移高优先级虚拟机。如果群集的资源不足,则仅迁移高优先级的虚拟机。

单击"确定"按钮。如果虚拟机正常运行,则会出现"待批准虚拟机的改变"对话窗口,单击"确定"按钮继续,如图 7-85 所示。

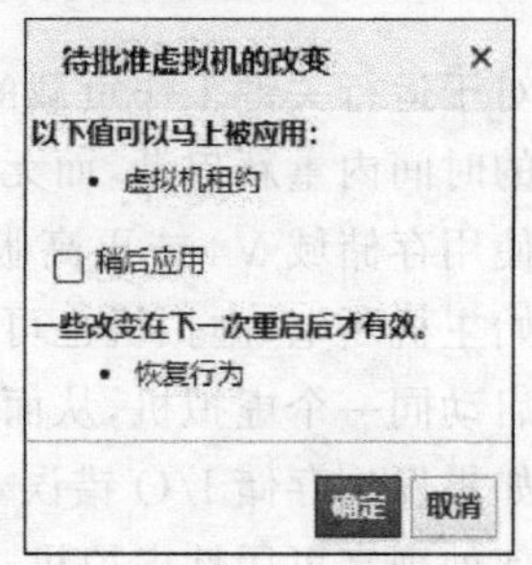


图 7-85 待批准虚拟机的改变

在实验中,配置完虚拟机的高可用性后,可以进行一个测试。通过 SSH 连接到此虚拟机的所在宿主机,使用 shutdown 命令关闭宿主机。oVirt 检测到宿主机故障后,会在其他的宿主机上重新启动虚拟机。可以通过查看事件日志获得更详细的信息,如图 7-86 所示。

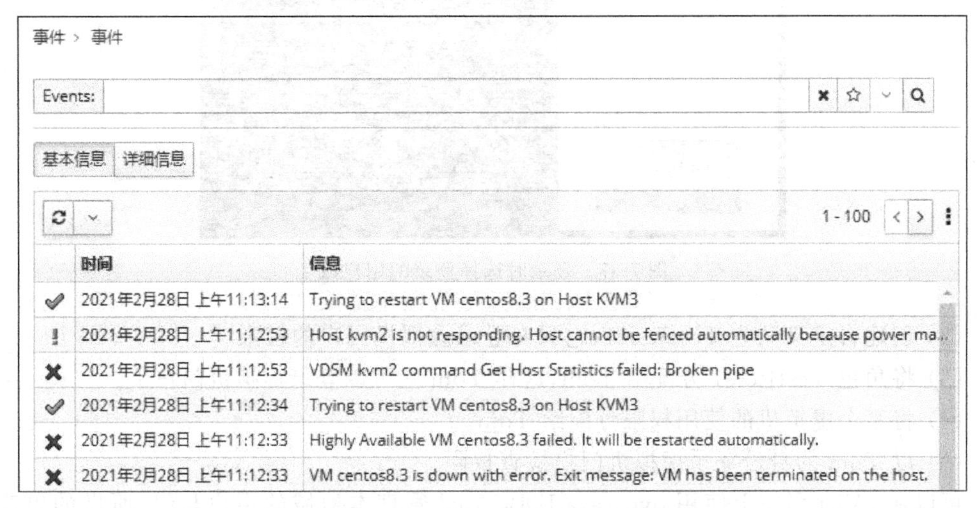


图 7-86 日志中的高可用虚拟机的重新启动信息

## 7.7 用户与权限管理

在 oVirt 中,有 2 种类型的用户域:本地域(Local Domain)和外部域(External Domain)。

在本地域中创建的用户账号被称为本地用户,它们的信息保存在 oVirt 的数据库中。在 oVirt 引擎的安装过程中,会创建一个默认的本地用户 admin。安装结束之后,还可以在 oVirt 引擎上使用命令行工具 ovirt-aaa-JDBC-tool 创建新的本地用户。

如果将外部目录服务器(例如 Red Hat Directory Server、Microsoft Active Directory、OpenLDAP)连接到 oVirt 环境,它们就是外部域。外部域中的用户账号被称为目录用户。用户在登录 oVirt 门户的时候,可以通过下拉列表框选择登录到外部域,如图 7-87 所示。

本地域和外部域都有用户组的概念。用户组是用户的集合,通过用户组可以简化用户的管理。使用 ovirt-aaa-JDBC-tool 可以管理本地域的用户组。

角色是预定义权限的集合。使用角色可以简化权限管理,可以为数据中心中不同级别的物理资源和虚拟资源设置访问和管理权限。角色主要有以下两种类型。

(1) 最终用户:可以使用和管理虚拟机门户中的资源。

(2) 管理员:可以使用管理门户来维护系统基础结构。

可以在管理门户中为本地用户和目录用户分配适当的角色和权限。

下面做一个简单的用户与权限管理的实验,主要包括以下几个步骤:

(1) 创建本地用户 Tom。

图 7-87　登录时选择登录的用户域

(2) 创建自定义角色 TestRole1,它只有登录虚拟机门户的权限。

(3) 将角色 TestRole1 分配给 Tom,这样 Tom 就可以登录虚拟机门户了。

(4) 将某个虚拟机的使用权限分配给 Tom。

(5) 以 Tom 的身份登录虚拟机门户检查权限。

可以在 oVirt 引擎上使用 ovirt-aaa-JDBC-tool 管理本地域的用户与组,所做的更改将立即生效。

查看此命令的联机帮助,示例命令如下:

```
[root@ovirt1 ~]# ovirt-aaa-JDBC-tool --help user
Usage: /usr/bin/ovirt-aaa-JDBC-tool [options] module ...
oVirt local user management command line tool.

Options:
  --db-config=[FILE]
    Path to the file which contains Database connection configuration.
    Default value: /etc/ovirt-engine/aaa/internal.properties

  --help
    Show help for this module.

  --log-level=[STRING]
    Log level of the tool.
    Valid values: FINEST|FINER|FINE|CONFIG|INFO|WARNING|SEVERE|ALL|OFF
    Default value: WARNING

  --version
    Show package version information.

Modules:
  user
```

```
  group
  group-manage
  query
  settings
  help
See: /usr/bin/ovirt-aaa-JDBC-tool [options] module --help for help on a specific module.
```

查看 user 子命令的联机帮助，示例命令如下：

```
[root@ovirt1 ~]# ovirt-aaa-JDBC-tool user --help
Usage: /usr/bin/ovirt-aaa-JDBC-tool [options] user module ...
Perform user related tasks.

Options:
  --help
    Show help for this module.

Modules:
  add
  edit
  delete
  unlock
  password-reset
  show
  help
See: /usr/bin/ovirt-aaa-JDBC-tool [options] user module --help for help on a specific user
module.
[root@ovirt1 ~]#
```

创建一个新的用户账号 tom，可选的--attribute 选项用于设置详细信息，示例命令如下：

```
[root@ovirt1 ~]# ovirt-aaa-JDBC-tool user add tom --attribute=firstName=Tom --
attribute=lastName=Chen
adding user tom...
user added successfully
Note: by default created user cannot log in. see:
/usr/bin/ovirt-aaa-JDBC-tool user password-reset --help.
```

为新用户设置密码。必须使用--password-valid-to 选项设置密码的到期时间，否则默认为当前时间，会导致新用户无法登录。日期格式为 yyyy-MM-dd HH：mm：ssX。在此示例中，使用中国时区，+0800 代表格林尼治标准时间加上 8 小时。示例命令如下：

```
[root@ovirt1 ~]# ovirt-aaa-JDBC-tool user password-reset tom --password-valid-to
="2030-01-01 12:00:00+0800"
Password:
```

```
Reenter password:
updating user tom...
user updated successfully
```

默认情况下,密码至少要有 6 个字符。更改密码时,无法再次使用最近使用过的 3 个密码。

查看新账号的信息,示例命令如下:

```
[root@ovirt1 ~]# ovirt-aaa-JDBC-tool user show tom
-- User tom(53ad5d47-b905-401d-949c-9746d08ea264) --
Namespace: *
Name: tom
ID: 53ad5d47-b905-401d-949c-9746d08ea264
Display Name:
Email:
First Name: Tom
Last Name: Chen
Department:
Title:
Description:
Account Disabled: false
Account Locked: false
Account Unlocked At: 1970-01-01 00:00:00Z
Account Valid From: 2021-02-28 13:17:46Z
Account Valid To: 2221-02-28 13:17:46Z
Account Without Password: false
Last successful Login At: 1970-01-01 00:00:00Z
Last unsuccessful Login At: 1970-01-01 00:00:00Z
Password Valid To: 2030-01-01 04:00:00Z
```

下面创建一个新角色。单击"管理"中的"配置",单击"角色"选项卡,会显示当前的所有角色、管理员角色和用户角色的列表,如图 7-88 所示。

单击"新建"按钮,输入新角色的名称和描述,选择"用户"作为账号类型。在权限选择中,仅选中"系统"下"配置系统"中的"登录权限",如图 7-89 所示。

单击"确定"按钮完成角色的创建。新角色将显示在角色列表中。

下面将添加 oVirt 用户并为其分配角色 TestRole1,这样此用户就拥有访问虚拟机门户的权限了。

单击"管理"中的"用户",会显示当前用户的列表,如图 7-90 所示。

单击"添加"按钮,会出现"添加用户和组"窗口,如图 7-91 所示。

选中"用户",在"搜索"下选择要搜索的域,在搜索文本字段中输入名称或名称的一部分,然后单击 GO 按钮。或者直接单击 GO 按钮查看所有用户和组的列表。

选中相应用户前面的复选框,单击"添加并关闭"按钮。

图 7-88 oVirt 角色列表

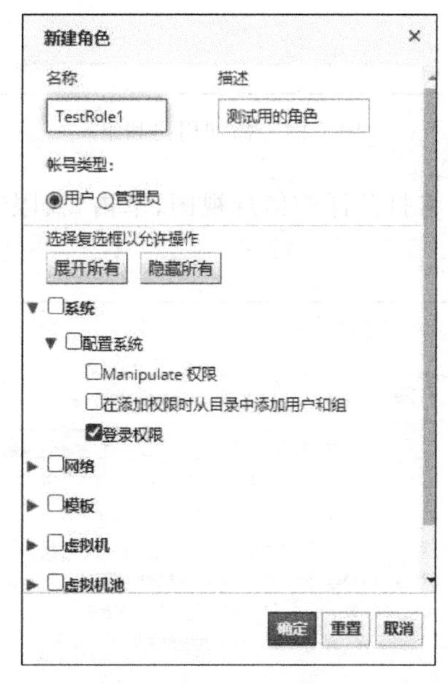


图 7-89 新建 oVirt 角色

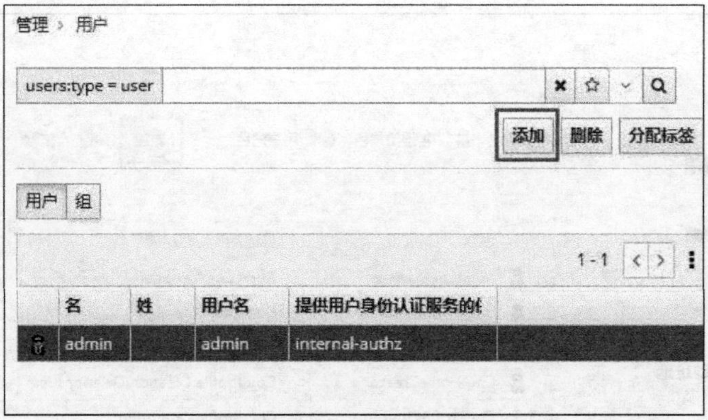

图 7-90 oVirt 用户列表

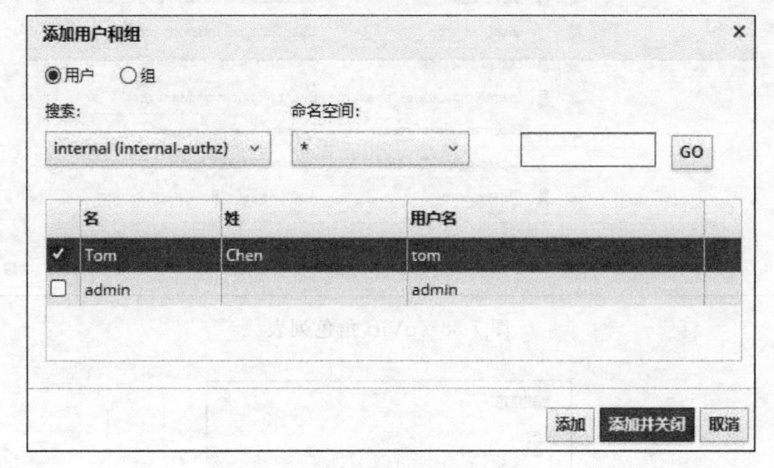


图 7-91 添加用户和组

单击新用户的用户名,这将打开详细信息视图,单击“权限”选项卡,会显示用户所拥有的默认角色,如图 7-92 所示。

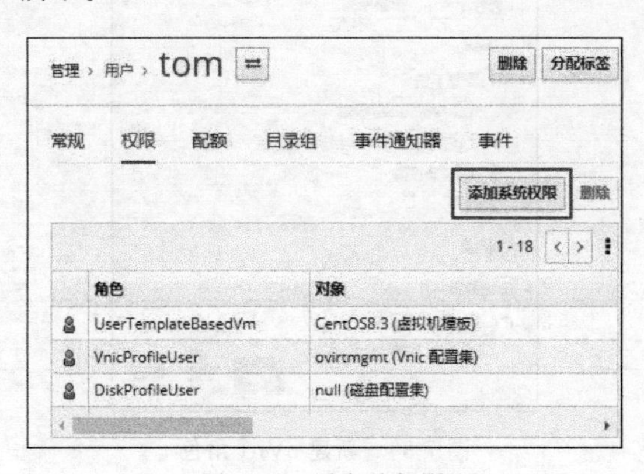


图 7-92 用户的权限

单击“添加系统权限”按钮，在新窗口中的“要分配的角色”下拉列表框中选择要添加的角色，单击“确定”按钮完成添加，如图 7-93 所示。

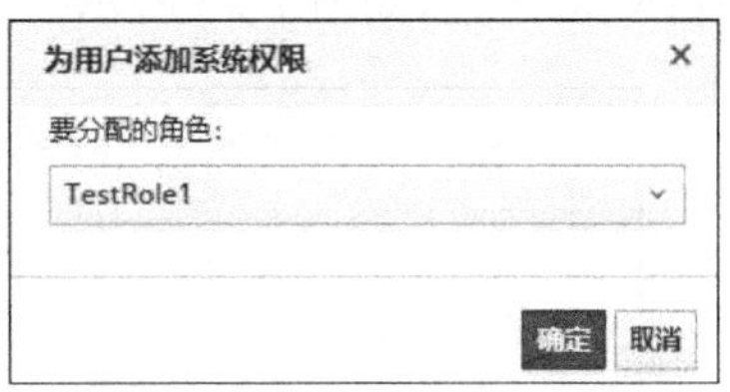

图 7-93　为用户添加系统权限

新打开一个浏览器，以 Tom 身份登录到虚拟机门户，如果可以成功登录，则说明具有登录权限。

但是 Tom 在虚拟机门户中没有任何虚拟机资源可供使用，这就需要为 Tom 分配权限。打开虚拟机的属性中的“权限”选项卡，如图 7-94 所示。

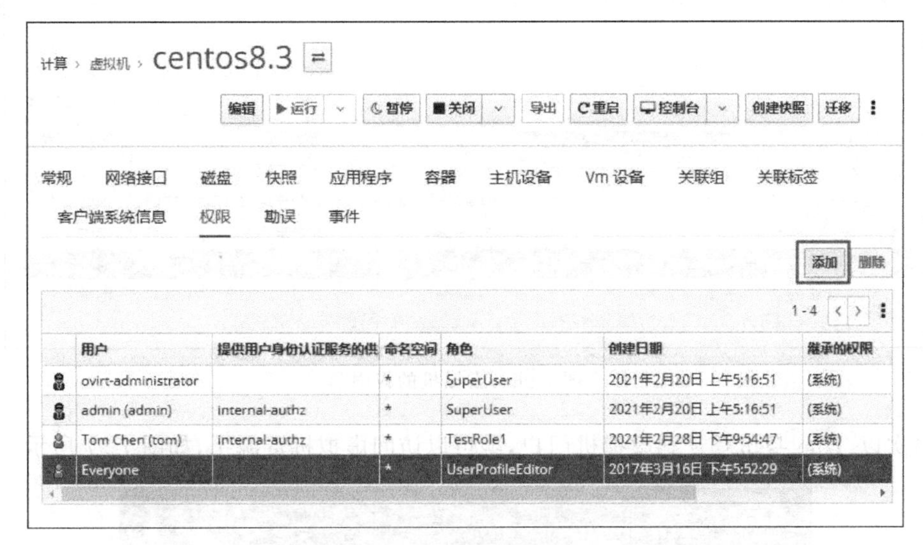

图 7-94　虚拟机的权限

单击“添加”按钮，会打开“为用户添加权限”窗口。最简单的方法是给 Tom 分配 UserRole 角色，如图 7-95 所示。

图 7-95　为用户添加权限

此时，用户 Tom 就拥有了从系统继承过来的角色 TestRole1，也拥有了来自本资源的 UserRole 角色，如图 7-96 所示。

图 7-96　虚拟机的权限

再次以 Tom 身份登录到虚拟机门户，就可以访问虚拟机资源了，如图 7-97 所示。

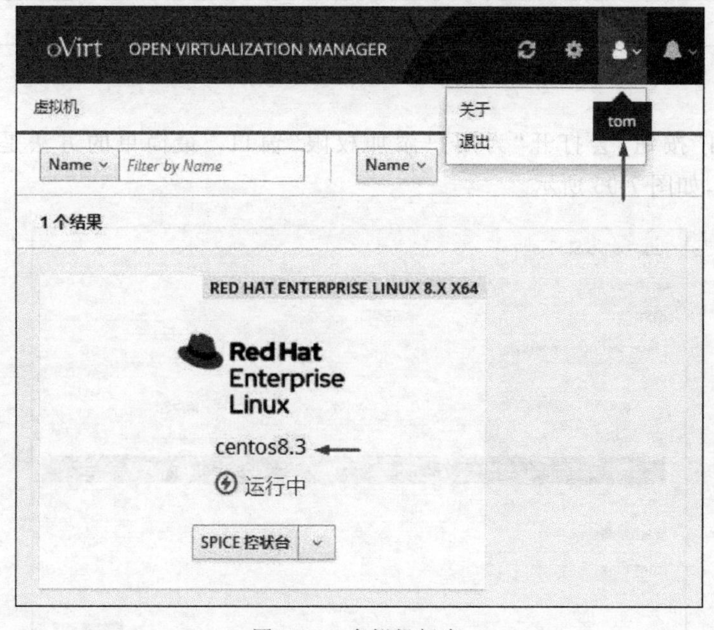


图 7-97　虚拟机门户

# 7.8　备份与恢复

oVirt 的备份工作包括 oVirt 引擎的备份和虚拟机的备份。

## 7.8.1　备份 oVirt 引擎

可以使用 engine-backup 工具定期备份 oVirt 引擎，它可以在不中断引擎服务的情况下将引擎的数据库和配置文件备份到单个文件中。

登录 oVirt 引擎，创建一个完整的备份，示例命令如下：

```
[root@ovirt1 ~]# engine-backup --scope=all --mode=backup --file=engbackup \
    --log=backuplog
Start of engine-backup with mode 'backup'
scope: all
archive file: engbackup
log file: backuplog
Backing up:
Notifying engine
- Files
- Engine Database 'engine'
- DWH Database 'ovirt_engine_history'
Packing into file 'engbackup'
Notifying engine
Done.
```

engine-backup 命令有两种操作模式：backup 和 restore。使用时，还需要通过额外的选项来指定备份范围、日志及引擎数据库的不同凭据。

1) --file 选项

在备份模式和还原模式下，这都是必需的选项。在备份模式下指定要备份目标的路径和文件名称，在还原模式下指定要读取的备份数据的路径和文件名称。

2) --log 选项

必需的选项，用于指定保存操作日志的文件。

3) --scope

指定备份或还原操作的范围。有 4 个选项：all 包括引擎数据库、数据仓库和配置文件，files 仅包括文件，db 仅包括引擎数据库，dwhdb 仅包括数据仓库。默认范围是 all。

备份文件是 gzip 的压缩打包文件。可以通过日志来获得与备份操作更多的信息。示例命令如下：

```
[root@ovirt1 ~]# file engbackup
engbackup: gzip compressed data, last modified: Mon Mar  1 06:15:11 2021, from UNIX, original
size 4444160

[root@ovirt1 ~]# head backuplog
2021-03-01 14:15:03 134386: Start of engine-backup mode backup scope all file engbackup
2021-03-01 14:15:03 134386: OUTPUT: Start of engine-backup with mode 'backup'
2021-03-01 14:15:03 134386: OUTPUT: scope: all
2021-03-01 14:15:03 134386: OUTPUT: archive file: engbackup
2021-03-01 14:15:03 134386: OUTPUT: log file: backuplog
2021-03-01 14:15:03 134386: OUTPUT: Backing up:
2021-03-01 14:15:03 134386: Generating pgpass
2021-03-01 14:15:03 134386: OUTPUT: Notifying engine
2021-03-01 14:15:03 134386: pg_cmd running: psql -w -U engine -h localhost -p 5432
engine -t -c SELECT LogEngineBackupEvent('files', now(), 0, 'engine-backup: Backup Started,
scope=files, log=/root/backuplog', 'ovirt1.tomtrain.local', '/root/backuplog');
```

### 7.8.2 恢复 oVirt 引擎

建议定期备份引擎的数据与文件，以应对物理错误和逻辑错误。在重要操作（如升级）之前也要进行备份，如果出现误操作，就可以从备份中还原以撤销更改。

如果是自托管引擎的架构，在恢复时，则需要先登录到自托管引擎所在节点，然后将其置于全局维护模式，示例命令如下：

```
[root@kvm1 ~]# hosted-engine --set-maintenance --mode=global
```

登录 oVirt 引擎，删除配置文件并清理与引擎关联的数据库，示例命令如下：

```
[root@ovirt1 ~]# engine-cleanup
[ INFO  ] Stage: Initializing
[ INFO  ] Stage: Environment setup
          Configuration files: /etc/ovirt-engine-setup.conf.d/10-packaging-jboss.conf,
/etc/ovirt-engine-setup.conf.d/10-packaging.conf, /etc/ovirt-engine-setup.conf.d/20
-setup-ovirt-post.conf
          Log file: /var/log/ovirt-engine/setup/ovirt-engine-remove-20210301144805-
tww339.log
          Version: otopi-1.9.2 (otopi-1.9.2-1.el8)
[ INFO  ] Stage: Environment packages setup
[ INFO  ] Stage: Programs detection
[ INFO  ] Stage: Environment customization
[WARNING] Failed to resolve ovirt1.tomtrain.local using DNS, it can be resolved only locally

          --== PRODUCT OPTIONS ==--

[ INFO  ] Stage: Setup validation
          During execution engine service will be stopped (OK, Cancel) [OK]:
```

```
[ INFO  ] Hosted Engine HA is in Global Maintenance mode.
          All the installed ovirt components are about to be removed, data will be lost (OK, Cancel) [Cancel]: ok <-- 输入 OK
[ INFO  ] Stage: Transaction setup
[ INFO  ] Stopping engine service
[ INFO  ] Stopping ovirt-fence-kdump-listener service
[ INFO  ] Stopping dwh service
[ INFO  ] Stopping vmconsole-proxy service
[ INFO  ] Stopping websocket-proxy service
[ INFO  ] Stage: Misc configuration (early)
[ INFO  ] Stage: Package installation
[ INFO  ] Stage: Misc configuration
[ INFO  ] Backing up PKI configuration and keys
[ INFO  ] Backing up Database localhost:engine to '/var/lib/ovirt-engine/backups/engine-20210301144851.z6vutbod.dump'.
[ INFO  ] Clearing Engine Database engine
[ INFO  ] Backing up Database localhost:ovirt_engine_history to '/var/lib/ovirt-engine-dwh/backups/dwh-20210301145025.176t4mb9.dump'.
[ INFO  ] Clearing DWH Database ovirt_engine_history
[ INFO  ] Install seLinux module /usr/share/ovirt-engine/seLinux/ansible-runner-service.cil
[ INFO  ] Removing files
[ INFO  ] Reverting changes to files
[ INFO  ] Stage: Transaction commit
[ INFO  ] Stage: Closing up

          --== SUMMARY ==--

          Engine setup successfully cleaned up
          A backup of PKI configuration and keys is available at /var/lib/ovirt-engine/backups/engine-pki-20210301144845za1tadny.tar.gz
          ovirt-engine has been removed
          A backup of the Engine Database is available at /var/lib/ovirt-engine/backups/engine-20210301144851.z6vutbod.dump
          A backup of the DWH Database is available at /var/lib/ovirt-engine-dwh/backups/dwh-20210301145025.176t4mb9.dump

          --== END OF SUMMARY ==--

[ INFO  ] Stage: Clean up
          Log file is located at /var/log/ovirt-engine/setup/ovirt-engine-remove-20210301144805-tww339.log
[ INFO  ] Generating answer file '/var/lib/ovirt-engine/setup/answers/20210301145140-cleanup.conf'
[ INFO  ] Stage: Pre-termination
[ INFO  ] Stage: Termination
[ INFO  ] Execution of cleanup completed successfully
```

engine-cleanup 命令仅清除引擎数据库中的数据，不会删除数据库和数据库的用户，所以在还原完整备份时不需要创建新的数据库或指定数据库凭据。示例命令如下：

```
[root@ovirt1 ~]# engine-backup --mode=restore --file=engbackup \
    --log=restorelog --restore-permissions
Start of engine-backup with mode 'restore'
scope: all
archive file: engbackup
log file: restorelog
Preparing to restore:
- Unpacking file 'engbackup'
Restoring:
- Files
- Engine Database 'engine'
  - Cleaning up temporary tables in engine Database 'engine'
  - Updating DbJustRestored VdcOption in engine Database
  - Resetting DwhCurrentlyRunning in dwh_history_timekeeping in engine Database
  - Resetting HA VM status
------------------------------------------------------------------------------
Please note:

The engine Database was backed up at 2021-03-01 14:15:08.000000000 +0800 .

Objects that were added, removed or changed after this date, such as virtual
machines, disks, etc., are missing in the engine, and will probably require
recovery or recreation.
------------------------------------------------------------------------------
- DWH Database 'ovirt_engine_history'
You should now run engine-setup.
Done.
```

参数--restore-permissions 是必需的，它指定恢复数据库用户的权限。

重新配置 oVirt 引擎。在配置的过程中，保留默认选项即可，示例命令如下：

```
[root@ovirt1 ~]# engine-setup
...略...
```

最后，登录到运行自托管引擎的宿主机，退出全局维护模式，示例命令如下：

```
[root@kvm1 ~]# hosted-engine --set-maintenance --mode=none

[root@kvm1 ~]# hosted-engine --vm-status
--== Host kvm1 (id: 1) status ==--

Host ID                            : 1
Host timestamp                     : 493230
```

```
Score                                    : 3400
Engine status                            : {"vm": "up", "health": "good", "detail": "Up"}
Hostname                                 : kvm1
Local maintenance                        : False
stopped                                  : False
crc32                                    : e94f7cfe
conf_on_shared_storage                   : True
local_conf_timestamp                     : 493230
Status up-to-date                        : True
Extra metadata (valid at timestamp):
    metadata_parse_version=1
    metadata_feature_version=1
    timestamp=493230 (Mon Mar  1 15:29:50 2021)
    host-id=1
    score=3400
    vm_conf_refresh_time=493230 (Mon Mar  1 15:29:50 2021)
    conf_on_shared_storage=True
    maintenance=False
    state=GlobalMaintenance
    stopped=False
```

这样就完成了 oVirt 引擎的恢复。

## 7.8.3　准备备份存储域

备份存储域用于虚拟机的备份和还原，可以在其中保存虚拟机和虚拟机模板，但是虚拟机不能在备份存储域上运行，即备份存储域上的所有虚拟机均处于关闭状态。

与其他类型的存储域一样，可以将备份存储域附加到数据中心或从数据中心分离，因此，除了存储备份外，还可以使用备份存储域在数据中心之间迁移虚拟机。

与导出域相比，备份存储域具有多个优势：

(1) 数据中心中只能有一个导出域，但可以有多个备份存储域。

(2) 使用备份存储域迁移大量虚拟机、模板或 OVF 文件要比导出域快得多。

(3) 备份存储域比导出域可以更有效地使用磁盘空间。

(4) 备份存储域支持文件存储(NFS 和 Gluster)和块存储(光纤通道和 iSCSI)，而导出域仅支持文件存储。

(5) 可以动态启用和禁用存储域的备份属性。

可以将某个现存的存储域设置为备份存储域。单击“存储”中的“域”，选择一个现有的存储域，然后单击“管理域”，这将打开“管理域”对话框，如图 7-98 所示。

在“高级参数”下选中“备份”复选框，单击“确定”按钮。该域现在就是一个备份存储域了。在存储域列表中，备份存储域会有一个小图标，如图 7-99 所示。

管理域

数据中心 Default (V5)
名称 bakup1
域功能 数据
描述
存储类型 NFS
注释
主机 kvm1

导出路径 192.168.1.235:/backup1
自定义连接参数
高级参数
警告级低磁盘空间 (%) 10
(5 GB)
严重的空间操作限制阈值（GB） 5
警告级低磁盘空间 (%) 10
格式 V5
删除后清理
备份
确定 取消

图 7-98　配置备份存储域

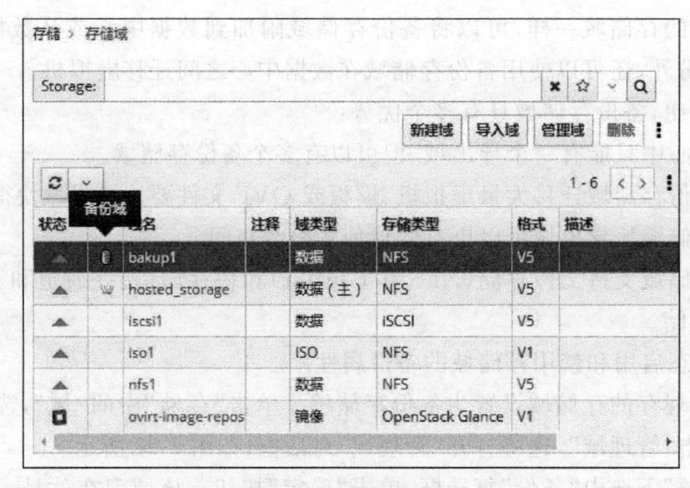


图 7-99　存储域列表

## 7.8.4 备份和还原虚拟机

有了备份存储域,就可以备份和还原虚拟机了。

要想备份虚拟机,首先需要克隆该虚拟机,如图 7-100 所示。

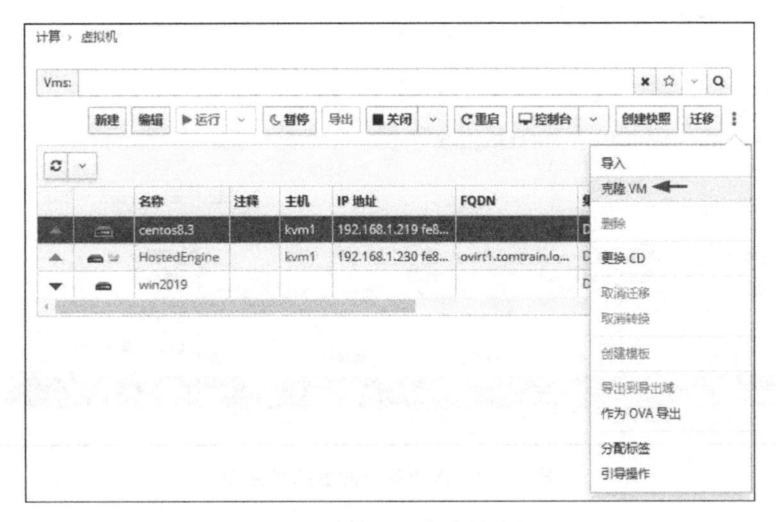


图 7-100 克隆需要备份的虚拟机

将克隆出的新虚拟机置于关闭状态,然后将它导出到备份存储域,这就完成了备份操作,如图 7-101 所示。

图 7-101 将虚拟机的克隆副本导出到备份存储域

还原虚拟机的操作,就是从备份存储域导入虚拟机的过程。

单击"计算"中的"虚拟机",选择保存在备份存储域中的虚拟机备份。单击"磁盘"选项卡，选择属于虚拟机的所有磁盘。单击更多操作标识符“⋮”，然后选择“移动”，如图 7-102 所示。

图 7-102　移动备份的虚拟机磁盘

在“目标”下，选择数据存储域，单击“确定”按钮，如图 7-103 所示。

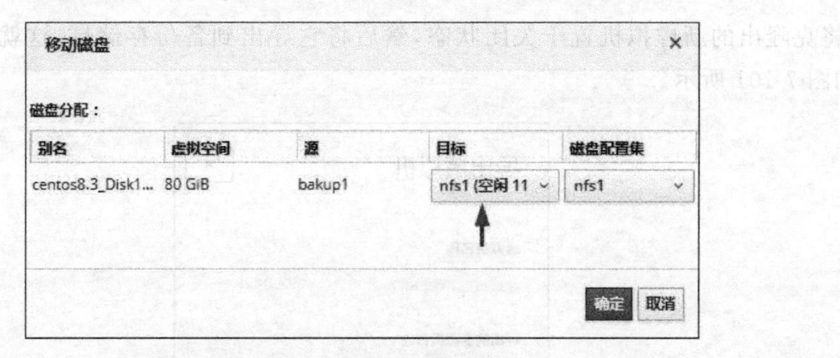


图 7-103　指定目标存储域

将当前备份存储域中的磁盘全部迁移到数据域之后，就完成了恢复操作。

**提示**：oVirt 提供了完整 API。通过编程语言或脚本调用 REST API 可以进行高效备份与恢复操作。

## 7.9　本章小结

本章通过实验讲解了 oVirt 管理操作，包括安装、数据中心、存储、宿主机、虚拟机、高可用、用户、权限及备份与恢复。更多的管理技术，建议阅读 oVirt 的官方文档。

**图书推荐**

| 书　　名 | 作　　者 |
| --- | --- |
| 鸿蒙应用程序开发 | 董昱 |
| 鸿蒙操作系统开发入门经典 | 徐礼文 |
| 鸿蒙操作系统应用开发实践 | 陈美汝、郑森文、武延军、吴敬征 |
| 华为方舟编译器之美——基于开源代码的架构分析与实现 | 史宁宁 |
| 鲲鹏架构入门与实战 | 张磊 |
| 华为 HCIA 路由与交换技术实战 | 江礼教 |
| Flutter 组件精讲与实战 | 赵龙 |
| Flutter 组件详解与实战 | [加]王浩然(Bradley Wang) |
| Flutter 实战指南 | 李楠 |
| Dart 语言实战——基于 Flutter 框架的程序开发(第 2 版) | 亢少军 |
| Dart 语言实战——基于 Angular 框架的 Web 开发 | 刘仕文 |
| IntelliJ IDEA 软件开发与应用 | 乔国辉 |
| Vue+Spring Boot 前后端分离开发实战 | 贾志杰 |
| Vue.js 企业开发实战 | 千锋教育高教产品研发部 |
| Python 人工智能——原理、实践及应用 | 杨博雄主编，于营、肖衡、潘玉霞、高华玲、梁志勇副主编 |
| Python 深度学习 | 王志立 |
| Python 异步编程实战——基于 AIO 的全栈开发技术 | 陈少佳 |
| Python 数据分析从 0 到 1 | 邓立文、俞心宇、牛瑶 |
| 物联网——嵌入式开发实战 | 连志安 |
| 智慧建造——物联网在建筑设计与管理中的实践 | [美]周晨光(Timothy Chou)著；段晨东、柯吉译 |
| TensorFlow 计算机视觉原理与实战 | 欧阳鹏程、任浩然 |
| 分布式机器学习实战 | 陈敬雷 |
| 计算机视觉——基于 OpenCV 与 TensorFlow 的深度学习方法 | 余海林、翟中华 |
| 深度学习——理论、方法与 PyTorch 实践 | 翟中华、孟翔宇 |
| 深度学习原理与 PyTorch 实战 | 张伟振 |
| ARKit 原生开发入门精粹——RealityKit+Swift+SwiftUI | 汪祥春 |
| HoloLens 2 开发入门精要——基于 Unity 和 MRTK | 汪祥春 |
| Altium Designer 20 PCB 设计实战(视频微课版) | 白军杰 |
| Cadence 高速 PCB 设计——基于手机高阶板的案例分析与实现 | 李卫国、张彬、林超文 |
| Octave 程序设计 | 于红博 |
| AutoCAD 2022 快速入门、进阶与精通 | 邵为龙 |
| SolidWorks 2020 快速入门与深入实战 | 邵为龙 |
| SolidWorks 2021 快速入门与深入实战 | 邵为龙 |
| UG NX 1926 快速入门与深入实战 | 邵为龙 |
| 西门子 S7-200 SMART PLC 编程及应用(视频微课版) | 徐宁、赵丽君 |
| 三菱 FX3U PLC 编程及应用(视频微课版) | 吴文灵 |
| 全栈 UI 自动化测试实战 | 胡胜强、单镜石、李睿 |
| pytest 框架与自动化测试应用 | 房荔枝、梁丽丽 |
| 软件测试与面试通识 | 于晶、张丹 |
| 深入理解微电子电路设计——电子元器件原理及应用(原书第 5 版) | [美]理查德·C. 耶格(Richard C. Jaeger)、[美]特拉维斯·N. 布莱洛克(Travis N. Blalock)著；宋廷强译 |
| 深入理解微电子电路设计——数字电子技术及应用(原书第 5 版) | [美]理查德·C. 耶格(Richard C. Jaeger)、[美]特拉维斯·N. 布莱洛克(Travis N. Blalock)著；宋廷强译 |
| 深入理解微电子电路设计——模拟电子技术及应用(原书第 5 版) | [美]理查德·C. 耶格(Richard C. Jaeger)、[美]特拉维斯·N. 布莱洛克(Travis N. Blalock)著；宋廷强译 |

# 图书资源支持

感谢您一直以来对清华大学出版社图书的支持和爱护。为了配合本书的使用，本书提供配套的资源，有需求的读者请扫描下方的“书圈”微信公众号二维码，在图书专区下载，也可以拨打电话或发送电子邮件咨询。

如果您在使用本书的过程中遇到了什么问题，或者有相关图书出版计划，也请您发邮件告诉我们，以便我们更好地为您服务。


**我们的联系方式：**

地　　址：北京市海淀区双清路学研大厦 A 座 714

邮　　编：100084

电　　话：010-83470236　010-83470237

资源下载：http://www.tup.com.cn

客服邮箱：tupjsj@vip.163.com

QQ：2301891038（请写明您的单位和姓名）


教学资源 · 教学样书 · 新书信息

人工智能科学与技术
人工智能|电子通信|自动控制

资料下载 · 样书申请

书圈

**用微信扫一扫右边的二维码，即可关注清华大学出版社公众号。**